W0245964

Supramolecular
Structure and
Function

Supramolecular Structure and Function

Edited by

Greta Pifat

Rudjer Bošković Institute
Zagreb, Yugoslavia

and

Janko N. Herak

University of Zagreb
Zagreb, Yugoslavia

PLENUM PRESS • NEW YORK AND LONDON

Library of Congress Cataloging in Publication Data

International Summer School in Biophysics (1981: Dubrovnik, Yugoslavia)
   Supramolecular structure and function.

   "Lectures given at the International Summer School in Biophysics, held September
16–25, 1981, in Dubrovnik, Yugoslavia"—T.p. verso.
   Includes bibliographical references and index.
   1. Molecular biology—Addresses, essays, lectures. 2. Macromolecules—Addresses,
essays, lectures. I. Pifat, Greta. II. Herak, Janko N. III. Title.
QH506.I49 1983   574.8′8   82-24604

   ISBN-13:978-1-4684-4480-3   e-ISBN-13:978-1-4684-4478-0
   DOI: 10.1007/978-1-4684-4478-0

Lectures given at the International Summer School in Biophysics,
held September 16–25, 1981, in Dubrovnik, Yugoslavia

© 1983 Plenum Press, New York
Softcover reprint of the hardcover 1st edition 1983

A Division of Plenum Publishing Corporation
233 Spring Street, New York, N.Y. 10013

All rights reserved

No part of this book may be reproduced, stored in a retrieval system, or transmitted
in any form or by any means, electronic, mechanical, photocopying, microfilming,
recording, or otherwise, without written permission from the Publisher

PREFACE

The molecular basis of life has been a rapidly growing field
of science. There is perhaps no other field where such diverse
profiles of scientists, ranging from applied mathematicians and
theoretical physicists to experimental biologists and medical
doctors(physicians),are compelled to communicate and even to col-
laborate. This diversity makes the exchange of information richer
but at the same time more cumbersome. One way to facilitate the
exchange of information and to overcome the barriers between the
different languages used by physicists, chemists and biologists
is to organize a meeting on a subject of common interest. A particularly suitable form of such a meeting for younger scientists is
a school at an undergraduate or postgraduate level.

This volume contains a collection of lectures presented
at the International Summer School in Biophysics, held under
the title "Supramolecular Structure and Function" in Dubrovnik,
Yugoslavia, in September 1981. The topics discussed at the school
were inter- and intramolecular interactions in biological systems,
and structure, organization and function of biological macromolecules and supramolecular structures. Although not all the lectures
could be prepared in a written form on time for publication, we
hope that the present volume contains valuable up-to-date information on various aspects of the molecular basis of life. We wish
to express our gratitude to the eminent authors and to state that,
having received so much valuable assistance from them, we as
editors can only attach our names to apologies for any errors
that may remain.

The School was sponsored by the International Union for Pure
and Applied Biophysics (IUPAB), the Yugoslav Biophysical Society
and the Croatian Biophysical Society as well as by the Scientific
Councils of Croatia and Yugoslavia. Financial aid granted by the

Councils has also been used in part for the preparation of this
volume.

                                        Greta Pifat

                                        Janko N. Herak

Zagreb, June 1982

CONTENTS

Potential Energy Functions for Structural
    Molecular Biology   1
  S. Lifson

Theoretical and Experimental Aspects of
    Protein Folding   45
  H.A. Scheraga

Some Aspects of the Macromolecular Chemistry
    of Carbohydrate Polymers   59
  V. Crescenzi

Hydration and Interactions in Aqueous Solutions
    of Ions and Molecules   95
  F. Franks

Biomembranes   127
  D. Marsh

X-Ray and Neutron Small-Angle Scattering on
    Plasma Lipoproteins   179
  P. Laggner

Structure and Dynamics of Human Plasma Lipoproteins   205
    L.C. Smith, J.B. Massey, J.T. Sparrow, A.M. Gotto, Jr.,
  and H.J. Pownall

Crystallographic Studies of the Protein
    Biosynthesis System   245
  A. Liljas and M. Leijonmarck

Hemoglobin Oxygen Binding, Erythrocyte Shape
    Transformations, and Modeling of Cell
    Differentiation as Examples of Theoretical
    Approaches in Studying the Structure-Function
    Relationship in Biological Systems   275
  S. Svetina

Evolution of Polynucleotides   309
  P. Schuster

List of Contributions   357

Subject index   359

POTENTIAL ENERGY FUNCTIONS FOR STRUCTURAL MOLECULAR BIOLOGY*

Shneior Lifson

Department of Chemical Physics
Weizmann Institute of Science
Rehovot, Israel

INTRODUCTION

Structural molecular biology is first and foremost an experimental science. This is quite understandable. As long as we did not know what is the structure of enzymes how could we ask questions on how do they obtain their structure or how is their structure determining their catalytic function? However, as our knowledge of facts about biological structures grows now at a tremendous rate, there is an ever-growing need to apply theory, and calculations based on theory, to supplement the experimental study of structural molecular biology.

In the following chapters we shall discuss the empirical methods of choosing the analytic forms of energy functions and of determining their constant coefficients (the so-called energy parameters). We shall enquire how the empirical methods and the quantum mechanical methods are related to each other, and shall find it

---

* The three lectures given here constitute a revised and enlarged version of lectures given in the NATO Advanced Study Institute/ FEBS Advanced Course No. 78, Maratea, S. Italy, May 3-16, 1981, published as Structural Molecular Biology. Methods and Applications, D.B. Davies, ed., Plenum Press, 1982.

similar to the ways by which inductive and deductive methods are
related to each other in all branches of the exact sciences.

The potential energy of intra- and inter-molecular interactions is one of the central concepts in a theoretical approach to
problems of structure and function in biology. What do we know about
the molecular potential energy in biomolecules and related organic
compounds? Where does this knowledge come from? How reliable or approximate is it? This is the subject of the first part of the presentation.

In the second part we will ask what are the potentialities and
limitations of the applications of energy functions for physical
chemistry in general and for molecular biology in particular. How
do we relate the potential energy to experimental data on structural, dynamic and thermodynamic properties of molecules? What are
the theoretical considerations and which are the computational algorithms available? Finally we shall point out the advantages of a
rigorous application of the inductive approach based on objective
search for "consistent force fields" of energy functions and energy
parameters, and of establishing a "bench-mark" for comparing alternative force fields.

The third part will be devoted to a discussion of the nature
of the hydrogen bond. Is it basically a weak chemical bond? Or is
it a strong non-bonded interaction? And why is this distinction
important for molecular biology? We shall show how an application
of the Consistent Force Field method can give a definite answer to
the above questions, and shall compare this answer with results obtained by quantum mechanical methods of calculating the hydrogen
bond and of partitioning the hydrogen bond energy into its various
components.

1. EMPIRICAL POTENTIAL ENERGY FUNCTIONS

The Basic Facts of Life of Molecules

We start our discussion of the use of potential energy functions in structural biology by a concise review of the quantum mechanical and the empirical basis of our qualitative understanding
and quantitative determination of the potential functions of molecular interactions. Such an introductory discussion is necessary because an intelligent and useful application of potential functions

in biology requires a feeling for the power as well as for the limitations of both the theoretical and the empirical approaches, and
for the complex and subtle nature of their inter-relations.

   The nature of molecular forces is in principle well understood, thanks to quantum mechanics, which "made theoretical chemistry
a branch of applied mathematics" according to the famous exagger-
ation by Dirac. Atoms and molecules are made of nuclei and electrons,
which carry positive and negative electrostatic charges, respectively. The only potential of interaction between nuclei and electrons
which is of relevance to our discussion is therefore the Coulomb
potential. Thus, the potential energy of interatomic and intermolecular interactions originates from the Coulomb (electrostatic)
interactions,

$$V_{Coulomb} = (1/2) \ \Sigma_{i,j} e_i e_j / r_{ij}, \tag{1}$$

between the nuclei and electrons which form the atomic and molecular
assemblies. The dynamic behavior of such assemblies is, however,
controlled by the laws of quantum mechanics, i.e. by the Schrödinger
equation.

   The solution of the Schrödinger equation is neither needed nor
generally possible for the molecules of interest to our subject of
discussion, namely, structural molecular biology. Yet an understanding of some basic properties of the laws of quantum mechanics incorporated in this equation is essential to the understanding of the
nature of the molecular forces which control all life processes.
Such understanding requires no mathematics, yet may be deep and
fundamental. Let us try to put it in the simplest terms compatible
with the purpose of our discussion.

   We shall start with an analysis of the nature of the total
energy of an assembly of nuclei and electrons which form a molecule.
The total energy is composed of an electrostatic potential energy
which is given by Eq. (1), and a kinetic energy, given by

$$T(p) = (1/2) \ \Sigma_i p_i^2 / m_i, \tag{2}$$

where m_i is the mass and p_i is the momentum (mass times velocity)
of the i-th particle. The total energy, when given as a function of
the coordinates and momenta of the nuclei and the electrons is the

Hamiltonian H, defined by

$$H = T + V \tag{3}$$

and introduced by Hamilton in the nineteenth century as a powerful
tool of classical mechanics. The quantum mechanical interpretation
of the Hamiltonian takes into account the wave-like character of all
elementary particles. Their motion becomes restricted in such a way
that the total energy can obtain only a set of restricted values,
called the eigenvalues, E_α , of the system, which correspond to the
wave-functions, or eigen functions, Ψ_α , of the Schrödinger wave
equation,

$$H \Psi_\alpha = E_\alpha \Psi_\alpha. \tag{4}$$

The eigen-values are discrete for any system of electrons and nuclei
confined in space, such as is any stable molecule, and are called
also energy levels. The transition from one level to another requires an adsorption or emission of an electromagnetic quantum. The
formalism by which H is converted from a function of coordinates and
momenta to a differential equation for the motion of wave-like
particles need not concern us at all, nor are we interested in
solving the Schrödinger equation for particular cases. Our interest
is to understand how does the quantum mechanical character of the
Hamiltonian determine the nature of molecular forces.

    The Born-Oppenheimer Approximation. The quantum mechanical
theory of molecular forces is based on an analysis of the
Schrödinger equation due to Born and Oppenheimer (1927), known as
the Born-Oppenheimer approximation. It is based on the fact that
electrons are lighter than nuclei by several orders of magnitude
(the mass of a proton is $\sim$ 2000 times the mass of an electron). The
Born-Oppenheimer approximation treats the Schrödinger equation in
two stages, or two levels: electronic and nuclear.

    At the first stage of the Born-Oppenheimer (BO) approximation,
which is the subject of all molecular orbital theories of quantum
chemistry, only the electrons are considered as dynamic variables
of the molecular Hamiltonian, while the atoms and nuclei are assumed
to be in fixed positions. The resulting eigenvalues represent the
electronic energy levels of a molecule as continuous functions of
the molecular coordinates, commonly known as Born-Oppenheimer energy

surfaces (BOES). Such energy surfaces were never derived in an analytical form, but are conceptually very important. The BOES of the
electronic ground state will be denoted by $V(\vec{r}^n)$, where $\vec{r}^n$ is a 3n-
dimensional vector which represents the Cartesian coordinates of
all atoms of an n-atom molecule. This is generally a very good approximation. This can be deduced from the fact that the frequencies
of electronic transitions are mostly orders of magnitude higher than
those of molecular vibrations. Consequently, the nuclei may be considered stationary on the time scale of electronic vibrations (Ex-
ceptional cases where this rule is not obeyed are known, but need
not be discussed here.).

The second stage of the Born-Oppenheimer approximation considers the quantum mechanical behavior of the nuclei. The molecular
Hamiltonian of the Schrödinger equation is now comprised of the
kinetic energy of the atoms as the differential operator, and of
the BO energy surface as the potential energy. Thus the BO potential
energy replaces the Coulomb energy as the potential of inter-atomic
interactions. In other words, an analytic representation of the BO
energy surface is necessary in order to obtain the quantum mechanical
dynamic behavior of the atoms in the molecules. The analytic representation of the BO surface is also necessary in order to obtain
the equilibrium structure of molecules, since it is that structure
for which the BO surface has a minimum value.

Quantum chemistry of semi-empirical and ab initio methods have
advanced during the years, improving the solutions of the Schrödinger
equation in the first stage of the BO approximation. They have become
applicable for medium size molecules of up to about 10-20 atoms,
with the aid of modern computers. By spending sufficient computer
time, it is possible now to obtain a numerical mapping of BO surfaces. However, quantum chemistry is incapable, at least in the
foreseeable future, to yield a mathematical analytical description
of BO surfaces of polyatomic molecules. Instead, the BO surface is
commonly represented by simple models, such as the particle in a
box, the rotating rigid top or the harmonic oscillator, for which
the Schrödinger equation can be solved.

Such solutions are applicable also for polyatomic molecules.
For example, the harmonic oscillator model gives basically correct,
though approximate, vibrational energy levels of a polyatomic molecule. It is based on the representation of the Born-Oppenheimer

energy surface of the polyatomic molecule around equilibrium by a
quadratic function of atomic coordinates. This simple representation
of the BO energy surface near its minimum point is used extensively
also in conformational analysis of polyatomic molecules. It assumes,
essentially, that the intramolecular forces near equilibrium resemble
elastic springs which keep the bond lengths and bond angles at their
equilibrium values. `

From this assumption stems the common use of the term "molecular mechanics" to describe the elementary use of empirical functions in conformational analysis. This similarity to macroscopic
bodies made of elastic springs created the common impression as if
empirical energy functions are "classical". In fact, empirical
methods of describing molecular energies are distinct from molecular
orbital methods of quantum chemistry not by being "classical", but
by their being empirical guesses not based on any deductive principle, whether classical or quantum mechanical.

The other distinction, which is of great importance for our
discussion is that while molecular orbital methods derive numerical
values for points on the BOES from first principles, empirical
methods offer an analytical description of the BOES as a whole,
based on physically plausible guesses. These are based on experimental information regarding static and dynamic properties of sets
of molecules of similar composition. Indeed, the low esteem in which
empirical functions are held in the eyes of some theoretical chemists
is not due to their being classical, but rather to their ad hoc and
arbitrary nature. This attitude is unjustified in general, since
empirical funtions have had many useful applications in conformational analysis (Ermer, 1976; Allinger, 1976; Dunitz and Burgi,
1975; Engler et al., 1973).

The possibility of representing the molecular energy, i.e. the
BOES, by empirical and semi-empirical functions of bond length, bond
angles, torsional angles and interatomic distances, rests on a
wealth of experimental data from most branches of chemistry, particularly from thermodynamics, spectroscopy and conformational analysis. We know that bond energies, bond lengths, bond angles etc.
have similar values in similar molecules, and these empirical facts
strongly indicate the following two closely linked general propositions. The first is concerned with additivity of potential energy
functions, the second with their transferability.

1. Empirical energy functions are additive. That is, the BOES (at least in the vicinity of the equilibrium point of minimum energy), can be expressed as the sum of a finite number of empirical functions of one, or at most very few, properly chosen variables, i.e.

$$V(\vec{s}) = \Sigma_i V(s_i) + \Sigma_{i,i'} V(s_i, s_{i'}) + \dots \qquad (5)$$

where $\vec{s}$ is a vector whose components, s_i, are all the bond lengths, bond angles, torsional angles and interatomic distances between non-bonded atoms.

2. Empirical energy functions are transferable. That is, the same atoms, bonds, bond angles, etc. have the same energy functions in different molecules of a sufficiently similar chemical structure.

Let us now see how the various energy potentials used in molecular calculations have been derived, either from theoretical quantum mechanical or from empirical considerations. We shall start with intramolecular potentials.

Intramolecular Potential Energy Functions

The Bond Potential. A notably successful guess of the analytic form of a BO surface of diatomic molecules is due to Morse (1929), who proposed the function now called the Morse potential, for the electronic ground state of a diatomic molecule as a function of its bond length b, over the whole range of inter-atomic distances, $0 < b < \infty$:

$$V_M(b) = D(\exp(-\alpha(b-b_o)) - 1)^2 - D, \qquad (6)$$

where D, α and b_o are adaptable constant parameters. $V_M(b)$ has a minimum, $-D$, at b_o and vanishes at infinity thus identifying D and b_o with the equilibrium bond energy and bond length, respectively. Morse solved the Schrödinger equation for the atomic nuclei (the second level of the Born-Oppenheimer approximation). He obtained, using only one more adaptable parameter, α , a good approximation for the whole molecular vibration spectrum of diatomic molecules, including its anharmonic overtones and combinations. It was a unique and conspicuous success of the merger of empirical and quantum mechanical considerations.

Although the Morse potential was derived for diatomic molecules only, it probably describes fairly well the potential of the chemical bond in polyatomic molecules as well. However, the variability of the chemical bonds has a very narrow range, of the order of 0.1 Å. Therefore, for applications to structural molecular biology, chemical bonds may be assumed as rigid, or else a quadratic (harmonic) approximation to the bond potential is amply sufficient,

$$V_{bond} = (1/2)K_b(b-b_o)^2, \qquad (7)$$

where K_b is related to the parameters of the Morse potential by $K_b = 2D\alpha^2$. This is not the case when vibrational spectra are concerned. Here the Morse potential is significantly superior over the harmonic potential (Hagler, Stern, Lifson, and Ariel, 1979; Lifson and Stern, 1982). We shall return to this point in the next chapter.

Bond-Bond Interactions. Bond angles variability is much more important to structural analysis of polyatomic molecules than the variability of bond lengths, but no theoretical basis for an analytical presentation of bond angle energy potentials has been offered as yet. However, quantum mechanics has given us a comprehensive conceptual insight into the nature of the forces which determine bond angles and their variability. Thanks to quantum mechanics we know that atoms tend to have closed valence electron shells; that bonds are formed by bond orbitals of valence electron pairs, one electron pair for each valence; that electrons in the closed valence shells which do not participate in chemical bond appear always in pairs, and that these "lone pair electrons" form "lone pair orbitals" which resemble to some extent bond orbitals. We know that the regions of lowest energy on the BO surface, which belong to the observed equilibrium structure of molecules, possess certain symmetry properties with respect to the bond orbitals, whose origin is the mutual repulsion between adjacent orbitals. We shall now see how these trends affect the bond angles and the torsional angles.

The angles between bonds connected to a given atom are essentially determined by the tendency of the bonds to be as far apart as possible. Thus, if only two bond orbitals are connected to a single atom, as in BeH_2, they will be collinear. Similarly, three orbitals connected to an atom are always coplanar, with bond angles of around $120°$. Each bond angle separately would tend to increase, but since it is impossible for one of the three bond angles in the

plane to increase without decreasing the others, the minimum of the
bending potential energy is reached when the three bond angles are
about equal. (The angles would be exactly equal for identical bonds,
as one would expect to observe in CH_3^+). Similarly, four bonds connected to an atom form a tetrahedral structure*, with all six angles
between the four bonds equal to the tetrahedral angle $109,47^{\circ}$ when
the four bonds are identical, as they are in CH_4. Deviations from
the tetrahedral angle occur when the four bonds are not identical,
as is the case in most tetrahedral carbon atoms. The deviations are
usually small, 2 to 3°. They become large if the molecule is strained
by other forces, as is the case in a number of ring molecules, like
cyclobutane, cyclodecane, and other cycloalkanes, where the closure
of the ring imposes significant deviations from the tetrahedral
angles, or in overcrowded molecules like tri-tertiarybutylmethane.
In the same way, when five bonds are connected to an atom, as in
pentavalent phosphorus compounds, three bonds are coplanar (equatorial) and two are perpendicular to the plane of the others (axial).
This is the highest symmetry available for such a structure, and the
fact that it is an energy minimum point can be derived from symmetry
considerations.

Deviations from these rules occur when lone pairs of electrons
form bond-like orbitals. For example, the two OH bonds in the water
molecule form an angle of 105°, but there are two lone pairs of
electrons which are connected to the oxygen atom and form "lobes"
which behave like chemical bonds in the sense that they repel each
other and the adjacent OH bonds. Thus, the two OH bonds and the two
lone pairs form a tetrahedral structure around the oxygen atom.
Similarly, the three NH bonds and the one lone pair of NH_3 form a
tetrahedral structure, so that NH_3 is pyramidal, not planar.

The Bending Potential. As we have seen, the qualitative nature
of the intramolecular bending energy may be considered to be well
understood. However, we are still unable to derive this energy quantitatively. Therefore, the common empirical representation of the
bending potential of a polyatomic molecule is given in terms of
the deviation of the bond angles θ from the "equilibrium value" or

---

*Tetrahedral structure means that the central atom is considered
  to be at the centre of a tetrahedron while the 4 atoms bonded to it
  are located at the vertices.

"reference value" θ_o, and is usually assumed to be quadratic, i.e. of the form

$$V_{angle}(\theta) = (1/2)K_{\theta}(\theta - \theta_o)^2, \qquad (8)$$

where K_{θ} is the "bending-force constant", a term taken from the theory of vibrational spectra, and θ_o is the reference angle. The reference angle θ_o is assumed to be 120° for coplanar trivalent structures and 109.47° for tetrahedral structures; alternatively, θ_o may be used as an adaptable parameter, like K_{θ}, where the values are adapted to give the best fit with the experimental data.

By definition, this empirical potential is good for small deviations from the equilibrium angle. For larger deviations, like those mentioned above for some cyclodecane molecules, it is not satisfactory. Unfortunately, various efforts to replace it by some better functions have not so far been particularly successful. It is, however, worth bearing these limitations in mind.

The Torsional Potential. When a bond connects two polyatomic groups in a molecule, as for example the C–C bond in CH_3–CH_3, it serves as an axis for the rotation of the two groups relative to each other. Such a rotation is called internal rotation, and the angle of internal rotation is called torsional angle (or also dihedral angle). The torsional potential is due to repulsive interactions between the orbitals of the two groups. Much is known about the torsional potential in hydrocarbons, mainly due to Pitzer (1951) in honor of whom it is often called "the Pitzer potential". The simple empirical form of the torsional potential is

$$V_{torsion}(\Phi) = (1/2)K_{\Phi}(1 + \cos n(\Phi - \Phi_o)), \qquad (9)$$

where Φ is the torsional angle, K_{Φ} is the height of the potential, n is its periodicity, and Φ_o is the reference angle where $V(\Phi)$ is maximum. For tetrahedral carbons, as for example, in the alkane chains ...–CH_2–CH_2–... occuring in many R groups of amino acids, K_{Φ} is about 3 kcal/mole and the periodicity is 3. The potential is very high, K_{Φ} = 40 kcal/mole and the periodicity is n = 2 for rotation around the double bond –C=C– as in ethylene and its derivatives. The peptide bond C´–N in proteins is known to have a partial double bond character and its torsional potential height is about 20 kcal/mole, with n = 2. On the other hand, the torsional potentials of the two other bonds of the backbone of proteins, N–C$^{\alpha}$ and C$^{\alpha}$–C$'$ are very low, because only the orbitals around C$^{\alpha}$ form a tetrahedral structure, while the orbitals around C$'$ and N are planar, the first because of the double bond C$'$=O, and second because of the partial double bond character of the C$'$–N bond. In fact, little is known about these torsional potentials, and even the periodicity and location of the minima are disputable.

   Out-of-Plane Torsion. When three bond orbitals are confined to be planar, as are the bonds around the ethylenic double-bonded carbons, or the bonds around the C$'$ and N which form the planar peptide unit, there is a force which resists the distortion of the planar structure. The out-of-plane torsion can be envisaged as the angle χ between the planes through the points (1,2,3) and (1,2,4) in the structure $1 = 2 < {}^{3}_{4}$, which intersect along the axis $1 = 2$. The potential of such out-of-plane distortion may be represented, for small angles, by

$$V_{\chi} = (1/2)K_{\chi}\chi^{2}. \qquad (10)$$

Spectroscopic analysis of ethylenes tells us that the stiffness of V_{χ} is about the same as that of V_{ϕ} for rotational torsion of the C = C double bond. It has been noted that the peptide bond in proteins is often distorted, i.e. values of the torsional angle ϕ of the peptide bond up to around 10° may occur as a result of internal strains in the molecule. Out-of-plane distortions of the peptide bond are as likely and as abundant as those of the torsional distortions.

   Cross Terms of Geminal Interactions. We have listed hitherto energy functions of single variables: bond lengths, bond angles, torsional angles and out-of-plane torsions of planar orbitals. However, as indicated by Eq. (5) which defined the additivity propositions, functions of single variables are not sufficient for a proper presentation of the BO energy surface. And indeed it was found that the vibrational frequencies of molecules are only crudely calculated by single variable terms, with errors mounting to few hundred of wave numbers (cm^{-1}). A big improvement is obtained when functions of two adjacent variables are added to the list, such as two adjacent bonds, or a bond and a bond angle, and the like. These functions represent the geminal interactions namely the interactions between adjacent orbitals, which are functions of the two bond

lengths and the angle between them. Such functions are supposed to
improve the representation of the BOES, but we have a very poor
theoretical information about their nature. Thus the only way left
is to assume for them a simple, though arbitrary, analytic form,
and fit their constant parameters to available spectroscopic data.
There are two alternative forms for these functions. One is a bi-
linear function of adjacent variables,

$$V_{geminal} = K_{xy}(x-x_o)(y-y_o),\tag{11}$$

where x and y represent any internal variables. The other form, due
to Urey and Bradley (1931), is to represent the geminal interaction
by a function of the distance between the geminal atoms. For
further comments on the geminal interactions see the next chapter.

   The Non-Bonded Intramolecular Potential. Interactions between
groups of atoms in a large molecule which are close to each other
but not linked by chemical bonds constitute the last contribution
in our list of intramolecular interactions. These are, however, not
different by their nature from the intermolecular interactions.
Indeed the study of intermolecular interactions in gases and crystals
has been a rich and reliable source of information about these interactions, since in the study of intermolecular interactions, the
molecular structure may be considered in many instances to be rigid
so that the intramolecular energy remains constant.

Intermolecular Potential Energy Functions

   Intermolecular interactions, like the intramolecular interactions, are of electrostatic origin. We distinguish between electrostatic interactions between charged or polar groups, which obey the
Coulomb law, and other interactions which exist always, even in
completely nonpolar molecules. The latter are often called van der
Waals interactions because van der Waals was the first to estimate
them quantitatively in his famous study of the deviations of real
gases from the Boyle-Mariotte law of ideal gases. Van der Waals interactions are repulsive at the short range of atomic contact, and attractive at long range.

   The Attractive Dispersion Force. The quantum mechanical theory
of the long range attraction was derived by London (1930), and

although its mathematical derivation is rather elaborate, its
physical basis and qualitative explanation are rather simple and
clear. Consider two electro-neutral atoms at a distance long enough
so that their electron clouds do not overlap. At first approximation
the electrostatic force between them is zero. However, the electrons
of both atoms are in constant motion and therefore possess instan-
taneous, fluctuating, dipole moments whose magnitude depends on the
atomic polarizabilities. The fluctuating dipoles of the two atoms
interact with each other, thus producing a net attractive force.
London showed that the energy function of this attraction at a
distance r has the form of the power series

$$V_{London} = -C/r^6 - C'/r^8 - C''/r^{10} - \ldots \, , \tag{12}$$

where the first term is the leading one at large enough distances.
An extensive literature exists on the so-called dispersive or London
attraction potential. Various theoretical derivations for the coefficient C have been proposed, and the one best fitting to experiments
on noble gases (within 15-20%) is that due to Slater and Kirkwood
(1931):

$$C_{a,b} = (3e\hbar/2 \, m_e^{1/2}) \, \alpha_a \alpha_b ((\alpha_a/N_a)^{1/2} + (\alpha_b/N_b)^{1/2})^{-1} \, , \tag{13}$$

where e, $\hbar$ and m_e are universal constants, a and b denote the kinds
of atoms, α their polarizabilities, and N their number of electrons
(preferably only p electrons) in the outer shell. London´s theory is
strictly speaking applicable to pairs of atoms with spherically
symmetric distribution of electrons. However, its importance goes
far beyond this limitation. Since dispersion forces are pair-wise
(approximately) additive, they are the basis of our understanding
of the intermolecular forces which hold molecules of solids and
liquids together, as well as the intramolecular forces between nonbonded atoms in polyatomic molecules. It has been consequently generally assumed, although never rigorously proved, that dispersion
interactions in polyatomic molecules are given by the sum of such
interactions between nonbonded atoms, both between the molecules and
inside each molecule. In the molecule atoms may be defined as nonbonded, for both the attractive and the repulsive interactions, if
they are separated by at least 3 or 4 consecutive bonds. The choice
between 3 and 4 is in fact arbitrary, however if it is 3 then its
contribution to the torsional barrier (Eq. (9)) has to be recognized
(see also next chapter).

From Eq. (13) it follows that the coefficients $C_{a,b}$ obey a combination rule which relates a-b interactions between atoms a and b to a-a interactions, i.e.

$$2 \alpha_a \alpha_b / C_{a,b} = \alpha_a^2 / C_{a,a} + \alpha_b^2 / C_{b,b} \cdot \tag{14}$$

The application of combination rules is essential when the coefficients are determined by empirical methods. Consider, for example, only the most frequent atoms in bio-molecules, H,C,N,O,S,P. Without a combination rule we would need 21 instead of 6 empirical parameters, and since these are not independent by their nature, they cannot be uniquely defined by empirical methods. The common practice in potential energy calculations has been to use a geometric mean combination rule

$$C_{a,b}^2 \doteq C_{a,a} C_{b,b} \cdot \tag{15}$$

Its practical advantage is that it avoids the use of the polarizabilities α ; its disadvantage is that it has no theoretical justification and that it is inaccurate. Kramer and Herschbach (1970) have compared the two combination rules with a set of accurate published calculations of 153 unlike coefficients. The root mean square deviation from experiment was 3.25% when Eq. (14) was used, but 73.5% when Eq. (15) was used! Whether the use of Eq. (15) is justified depends on estimates of the overall accuracy of the various contributions to the total set of potential energy functions, a subject to which we shall return in the next chapter.

The Repulsive Force. Short range repulsion forces arise whenever atoms or molecules come so near each other that their electron clouds interpenetrate. Quantum mechanics offers a satisfactory explanation for such repulsions in terms of Pauli´s exlusion principle and of perturbation theory, but no formula of general applicability is available. It is commonly accepted that the leading factor implied by theory is exponential, so that a simple form for the repulsive energy function is

$$V_{repulsive} = A \exp(-br). \tag{16}$$

The constants A and b are, however, not derivable in terms of atomic properties, and are therefore indeterminated.

Moreover, since the repulsive potential is very steep, and it is very difficult to isolate it from other interactions experimentally, the determination of A and b are not a simple matter. Lennard-Jones (1924) avoided this difficulty in his classical study of the van der Waals interactions, by proposing the empirical formula for such interactions, now known as the Lennard-Jones potential,

$$V_{LJ}(r) = A/r^n - C/r^6, \tag{17}$$

where A and n are empirical constants, and C is the dispersion coefficient. Lennard-Jones examined various values of n by fitting A and C to experimental data, and found no distinction between different values of n, or between A/r^n and the exponential form of Eq. (16), provided the calculated repulsion is sufficiently steep. Eventually, n = 12 was chosen (Hirschfelder et al., 1964) for computational convenience, and this form of the potential became the most common in energy calculations. The common combination rule for A is similar to that given in Eq. (15) for C, namely the geometric mean,

$$A_{a,b}^2 = A_{a,a} A_{b,b}. \tag{18}$$

The heuristic nature of this potential is evident in another way. It is not rigorously correct just to simply take C/r^6 for the attractive term, since this is the leading term in Eq. (12) for dispersion interactions only for large distances where A/r^n is negligible, while at contact distaces, where the interaction energy is highest, other dispersion terms might be significant. Nevertheless, Eq. (17) has been proven to be extremely useful in a wide range of applications, particularly when A and C are considered empirically adaptable parameters, since it represents faithfully the essential features of nonbonded interactions, namely steep repulsion at short distances, minimum energy at an intermediate distance, and r^{-6} behaviour at long distances. If we denote by r^* the distance for which V_{LJ} is at minimum, and by $-\varepsilon$ the value of this minimum, the Lennard-Jones potential takes the form

$$V_{LJ}(r) = \varepsilon (n-6)^{-1} (6(r/r^*)^{-n} - n(r/r^*)^{-6}), \tag{19}$$

which is less handy for computations but has the advantage that its parameters r^* and ϵ are more directly related to the observable properties of molecular systems, and therefore their numerical values are more meaningful. Also the combination rule $r^*_{a,b} = (r^*_{a,a} + r^*_{b,b})/2$, in conjunction with Eq. (14), is perhaps more physically justified than the geometric average for A (Eq. (18)), although this suggestion is still awaiting an empirical test.

The Three-Body Force. Intermolecular forces are commonly believed to be pair-wise additive. That is, the energy of an assembly of atoms or molecules is considered equal to the sum of the interaction energies of all pairs of particles. However, this assumption has been challenged. It is certainly not strictly valid for condensed systems, namely also for systems of interest in molecular biology. The reason is that when an atom is polarized simultaneously by two other neighbouring atoms, its polarization is proportional to the vectorial sum of the electrostatic forces. Thus the dispersion energy of three neighbouring atoms must depend not only on the three interatomic distances but also on the three angles between them. According to Axilrod and Teller (1943), this three body interaction potential, namely the correction to the pairwise additive potential energy of the dispersion interaction, is

$$V_3 = C_{abc}(1 + 3\cos\theta_a \cos\theta_b \cos\theta_c)(r_{ab}r_{ac}r_{bc})^{-3}, \qquad (20)$$

where C_{abc} is a constant and the r's and θ's are the sides and angles of the triangle formed by the atoms a,b and c. While it is common practice to neglect this term in calculations of nonbonded interactions, we should realize that they are by no means always negligibly small. For example, Kestner and Sinanoglu (1963) concluded that three body forces reduced the dispersion interaction between base pairs in the DNA double helix by 28%.

Electrostatic Interactions. Most chemical bonds are not purely covalent, but possess a partial electrostatic character. Whenever the electronegativities of two bonded atoms are different, the charge distribution of the bonding electrons is shifted partly towards the more electronegative atom, thus making the total charge (nucleus and electrons) in the vicinity of this atom equivalent to a partial negative charge, while the less electronegative atom obtains a partial positive charge. The detailed charge distribution may be obtained from the wave functions of the Schrödinger equation. However, the calculation of electrostatic interactions from con-

tinuous charge distributions is extremely cumbersome, involving
multidimensional integrals. Since the available solutions to the
Schrödinger equation are approximate and so are the charge distributions, there is no reason to attempt such calculations. A major
simplification of the problem is obtained if the continuous charge
is represented approximately by point charges located on the atoms.
This is done, for example, by the so-called Mulliken population
analysis (Mulliken, 1955), where the charges in a given molecular
orbital are assigned to the individual atoms, according to the extent to which the atomic orbitals of each atom contribute to the
molecular orbital. There is a vast literature on the Mulliken analysis of various molecules, and also on other similar methods. Unfortunately, the partial charges thus obtained depend heavily on
the method or the type of approximation employed. For example,
Mulliken population analysis of ab initio calculations yields widely
different values of the partial charges when different basis sets
are used in amides (Hagler and Lapiccirella, 1976) and carboxylic
acids (Hagler and Lapiccirela, 1976; Lifson et al., 1979). They all
indicate, however, that electrostatic interactions between partial
charges are an important part of the nonbonded interactions, in
particular in molecules such as are the subject of structural molecular biology, in which polar bonds and groups are abundant.

The electrostatic potential energy of interaction between the
partial charges q_i is given by the Coulomb law:

$$V_{electrostatic} = (1/2) \, \Sigma_{i,j} q_i q_j / r_{ij} , \qquad (21)$$

where r_{ij} is the distance between atoms i and j. For intermolecular
interactions the summation is obviously over the atoms i of one molecule and j of the other. For intramolecular interactions one is
confronted with a difficulty: If the summation would extend over all
pairs, i,j (i≠j), the electrostatic forces would be considered as
if they acted even between bonded atoms, over and above the bond
potential. On the other hand, if the definition of non-bondedness
for van der Waals interactions is adopted also for electrostatic
interactions, namely, the summation extends over atoms separated by,
say at least 3 consecutive bonds, then the electrostatic attractions
between opposite charges and repulsions between similar charges, do
not balance properly (for example this rule would retain only H...H
repulsions in ethane). There is not as yet, an agreed solution for
this difficulty.

Equation (21) represents electrostatic interactions in vacuum, i.e. the dielectric constant has been given the value D = 1. Some authors have used D > 1, arguing that interactions in a polarizable medium are weaker than in vacuum. This argument seems, however, to be based on a misconception. When the charge-carrying atoms constitute the medium, then their interaction must be considered as interaction in vacuum. Thus intramolecular interactions or intermolecular interactions in crystals must be represented by Eq. (21) with D = 1. On the other hand intermolecular electrostatic interactions between molecules dissolved in a solvent of dielectric constant D must be reduced by a factor D if (and only if) the solvent is considered as a continuous medium. The problem of the meaning of a dielectric medium at microscopic distances is very complex and could not be discussed here.

The following "Gedanken Experiment" will support this view: consider a molecular crystal whose molecules carry atomic partial charges q_i; assume that the molecules have no permanent dipole moments. The dielectric constant D of the crystal is then determined only by the polarizability of the molecules.

Imagine now that it would be possible to freeze the polarizability and then release it at will. When it is frozen then D = 1 and when it is released then D > 1. In the first case Eq. (21) is certainly valid. In the second case the electrostatic energy would be divided by D > 1. It would then be smaller if the total sum would be positive, and larger (smaller in absolute value) if the sum would be negative. However, when the frozen polarizability is released, the crystal adapts spontaneously to the gained freedom, and may respond only by decreasing the total electrostatic energy in both cases!

2. EMPIRICAL FUNCTIONS AND OBSERVED MOLECULAR PROPERTIES

Molecular Systems: Additivity and Transferability

Summing up the above discussion of the various intra- and intermolecular energy potentials we shall state now the important though seemingly self-evident assumption, that the total potential energy of a molecular system is the sum of all the various potential energy functions reviewed above, namely that

$$V_{total} = V_{bond} + V_{angle} + V_{torsion} + V_{out-of-plane} +$$

$$V_{geminal} + V_{LJ} + V_{electrostatic} + \ldots , \qquad (22)$$

where the summations are extended over all bond lengths, bond angles, torsional and out-of-plane torsion angles, and over all inter-atomic distances of nonbonded atoms.

In fact the additivity of these single variable or two variables potential functions is only approximate, otherwise we would not insert the three dots, which cover-up the shortcomings of the assumption of additivity. These dots indicate the possibility of adding more functions to the list, such as the energy due to atomic polarizabilities or three body forces, but such additions complicate an already complicated situation of having to deal with a large number of empirical functions with adaptable, or indeterminate parameters.

As we shall see below, the choice of functions which comprize the total energy V_{total} must depend on the experimental resources from which potential functions are determined empirically, as well as on the purpose to be served by using V_{total}. In as much as our interest is focussed on using V_{total} for the study of biological macromolecules, it is my personal prejudice that neither polarizabilities nor three body forces can be effectively and usefully incorporated.

The concept of transferability of potential functions is closely linked with that of additivity, as already noted in the preceding chapter. Transferability means that the energy functions which give a satisfactory description of the physico-chemical properties of one molecule may be used with confidence to describe the properties of any other molecule of similar structure. Thanks to the transferability of potential functions it is possible to determine them by studying simple, small-size molecules, and then use them in complex, large molecules such as proteins and nucleic acids. However, in doing so we should always remember that transferability, like additivity, is an empirical and approximate rule, valid only if the concept of similar structure is used properly.

The Calculation of Observable Properties

The first major application of inter-molecular energy functions
was related to the statistical mechanics of real gases. It was one
of the great break-throughs of statistical mechanics that the
"second virial coefficient" was derived in terms of inter-molecular
interaction functions, such as the Lennard-Jones potential. A large
literature has been devoted to this subject which was focussed, however, only on the simplest molecules, mainly noble gases. The reason
was that intermolecular distances in fluids (gases and fluids) are
in constant dynamic change, and therefore the inter-molecular forces
are related to properties of fluids by a statistical averaging
process which erases the details of the functional form of the intermolecular or interatomic interactions. Thus potential energy functions are suitable for use in statistical mechanical calculations,
however their validity must be judged by other observable properties,
more directly related to the functional form of the potentials. Such
observables are best introduced and discussed by considering the
main features of the total potential energy function V_{total} (Eq.
(22)) as a whole, ignoring at this level the details of its components.

V_{total} of any molecular system is naturally a function of its
structure. The structure may be represented in various ways. One
way, very useful for computer calculations, is to specify the
Cartesian coordinates of all the atoms which comprize the system.
If the system is a single molecule, whether small or large, the
positions of its n atoms are specified by 3n Cartesian coordinates,
which we shall denote simbolically by a single vector $\vec{r}$. If the
system is, say, a crystal, then $\vec{r}$ may be considered as representing
the atomic coordinates of the atoms in one unit cell together with
the translation vectors of the unit cell. Another way to represent
the molecular systemm is by specifying the internal coordinates
(bond lengths, bond angles, torsional angles), from which all interatomic distances may be calculated by standard mathematical devices.
Whatever representation we choose, $\vec{r}$ represents the detailed structure of the molecular system. Now consider the molecular system to
be at equilibrium. Then its vector $\vec{r}$ has an equilibrium value, $\vec{r}_o$,
for which the total energy function $V_{total}(\vec{r}_o)$ is at its minimum.
Let us now develop V_{total} in a Taylor series around $\vec{r}_o$:

$$V_{total}(\vec{r}) = V_{total}(\vec{r}_o) + \Sigma_i (\partial V_{total} / \partial r_i)_o \delta r_i +$$

$$(1/2) \Sigma_i \Sigma_j (\partial^2 V_{total} / \partial r_i \partial r_j)_o \delta r_i \delta r_j + \dots . \qquad (23)$$

It is now common to write Eq. (23) in the notation of vectors and matrices, which is convenient because it is compact and avoids the cumbersome multiple summation notation:

$$V_{total}(\vec{r}) = V_{total}(\vec{r}_o) + grad\ V_{total}(\vec{r}_o)\ \delta\vec{r} +$$

$$(1/2)\ \delta\vec{r}'\underset{\sim}{F}\ \delta\vec{r} + \dots . \qquad (23')$$

In the above equations r_i and δr_i are the components of the vector $\vec{r}$ and of small deviations from $\vec{r}_o$, respectively; grad V, or the gradient of V_{total} is a vector whose components are $\partial V_{total} / \partial r_i$; and $\underset{\sim}{F}$ is the so called Hessian matrix, whose elements are the second derivatives of V_{total}, i.e.

$$F_{ij} = \partial^2 V_{total} / \partial r_i \partial r_j .$$

A close examination of this expression will show a whole world of applications to energy calculations of molecular properties.

Molecular and Crystal Energies and Thermodynamic Functions. $V_{total}(\vec{r}_o)$, the first term in the Taylor expansion (Eq. (23)), is the molecular energy at equilibrium. The corresponding experimental, measurable property is the heat of atomization (or some related functions such as the heat of formation, or the heat of combustion) of the system. In comparing the two, we have to take into consideration their differences. The heat (or enthalpy) of atomization is a thermodynamic function related to thermal agitation and temperature, while $V_{total}(\vec{r}_o)$ represents the energy of a single microscopically defined state. The heat of atomization's major component is the quantum mechanical energy of formation (or dissociation) of the chemical bonds, which is represented by Morse potentials for the various types of bonds (Eq. (2)) as well as by other terms of intra-molecular interactions included in V_{total}. It includes, however, other contributions which must be taken into account. First, there is the zero-point energy, which is the vibrational energy of the lowest vibrational quantum state corresponding to zero absolute temperature. Then there are the energies of molecular translations,

rotations and vibrations in the gas phase, or the molecular and lattice vibrations in the solid phase. These quantities are functions
of thermodynamic variables such as temperature. They can be calculated approximately from V_{total}, as we shall see below.

In many instances we are interested in the changes in energy
accompanying conformational or structural changes in the molecular
system. Examples are the cis-trans difference of the peptide bond,
the chair-half-chair-boat conformational transition of sugar rings,
or the trans-gauche transition in alkanes. In such cases, the zero-
point energy and the other thermodynamic contributions may cancel
out to a good approximation, and the correspondence of such transitions as calculated from differences in $V_{total}(\vec{r}_o)$ to experiment
is then simpler.

When intermolecular interactions are included in V_{total}, it is
possible to consider $V_{total}(\vec{r})$ as representing macroscopic condensed
phases, i.e. solids, liquids and solutions. A very useful and simple
example is the lattice energy of crystals, or the sublimation energy,
i.e. the energy of crystal-to-gas transition. For rigid molecules,
which maintain the same structure in gas and solid phases, the sublimation process involves work against inter-molecular forces only.
The intra-molecular energy and the enthalpy of molecular vibrations
remain invariant. Therefore, the heat of sublimation is an important
source of information on inter-molecular energy potentials. Note,
however, that here again, when energy calculations of crystal packing
are compared with experimental heat of sublimation, enthalpy corrections, for translations, rotations and the pV term in the gas
and for lattice vibrations in the solid, are required. Such corrections may be estimated satisfactorily by classical thermodynamics
(Hagler, Huler and Lifson, 1974).

Equilibrium Structure of Molecules and Crystals. The equilibrium
structure of molecules, as well as that of molecular crystals is
calculated from the second term of the Taylor expansion of $V_{total}(\vec{r})$,
equation (23). Since the equilibrium structure is the structure for
which the energy function V_{total} obtains a minimum value, all derivatives of V_{total} with respect to the components of $\vec{r}$, namely the
gradient of V_{total} (denoted grad V_{total}) must vanish at $\vec{r}_o$. The set
of equations

$$\partial V_{total}(\vec{r})/\partial r_i = 0 \qquad i = 1,\ldots,3n\ , \tag{24}$$

or in short gradV = 0 may therefore be used to solve for $\vec{r}_o$. Computer programs are available to obtain numerical solutions of Eq. (24), and various algorithms for such solutions have been studied extensively. The general idea of all such algorithms is the same: One starts at any non-equilibrium value $\vec{r}$ and calculates grad V_{total}. If the gradient does not vanish, $\vec{r}$ is changed to $\vec{r} + \delta \vec{r}$, where $\delta \vec{r}$ is calculated in such a way as to move $\vec{r}$ in the direction of $\vec{r}_o$. The process is iterated until finally the minimum is reached to the desired precision.

The choice of the appropriate minimization method is very important in energy calculations. The simplest method of solving Eq. (24) is the "steepest descent" method. At each iteration the vector grad V_{total} is calculated, and a "line search" is performed along the direction of the gradient. That is to say that the value of V_{total} is calculated at several points along the direction of the gradient, and a minimum point is obtained by interpolation between these points. The method is "stable", in the sense that each iteration leads to a lower value of V_{total}. However it usually gets stuck by progressively slow convergence.

Fast convergence is obtained if one considers the expansion of grad V_{total} in a Taylor series around the (yet unknown) point of minimum $\vec{r}_o$. In matrix notation such expansion is written as

$$\text{grad } V_{total}(\vec{r}_o) = \text{grad } V_{total}(\vec{r}) + \underset{\sim}{F}(\vec{r})\delta \vec{r} + \dots , \qquad (25)$$

where, just as in Eq. (23'), $\delta \vec{r} = \vec{r}_o - \vec{r}$ and $\underset{\sim}{F}$ is a matrix whose elements are $\partial^2 V_{total}(\vec{r})/\partial r_i \partial r_j$. Since grad $V_{total}(\vec{r}_o)$ is zero by definition of $\vec{r}_o$ as the point of minimum, one obtains an equation for $\delta \vec{r}$:

$$\delta \vec{r} = -\underset{\sim}{F}^{-1}(\vec{r})\text{grad } V(\vec{r}) . \qquad (26)$$

This method, called the Newton–Raphson method, yields the minimum in one step, provided the higher terms in the Taylor series are negligible, which is approximately true for points near the minimum of any well behaved function. Therefore, starting near the minimum, a few iterations of Eq. (26) converge progressively to the minimum. Such a convergence is called "quadratic convergence" since for a quadratic function the higher terms are precisely zero and the exact minimum is reached by Eq. (26) in one step. There are however dif-

ficulties with this algorithm. First, when $\bar{r}$ is far away from the
minimum the method may be unstable, namely $\bar{r}$ derived by Eq. (26)
may move $\bar{r}$ away from minimum. Second, the inverse of $\bar{F}$ does not
exist if $\bar{F}$ is singular, so that $\underset{\sim}{F}^{-1}$ in Eq. (26) must be replaced by
$\underset{\sim}{F}^{+}$, the "generalized inverse" matrix (Fletcher, 1968; Warshel and
Lifson, 1970). Third, when V_{total} is a function of a great number
of variables, as is the case for V_{total} of biological macromolecules
the calculation, storage and manipulation of the matrix $\underset{\sim}{F}$ becomes
technically unmanageable.

To overcome these difficulties, new methods have been developed
which combine the advantages of the steepest descent and the Newton-
Raphson methods but avoid their pitfalls. Such methods are both
stable and fast, namely quadratically, converging. Among these the
most suitable for energy calculations in molecular biology is the
so-called "conjugate gradient" method (Fletcher and Reeves, 1964).
Its great advantage is that like the steepest descent method its requirements for computer memory space is proportional to the number
of variables (3n for a molecule of n atoms), while other methods,
like the Newton-Raphson, require memory space proportional to the
square of the number of variables. We shall explain here the basic
ideas behind the "conjugate gradient" method, while skipping the
mathematical details.

The method starts at the first iteration like the steepest
descent, by a "line search" along the direction of the gradient.
At the second, third, or in general at the i´th iteration, the new
direction of the line search is determined by a linear combination
of the gradient at the present point (i) and the direction of line
search at the previous point (i-1). The coefficients of the linear
combination are chosen, very cleverly, such that if the function
to be minimized is quadratic to a good approximation then two major
goals are achieved. 1) Each new direction of search is linearly independent of all previous directions of search. 2) The minimum along
the direction of the line search at each iteration is also the
minimum of the function in the whole subspace spanned by all previous
directions. That is to say, after i iterations we obtain the minimum
of the function in the subspace spanned by the first i directions.
Therefore, if V is a quadratic function of N variables, the total
minimum must be reached in exactly N steps. The fast convergence
and modest requirements for memory space make the conjugate gradient
method a best choice for minimizing the energy of proteins and other
biopolymers.

Vibrational Modes in Molecules and Crystals. We come to the
third term in the expansion of the energy in a Taylor series around
equilibrium, Eq. (23). It represents the energy of the system due
to small deviations of the (atomic or internal) coordinates from
their equilibrium values. Such deviations must exist, because the
uncertainty principle of quantum mechanics implies that molecules
possess a vibrational energy even at absolute zero temperature (zero
point energy).

The derivation of vibrational energy of molecules and crystals
is linked to the concept of normal modes of vibrations. Consider a
single polyatomic molecule in a gas, composed of n atoms. Each atom
has three degrees of freedom, therefore the molecule has 3n degrees
of freedom. Molecular translations and rotations use up 3 degrees
of freedom each. The remaining 3n-6 are vibrational degrees of freedom, in which the atoms participate collectively, since an atom
cannot move within the molecule without exerting forces on its
neighbours. Thus to each vibrational degree of freedom there corresponds a collective, "normal", or "fundamental" mode of vibration
with a characteristic frequency.

The calculation of these normal modes can be performed in the
"harmonic approximation" by using the third (last) term of Eq. (23).
The higher, neglected terms would then supply the "anharmonic corrections" to the harmonic normal modes.

There is an extensive literature on the normal mode analysis
of polyatomic molecules. The classical treaties by Herzberg (1945)
and by Wilson et al. (1955), make use of symmetry properties of
polyatomic molecules and the theory of group representations. However, the application of potential functions to normal mode analysis (Lifson and Warshel,1968) is made much simpler and straightforward, mostly thanks to the availability of fast computers with
large memory space. Symmetry properties of the system are neither
needed nor assumed. Rather, they are derived for both the equilibrium structure and the normal vibrations around equilibrium.

We give here a brief summary of the derivation of normal modes
from potential energy functions in its simplest form. The reader is
referred to Lifson and Warshel (1968) for further details. The potential
energy of vibrations around the equilibrium conformation $\vec{r}_0$ is taken
from Eq. (23), omitting the second term since it vanishes at equi-

librium. It is presented in matrix notation by

$$V_{total}(\bar{r}_o + \delta\bar{r}) - V_{total}(\bar{r}_o) = (1/2)\delta\bar{r}'\underset{\sim}{F}(\bar{r}_o)\delta\bar{r} + \dots . \quad (27)$$

The vector $\bar{r}_o$ of the equilibrium structure is represented here in
Cartesian coordinates. Its 3n components are the x,y and z coordinates of the n atoms of the molecule, chosen in any convenient
order; since $\delta\bar{r}$ represents the deviations from equilibrium during the
vibrational motion, it is a function of time. The row vector $\delta\bar{r}'$
is the transpose of the column vector $\delta\bar{r}$; $\underset{\sim}{F}(\bar{r}_o)$ is the matrix of
the second derivatives of V_{total} at $\bar{r}_o$. Cartesian coordinates have
to be used here because only in these coordinates is the kinetic
energy given in the simple form

$$T(\delta\dot{\bar{r}}) = (1/2)\delta\dot{\bar{r}}'\underset{\sim}{M}\delta\dot{\bar{r}} , \quad (28)$$

where $\dot{\bar{r}} = d\bar{r}/dt$ is the vector of atomic velocities. $\underset{\sim}{M}$ is a diagonal
matrix of the atomic masses, each atomic mass appearing thrice, once
for each Cartesian coordinate of that atom.

From the potential and kinetic energies, the simultaneous
equations of motion for all atoms in the molecule are readily obtained:

$$\underset{\sim}{M}\cdot\delta\ddot{\bar{r}} + \underset{\sim}{F}(\bar{r}_o)\delta\ddot{\bar{r}} = \bar{0} \quad (29)$$

where $\delta\ddot{\bar{r}}$ is the vector of atomic accelerations. This is in fact a
very simple form of Newton's equations of motion. It says that for
each atom i the acceleration $\delta\ddot{r}_i$ times the mass M_i is equal to the
force $-\Sigma_j F_{ij}\delta r_j$ excerted by all other atoms. A normal mode of vibration with a frequency ν and a vector amplitude $\delta\bar{r}_o$, whose components are the amplitudes of the atoms participating in the normal
mode, is expressed in vector notation by $\delta\bar{r} = \delta\bar{r}_o\exp(2\pi i\nu t)$. The
corresponding acceleration vector is $\delta\ddot{\bar{r}} = -(2\pi\nu)^2\delta\bar{r}_o\exp(2\pi i\nu t)$.
When these expressions are inserted in Eq. (29), a set of algebraic
equations

$$(\underset{\sim}{F}(\bar{r}_o) - (2\pi\nu)^2\underset{\sim}{M})\delta\bar{r}_o = \bar{0} \quad (30)$$

is obtained for the amplitudes of the normal modes and their corresponding frequencies. In the standard mathematical language this
is a typical "eigen value problem". Since the order of the matrices

is 3n there must be 3n solutions. Each solution yields a characteristic vector, or "eigen vector", which is the vector-amplitude of
a normal mode. To each "eigen vector" there belongs an "eigen value"
representing the square of the characteristic frequency of this
normal mode. Six of the 3n frequencies are zero, corresponding to
the 3 transitions and 3 rotations of the molecule, and the remaining
3n-6 are the molecular harmonic frequencies. Once the eigen vectors
are derived in Cartesian coordinates, they can and should be transformed to internal coordinates where the symmetry properties of the
normal modes are easily recognized.

This formalism for calculating vibrational normal modes is
particularly suitable for large computers. It requires one and the
same program for any molecule, provided sufficient computer space
and time are available. It is, however, much more than a convenient
formalism. Its main innovation is that it links together the calculation of the static properties (the equilibrium energy and the
equilibrium structure), and the dynamic properties (the vibrational
modes), and brings out their intrinsic interdependence.

Molecular vibrations contain much information about the molecular energy surface in the whole vicinity of the equilibrium conformation, since they depend on the curvature of the energy surface
around equilibrium. Furthermore, the high accuracy of infrared and
Raman spectra of polyatomic molecules reflects itself in high accurate information on the changes of the BO energy surface with the
vibrational motion of the atoms. Thus a common accuracy of a measured frequency to within, say, 1 cm^{-1}, correspond to an energy
change of only 3 cal/mole. Therefore, the representation of V_{total}
as a sum of functions of single variables, which may do for conformational calculations, must be supplemented for the calculation of
vibrational spectra by functions of adjacent internal coordinates.

We discussed hitherto the variables of single molecules. By extending the potential energy V_{total} to include the lattice energy
of molecular crystals, the vibrations of molecules in a lattice are
obtained (Warshel and Lifson, 1970). The 6 degrees of freedom of molecular translations and rotations in the gas phase are replaced in
condensed phases by lattice vibrations and molecular librations.
Further, some molecular vibrations are modified by interactions with
neighbour molecules, and frequencies of degenerate modes are split
by the asymmetric environment of the molecules in the crystal.

Thermodynamic Functions. The thermodynamic energy, or enthalpy, of a system in the gas phase contains, as noted already, V_{total} as well as the energies of molecular translations, rotations and vibrations. In condensed systems, from solid to living cells, the vibrational energy is the main component of the energy which varies with temperature, and the main contributor to the thermodynamic functions such as enthalpy, free energy, specific heat and thermal expansion.

According to the equipartition law of classical statistical thermodynamics, the kinetic and potential energies of translation, rotation and vibration are equally distributed over all degrees of freedom of a system, each obtaining kT/2 of energy at a temperature T. Purely kinetic degrees of freedom like molecular translations, molecular rotations and free (unhindered) internal rotations obtain kT/2, while molecular vibrations, including hindered rotations, obtain 1 kT (half for kinetic and half for potential energy). For example, alanine has 13 atoms, i.e. 39 degrees of freedom. Assuming free rotation around Φ and Ψ and hindered rotation around the $C^{\alpha}-C^{\beta}$ bond, 8 degrees of freedom are purely kinetic (3 translations, 3 rotations and 2 internal rotations) and 31 degrees of freedom are vibrational. Therefore, alanine has, by classical theory, an energy of (4 + 31) kT per molecule in the gas phase due to translations, rotations and vibrations, namely over and above its equilibrium potential energy $V_{total}(\vec{r}_o)$. According to classical statistical thermodynamics, the energy of a system is independent of its vibrational frequency distribution. However, the law of equipartition is not supported by experiment. The recognition of this fact was strongly linked to the origin of the quantum theory. It led Planck to suggest his theory of black body radiation, and it led Einstein and Debye to produce the theory of specific heats of solids, all of which came to replace the equipartition law.

According to quantum statistical thermodynamics (Hill, 1960), the contribution of each normal mode of frequency ν to the thermodynamic functions depends on the ratio of $h\nu$ to kT, and the contributions are additive. For example the vibrational energy is given by

$$h\nu/2 + h\nu/(\exp(h\nu/kT) - 1) \, , \tag{31}$$

which tends to the classical limit kT either at high temperatures of at low frequencies ($h\nu \ll kT$), and to the zero-point energy limit $h\nu/2$ for $h\nu \gg kT$. The vibrational free energy is given by

$$h\nu/2 + kT \ln(1-\exp(-h\nu/kT)) \, , \tag{32}$$

while the vibrational entropy is obtained from the difference between energy and free energy.

Choice of Energy Parameters

  Theoretical quantum mechanical considerations guided us generally in finding the functional form for many of the potential energy functions, but not the numerical values of their coefficients, or energy parameters. These have to be determined, as a rule, by fitting calculated results to experimental ones. The number of parameters to be fitted this way for all atoms and groups of atoms involved in biological molecules is rather large. The experimental data on which the parameter fitting rests is varied. In some cases there are not enough data, in other cases they are not sufficiently accurate. Sometimes different authors derive different parameters from the same data, which is understandable because the research objectives as well as personal biases may affect the method by which the parameters are determined. Consequently, there is not yet a generally agreed "force field" for molecular biology. (The term "force field" denotes a set of energy functions and their parameters, fitted for a given family of molecules). It is to be expected, however, that force fields will improve as their application in chemistry and molecular biology will advance, and that Darwinian natural selection and survival of the fittest will lead to better force fields which will be gradually accepted according to agreed-upon standards.

  One useful way to further such a purpose is to establish a "bench-mark" for an objective comparison of force-fields, namely an extensive data-base of high quality experimental results, with efficient computer programs by which the corresponding theoretical results from different force-fields can be calculated and compared. Such a bench-mark for amides and carboxylic acids has been published recently (Hagler, Lifson and Dauber, 1979), and may serve as a first step in this direction. It was a natural outcome of the efforts of

our group in obtaining a "consistent force field" for organic molecules and biopolymers.

The Consistent Force Field (CFF) Method. This method consists
of a systematic selection of energy functions, and an objective empirical determination of their parameters. It is based on the reciprocal relations which exist between observable properties and
energy functions. If the energy functions are known, then the observable properties can be calculated. Conversely, if sufficient
independent experimental data are available, energy functions can
be determined which will yield calculated values for the observables
to fit their corresponding experimental values.

In order to obtain progressively better functions, scientific
methodology, sometimes called the inductive process, dictates that
a procedure made of the steps enumerated below should be carefully
followed:

1.   Choose your initial set of functions either from available
literature, or from common sense and intelligent guessing, or from
theoretical considerations. Collect all experimental data pertaining
to those molecular properties which you can and wish to calculate
from V_{total} and its derivatives. Prepare the necessary set of computer programs to calculate these molecular properties. It is essential to make sure that the number of independent data exceeds
the number of adaptable parameters, not only for V_{total}, but for
each one of the empirical functions which comprize V_{total} according
to Eq. (22). Sufficient independent experimental data are essential
for an unambiguous determination of the empirical values of the parameters of the energy functions.

2.   Test your calculated data against the corresponding experimental data, using the "residual" pertaining to step 1. The residual
R is defined as the square root of the sum of squares of all differences between the calculated observables y_{cal} and the experimental ones, y_{exp}, properly weighted, namely

$$R^2 = \Sigma_i (w \Delta y)_i^2 . \tag{33}$$

Here $\Delta y = y_{cal} - y_{exp}$ and w is a weight given to Δy, which should be
proportional to the reliability of the experimental data, namely
inversely proportional to the experimental error of y_{exp}. The

index i enumerates all experimental data participating in the formation of the residual. The residual serves as an objective test of
how good was the choice of functions in step 1. The smaller the
residual the better is the overall fit of calculation to observation.

(Here is perhaps the place to add a corrolary to the scientific
method: Expect disappointing results at this stage, but don´t be
discouraged, rather, try step 3).

3.   Consider the residual R as a function of the energy parameters,
namely the adaptable constants in all the energy functions introduced in step 1. By varying these constants you vary y_{cal}, the calculated values of the observables. You may therefore seek the set of
energy parameters which will reduce R^2 to a minimum, namely give a
best fit.This is achieved by a minimization algorithm called "least
square optimization" (Marquardt, 1963; Lifson and Warshel, 1968;
Warshel and Lifson, 1970; Hagler and Lifson, 1974; Lifson and Levitt,
1979), particularly constructed to minimize residuals. The resulting
energy parameters are called optimal or optimized parameters, and
the set of energy functions with optimized parameters is what we
call a "consistent force field" (CFF), the term "consistent" representing the fact that the force field has been optimized to yield
consistently best results for all available data, energies, geometric or structural data, vibrational spectra etc.

4.   Examine the following two questions:
a.   Are you satisfied with the final fit of calculated to experimental data, or in other words, are the optimized functions good
enough?

b.   If not – why? or in other words, which are the data for which
the results are particularly unsatisfactory, and which are the
energy functions which could be blamed for the failures?

5.   Put your energy functions to a "Darwinian selection", replacing
inferior functions by hopefully better ones, and repeat steps 3-4-5
until calculated values fall within experimental error.

The above set of steps is a schematic and somewhat idealized
description of the CFF. In reality it is not practical to include
"all" available data at once. One starts with a limited set of experimental data pertaining to one family of molecules, and a cor-

respondingly limited set of energy functions. The resulting force
field for this family may be consequently extended to larger families
of molecules, and to new classes of observables.

    We started with a force field for conformations, vibrations and
enthalpies of cyclic and normal alkanes (Lifson and Warshel, 1968),
and extended it to crystal structures, sublimation energies, lattice
vibrations and thermal expansions of alkanes crystals (Warshel and
Lifson, 1970). It was then further extended to include conformations,
vibrations, and heats of hydrogenations of non-conjugated olefins
(Ermer and Lifson, 1973); conformations and vibrations of amide-
alkane rings (lactams) (Warshel et al., 1970); heats of sublimation
and crystal structures of hydrogen bonded crystals of amides (Hagler,
Huler and Lifson, 1974) and carboxylic acids (Lifson et al., 1979),
as well as some of their dipole moments (Hagler, Huler and Lifson,
1974; Lifson et al., 1979), and enthalpies of dimerization in the
gas phase (Hagler, Dauber and Lifson, 1979).

    The families of alkanes, amides and carboxylic acids contain
the constituents which comprise most of the amino acid residues of
protein chains, and indeed the major motivation for deriving a consistent force field for these families was its potential application
for biopolymers. However, a comprehensive consistent force field for
biological molecules should include more groups e.g. aromatic and
heterocyclic rings, carbohydrate, phosphate and sulfur groups; it
should also extend considerably the consistent analysis of intramolecular energy parameters pertaining to polypeptide and polynucleotide chains. This could be done fully only if more experimental data
could be accumulated to supply the specific needs of the consistent
force field analysis.

3. HYDROGEN-BONDED CRYSTALS AND THE NATURE OF THE HYDROGEN BOND

    The hydrogen bond (Schuster, 1976) is a special case of inter-
atomic interactions. It is special in several ways. First, it is
ubiquitous in biological structures, on the molecular level as well
as on higher levels of structural biology, and plays a very important
role in controlling biological specificity and life processes.
Second, it is very weak, if considered as a chemical bond, but one
of the strongest, if considered as a nonbonded interactions, and if
strength of interaction is to be evaluated by the ratio of attrac-

tive energy to number of atoms involved. It is comparable in strength
with the salt bridge which is, however much less abundant. Third, it
is special in that, although it has been recognized about 80 years
ago, the question whether it is merely a strong nonbonded interaction
or whether it is more like a weak chemical bond has been the subject
of some controversy or soul searching until recently. Finally, it is
special in that it gives us an excellent opportunity to demonstrate
the power of the consistent force field (CFF) method. We shall see
how its application to the study of hydrogen bonded crystals offered
a definitive answer to the question: "What is the hydrogen bond"?

It is for these reasons that we shall discuss it in detail here.
From the point of view of potential energy calculations, the question
has been whether a special potential function is required to represent the hydrogen bond, which function should it be, and whether
it should be added to or replace the functions of van der Waals and
electrostatic interactions. In the past, various functions were introduced to represent the hydrogen bond in terms of the "bond" distances and/or "bond" angles, which we shall not review here. Instead, it would be instructive to see how diverse are these various
functions. Figure 1 represents the hydrogen bond energy as a function of the H...O distance for a number of such potentials (Hagler,
Lifson and Huler, 1974) and it is obvious that not all of them could
be reliable estimates of the hydrogen bond energy. One of the potential
curves (#3) is the result of our CFF analysis to be described below.

We started our CFF analysis of hydrogen bonded crystals (Hagler,
Huler and Lifson, 1974) about a decade ago, after we concluded a
CFF analysis of intra- and intermolecular potentials for alkanes.
The nonbonded interactions in the alkane force field were selected
to be of the "n-6-1" type, namely composed additively from a
Lennard-Jones ("n-6") potential, Eqs. (17) or (19), and a Coulomb
("1") potential (Eq. (21)). Atoms of the same molecule were considered as nonbonded if they were separated by at least 3 consecutive
bonds. The Lennard-Jones (LJ) potential is commonly considered to
be a "12-6" potential, that is, the exponent n in the repulsive
term Ar^{-n} is taken to be n = 12 (see above). Examining the LJ potential for alkanes, we found that the LJ parameters optimized for
intramolecular interactions were too low for intermolecular interactions, while the LJ parameters optimized for intermolecular interactions were too high for intramolecular interactions. These trends
indicated that the r^{-12} dependence is too steep. Trying various

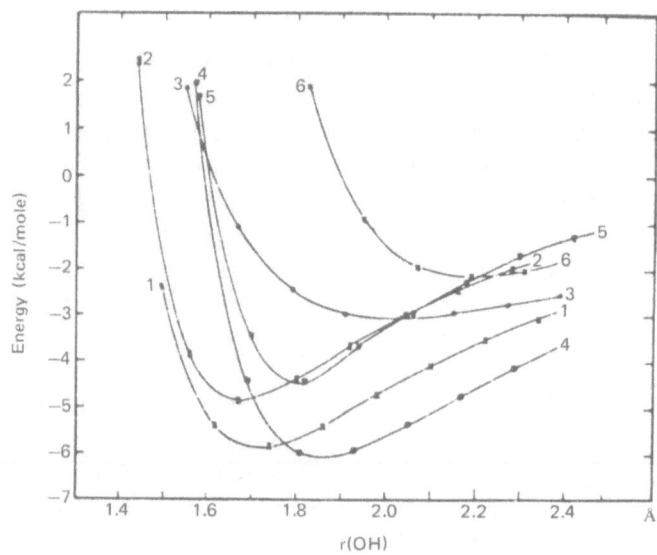

Fig. 1. The total interaction energy of the N-H and C=O groups as
a function of r(OH) for the different potential functions
used in conformational analysis: (1) Ooi et al. (1967);
(2) Brant (1968); (3) Hagler, Huler and Lifson (1974);
(4) Liquori (1969); (5) Balasubramanian et al. (1970);
(6) Popov et al. (1968).

values of n we found that the 9-6-1 potential (i.e. n = 9) was the
best for alkanes (Warshel and Lifson, 1970).

As we moved on to hydrogen bonded crystals, we were confronted
with the question, to what extent does the hydrogen bond have a covalent character? The answer to this question was of major importance for the application of an amide force field for condensed
systems in general and to molecular biology in particular. If two
atoms are covalently bound, then they cannot interact covalently
with other bonds unless their bond is broken. Thus if the hydrogen
bond would be purely covalent, then the energy of n hydrogen bonds
would be n times the energy of a single bond. On the other hand,
nonbonded interaction potentials, including the electrostatic potential, are pair-wise additive. Therefore, if the hydrogen bond

would be purely of nonbonded character, the number of energy terms representing n hydrogen bonds would be of the order of n^2. Thus finding the right balance between these two kinds of interactions is very relevant to all considerations of the energy of systems in which hydrogen bonds are abundant and play a major role, including essentially all biomolecules and biomolecular assemblies.

Having no a priori reason to choose between these two extreme alternatives, we assumed that both a covalent energy of the Morse type, modified by an angle dependent factor, and nonbonded interactions of the n-6-1 type, participate in the formation of the hydrogen bond. This assumption was supported in part, at that time, by an ab initio quantum mechanical study of the hydrogen bond by Dreyfus and Pullman (1970). They "partitioned" the hydrogen bond energy of the formic acid dimer $(HCOOH)_2$, and found that the total energy, -8 kcal/mole, was composed of +7, -10 and -5 kcal/mole for electronic cloud overlap repulsion, electrostatic attraction and covalent "charge-transfer" binding, respectively. Similar results were obtained at about the same time by Morokuma (1971) for water dimers, where the total hydrogen bond energy of -7 kcal/mole was partitioned into about 10, -8, -8 and -1 for repulsion, electrostatic, charge transfer and polarization components, respectively.

We used a very large data set for the optimization of the intermolecular energy parameters for the amide crystals. It included the structural data (unit cell parameters and the orientation of the molecules in the unit cell) for 10 amide crystals, the heat of sublimation of 6 amide crystals, and the dipole moments of urea, formamide and N-methyl-acetamide. The experimental data on the geometry of the hydrogen bond were rather varied, including O...H distances in the range of 1.85-2.31 Å, H...O=C angles in the range $106°$ - $155°$, and deviation of the angle N-H...O from $180°$ by up to $41°$. Thus, there was in the data base sufficient independent information to determine the parameters of the energy functions for amide crystals.

Following the steps 1-4 described in the previous chapter, we found ourselves in what was then to us a big surprise. The parameters of the Morse potential for the hydrogen bond appeared erratic. The potential well -D was small, and the standard deviations for α and b_0 were larger than the optimized values themselves. There was only one possible conclusion: The Morse potential offered no significant contribution to the residual (Eq. (33)) of the amide force field. In other words: There is no significant covalent contribution to the amide hydrogen bond. Indeed, when the N-H...O=C interaction was described, like all other nonbonded interactions, by the 9-6-1 or alternatively by the 12-6-1 potentials (we had optimized both alternatives for the sake of comparison), the overall agreement with the experimental data was as good as when the Morse potential was formally included although practically contributing negligibly. It appeared that the main attraction was due to the balance of all electrostatic forces, since the partial charges on the four atoms N-H...O=C were quite large, while the repulsion was due to the repulsive term in the Lennard-Jones potential. The atomic radii (see Eq. (19)) of nitrogen (r^*_{NN} = 4.01 Å) and oxygen (r^*_{OO} = 3.65) yielded, by the combination rule $r^*_{NO} = 1/2(r^*_{NN} + r^*_{OO})$, a van der Waals contact distance larger than their experimental hydrogen bond distance.

Another surprise was that the Lennard-Jones parameters of the hydrogen H_N of the NH bond were negligibly small. However, this also was understandable in the light of the large van der Waals contact distance of the two heavy atoms N and O, which left no distinguishable role for the van der Waals interaction of H_N with O. Putting these results together, an interesting explanation of the nature of the hydrogen bond emerged. Both the NH and the C=O bonds are strongly polar, due to the electronegativity of N and O relative to H and C, respectively. The electronegativity of the NH bond has the particular effect of shifting the pair of bonding electrons towards the N, thus exposing the H to a short van der Waals contact. As a result, the carbonyl oxygen can approach the NH bond closer than, say, an CH bond, getting into the range of N...O van der Waals repulsion. The electrostatic attraction of the N-H and O=C dipoles is then strong enough to compress the H...O distance to the range of ∼ 2 Å, which is the observed range of the "hydrogen bond length". Accordingly, the 4 atoms in the set N-H_N...O=C have, like the C and H_C atoms of alkanes, an n-6-1 type potential of pair-wise nonbonded interactions, which determine their interactions among themselves and with other atoms according to the rule of pair-wise additivity.

These results and conclusions contradicted the results obtained, and concepts adhered to, by quantum chemists, and were consequently either questioned (Schuster, 1976) or ignored by them. Even within the conceptual framework of the consistent force field methodology, one might argue that the amide hydrogen bond is the

weakest of all, and therefore its covalent character is minimal.
The O-H...O=C hydrogen bond in carboxylic acids is known to be significantly stronger, and its bond length (≈ 1.65 Å) significantly
shorter. Maybe a CFF analysis would prove the covalent "charge
transfer" to be significant for hydrogen bonds of carboxylic acids?
So we went ahead to extend the CFF to carboxylic acid crystals.

The data base of carboxylic acid crystals included 14 mono-
and dicarboxylic acids, of which 11 crystal structures, 7 sublimation energies and 4 dipole moments were included in the optimization of the energy parameters, while the rest were used to test
the resulting CFF on data not included in the optimization. The experimental hydrogen bond data included H...O distances in the range
of 1.6 - 1.8 Å, H...O=C angles between 114° and 130°, and deviations
of O-H...O angles from 180° by up to 33°.

Here again we were up to a number of surprises. First, the
Morse potential was again rejected by the least squares optimization
of the energy parameters, forcing us to the conclusion that the
O-H...O=C hydrogen bond, like the N-H...O=C one, is no more than a
nonbonded, pair-wise additive, n-6-1 interaction. Second, although
we optimized the parameters of the n-6-1 potential for the atoms of
the carboxylic group independently of the corresponding parameters
for the amide groups, we obtained the same parameters for the atoms
O, H and C. Thus the amide force field could be used on carboxylic
acids without any modification! But how could the hydrogen bond
energy be larger and the hydrogen bond distance shorter? Simply because the atomic radius r_{OO}^{*} of oxygen is smaller than that of nitrogen r_{NN}^{*}: the van der Waals contact distance O...O is smaller than
the contact distance N...O; therefore the electrostatic interaction
of the O-H and O=C bonds is stronger: therefore the H...O distance
in O-H...O=C is compressed to its equilibrium distance of ≈ 1.65 Å,
i.e. by about 0.3 Å smaller than the equilibrium distance N-H...O=C.

These results indicate convincingly that the covalent character
of the hydrogen bond is indeed negligible or at least less significant than its nonbonded character in carboxylic acids as well as in
amides. Is the covalent contribution still significant in water?
We did not apply the CFF method to the study of the properties of
water, but the answer came from the quantum chemists. In a more
recent study, with extended basis set, Umeyuma and Morokuma (1977)
have revised their results. Consequently, the covalent "charge
transfer" for water dimers has been reduced from -8.2 kcal/mole to

−1.8 kcal/mole. Bayer, Lischka and Schuster (see Schuster, 1976), using a different basis set, obtained for the same quantity the value of −1.2 kcal/mole. Thus the CFF conclusions were essentially corroborated ultimately by quantum mechanics.

The essence of the CFF method as applied to nonbonded interactions in carboxylic acids and amides, may be illustrated by Table 1 and Figs. 2-3 from the "bench-mark" paper by Hagler, Lifson and Dauber (1979). The table represents the root mean square deviation (rms dev) between calculated and measured observables for three force fields: CFF "12-6-1" (i.e. "12-6" for LJ and "1" for Coulomb).

Table 1. Root Mean Square Deviations of Properties Calculated for Carboxylic Acids and Amides by Three Force Fields

Property	Units	No. of terms	rms dev.		
			12-6-1	9-6-1	MCMS
		Acids			
energy	kcal/mol	12	2.486	2.053	2.118
UCV length	$\mathring{A}$	42	0.489	0.307	0.604
UCV angle	deg	17	3.456	2.856	4.465
volume	$\mathring{A}^3$	14	15.911	16.772	18.876
d < 4	$\mathring{A}$	14	0.247	0.190	0.322
H...O dist	$\mathring{A}$	16	0.062	0.072	0.058
O...O dist	$\mathring{A}$	16	0.047	0.071	0.041
C-O...O angle	deg	16	11.071	9.881	14.048
O...O=C angle	deg	16	7.843	7.760	11.786
H...O=C angle	deg	16	12.362	12.144	17.985
180°-O-H...O	deg	16	8.491	7.732	11.710
		Amides			
energy	kcal/mol	6	1.574	1.930	8.446
UCV length	$\mathring{A}$	36	0.208	0.235	0.261
UCV angle	deg	14	1.824	1.261	2.385
volume	$\mathring{A}^3$	12	7.057	17.797	13.951
d < 4	$\mathring{A}$	12	0.145	0.145	0.164
H...O dist	$\mathring{A}$	30	0.049	0.059	0.056
N...O dist	$\mathring{A}$	30	0.055	0.055	0.076
C-N...O angle	deg	22	3.337	3.575	4.071
N...O=C angle	deg	22	5.931	5.502	9.257
H...O=C angle	deg	30	5.830	5.609	7.329
180°-NH...O	deg	30	4.396	3.894	4.093

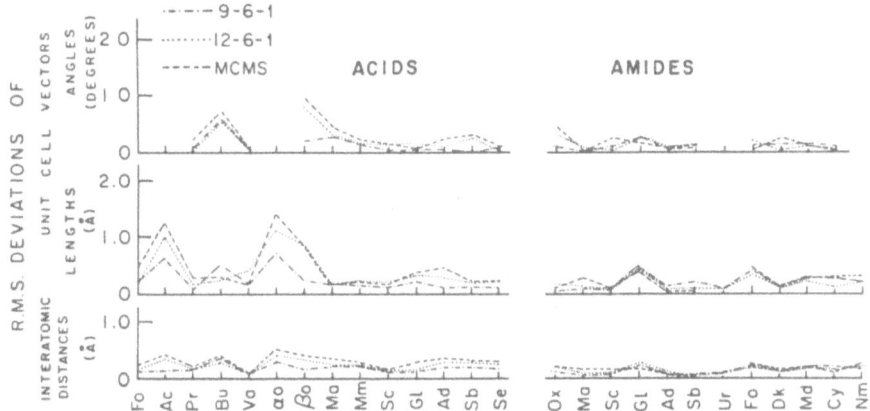

Fig. 2. The root mean square (rms) deviations of structural properties of acids and amides. These properties include interatomic distances less than 4 Å, unit cell vector lengths,
and angles.

Abbreviations for acids: formic (Fo), acetic (Ac), propi-
onic (Pr), butyric (Bu), valeric (Va), α-oxalic (α o), β -
oxalic (β o), malonic (Ma), methylmalonic (Mm), succinic
(Sc), glutaric (Gl), adipic (Ad), suberic (Sb), and sebacic
(Sc).

For amides: oxamide (Ox), malonamide (Ma), succinamide
(Sc), glutaramide (Gl), adipamide (Ad), suberamide (Sb),
urea (Ur), formamide (Fo), diketopiperazine (Dk), dimethyl
diketopiperazine (Md), cyclopropanecarboxamide (Cy), and
N-methylacetamide (Nm).

CFF "9-6-1" and MCMS (Momany et al., 1974· Dunfield et al., 1978).
a semi-empirical force field which uses a special function for the
hydrogen bond, and theoretical values for dispersion parameters and
partial charges. The figures are a concise summary of calculated vs.
experimental data for the individual molecules.

A close examination of the figures and the table, and of the
detailed tables in the "bench-mark" paper by Hagler, Lifson, Dauber
(1979) can tell us not only which force field is better and which
observables are better represented. One can look for systematic
trends of rms deviations, search for their cause or origin, and indicate the way to replace poorer potentials by better ones. Thus,

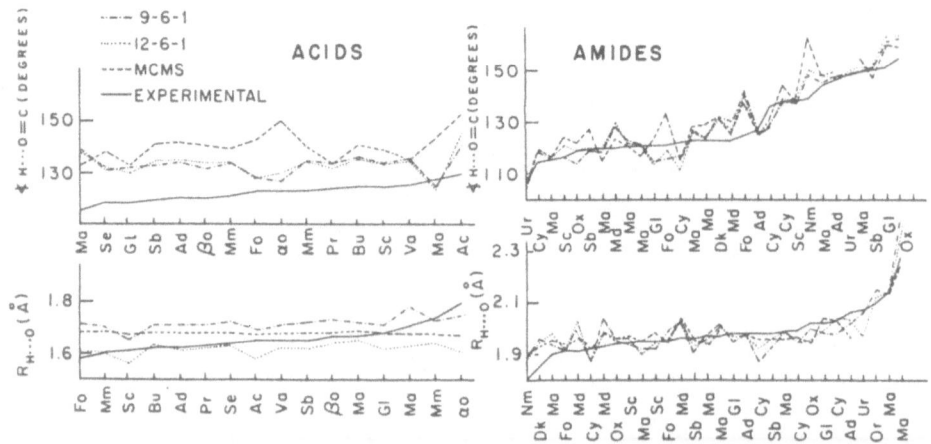

Fig. 3. The hydrogen bond geometry (H...O distance and H...O=C angle)
in acids and amides. Abbreviations for compounds as in Fig.
2. Molecules were ordered according to increasing observed
value of the property.

for example, it is noted that rms deviations are smaller in amides
than in carboxylic acids in both 12-6-1 and 9-6-1. The reason could
be traced to two facts: Amide crystals contain twice as many hydrogen bonds than carboxylic acids (there are 2 NH bonds per amide vs
one OH bond per carboxyl), and hydrogen bond distances are better
fit to experiment than alkane contact distances (most terms in
"d < 4"). The indication is that the alkane nonbonded interactions
need a reevaluation. Indeed we are trying out some ideas to improve
the van der Waals potential for alkanes. Such an improvement may be
important for a better representation of hydrophobic, non-polar
groups. Hydrophobic regions, typical of the interior of globular
proteins, are most suitable for potential energy calculations related
to protein structure, since they are not involved in interactions
with the solvent. Therefore, reducing the errors in the van der
Waals potential to the level of that of the polar interactions is a
desirable goal from the point of view of potential energy calculations in structural molecular biology.

REFERENCES

Allinger, N.L., 1976, Calculation of molecular structure and ener-
    gy of force-field methods, Adv. Phys. Org. Chem., 13:1.
Axilrod, B.M., and Teller, E., 1943, Interaction of the van der
    Waals type between three atoms, J. Chem. Phys., 11:299.
Balasubramanian, R., Chidamboram, R., and Ramachandran, G.N., 1970,
    Potential functions for hydrogen bond interactions.II. Formu-
    lation of an empirical potential function, Biochim. Biophys.
    Acta, 221:196.
Born, M., and Oppenheimer, R., 1927, Zur Quantentheorie der Mole-
    kulen, Ann. Physik, 84:457.
Brant, D.A., 1968, Conformational energy estimates for helical
    polypetide molecules, Macromolecules, 1:291.
Dreyfus, M., and Pullman, A., 1970, A non-empirical study of the
    hydrogen bond between peptide units, Theoret. Chim. Acta,
    19:20.
Dunfield, L.G., Burgess, A.W., and Scheraga, H.A., 1978, Energy
    parameters in polypetides. 8. Empirical potential energy
    algorithm for the conformational analysis of large molecules,
    J. Phys. Chem., 82:2609.
Dunitz, J.D., and Burgi, H.B., 1975, Nonbonded interactions in or-
    ganic molecules, Int. Rev. Sci. Phys. Chem. Ser. Two, 11:81.
Engler, E.M., Andose, J.D., and Schleyer, P.v.R., 1973, Critical
    evaluation of molecular mechanics, J. Amer. Chem. Soc.,
    95:8005.
Ermer, O., 1976, Calculation of molecular properties using force
    fields. Application in organic chemistry, Struc. Bonding
    (Berlin), 27:161.
Ermer, O., and Lifson, S., 1973, Consistent force field calculations
    III. Vibrations, conformations, and heats of hydrogenation
    of nonconjugated olefins, J. Amer. Chem. Soc., 95:4121.
Fletcher, R., 1968, Generalized inverse methods for the best least
    squares solution of systems of non-linear equations, Computer
    J., 10:392.
Fletcher, R., and Reeves, C.M., 1964, Function minimization by co-
    njugated gradients, Computer J., 7:149.
Hagler, A.T., Dauber, P., and Lifson, S., 1979, Consistent force
    field studies of intermolecular forces in hydrogen bonded
    crystals   3. The C=O...H-O hydrogen bond and the analysis
    of the energetics and packing of carboxylic acids, J. Amer.
    Chem. Soc., 101:5131.

Hagler, A.T., Huler, E., and Lifson, S., 1974, Energy functions
    for peptides and proteins. I. Derivation of a consistent force
    field including the hydrogen bond from amide crystals, J. Amer.
    Chem. Soc., 96:5319.

Hagler, A.T., and Lapiccirella, A., 1976, Special electron distri-
    bution and population analysis of amides, carboxylic acids and
    peptides, and their relation to empirical potential functions,
    Biopolymers, 15:1167.

Hagler, A.T., and Lifson, S., 1974, A procedure for obtaining energy
    parameters from crystal packing, Acta Crystallogr. B, 30:1336.

Hagler, A.T., Lifson, S., and Dauber, P., 1979, Consistent force
    field studies on intermolecular forces in hydrogen bonded
    crystals. 2. A benchmark for the objective comparison of al-
    ternative force fields, J. Amer. Chem. Soc., 101:5122.

Hagler, A.T., Lifson, S., and Huler, E., 1974, The amide hydrogen
    bond in energy functions for peptides and proteins, in: "Pep-
    tides, Polypeptides and Proteins", E.R. Blout, F.A. Bovey,
    M. Goodman, and N. Lotan, eds., Wiley, New York, p. 35.

Hagler, A.T., Stern, P.S., Lifson, S., and Ariel, S., 1979, Urey-
    Bradley force field, valence force field, and ab initio study
    of intramolecular forces in tri-tert-butylmethane and isobu-
    tane, J. Amer. Chem. Soc., 101:813.

Herzberg, G., 1945, "Molecular Spectra and Molecular Structure.
    Infrared and Raman Spectra of Polyatomic Molecules", Vol. II,
    Van Nostrand Co., New York.

Hill, T.L., 1960, "An Introduction to Statistical Thermodynamics",
    Addison-Wesley Inc., Reading, Mass.

Hirschfelder, J.O., Curtiss, C.F., and Bird, R.B., 1964, "Molecu-
    lar Theory of Gases and Liquids", J. Wiley and Sons, New York,
    2nd edition.

Kramer, H.L. and Herschbach, D.R., 1970, Combination rules for van
    der Waals force constants, J. Chem. Phys., 53:2729.

Kestner, N.R. and Sinanoglu, O., 1963, Effective intermolecular
    pair potentials in nonpolar media, J. Chem. Phys., 38:1730.

Lennard-Jones, J.E. 1924, On the determination of molecular fields,
    Proc. Roy. Soc. London A, 106:463.

Lifson, S., Hagler, A.T., and Dauber, P., 1979, Consistent force
    field studies of intermolecular forces in hydrogen bonded
    crystals. 1. Carboxylic acids, amides and the C=O...H - hydro-
    gen bonds, J. Amer. Chem. Soc., 101:5111.

Lifson, S. and Levitt, M., 1979, On obtaining energy parameters
    from crystal structure data, Computers and Chem., 3:49.

Lifson, S., and Stern, P.S., 1982, Born-Oppenheimer energy surface
   of similar molecules: interrelations between bond-lengths,
   bond angles and frequencies of normal vibrations in alkanes,
   J. Chem. Phys., in press.

Lifson, S., and Warshel, A., 1968, Consistent force field for calculations of conformations, vibrational spectra and enthalpies
   of cycloalkane and n-alkane molecules, J. Chem. Phys., 49:5116.

Liquori, A.M., 1969, Stereochemical code of amino acid residues in
   polypeptides and proteins, in: "Symmetry and Function of
   Biological Systems at the Molecular Level", A. Engstrom and
   B. Strandbey, eds., Wiley, New York, p. 101.

London, F., 1930, Ueber einige Eigenschaften und Anwendungen der
   Molekularkraefte, Z. Physik. Chem. B, 11:222.

Marquardt, D.W., 1963, An algorithm for least-squares estimation
   of non-linear parameters, J. Soc. Industr. Appl. Math ,
   11:431.

Momany, F.A., Carruthers, L.M., McGuire, R.F., and Scheraga, H.A.,
   1974, Intermolecular Potentials from crystal data. III. Determination of empirical potentials and application to the
   packing configurations and lattice energies in crystals of
   hydrocarbons, carboxylic acids, amines, and amides, J. Phys.
   Chem., 78:1595.

Morokuma, K., 1971, Molecular orbital studies of hydrogen bonds.
   III. C=O...H-O hydrogen bond in $H_2CO...H_2O$ and $H_2CO...2H_2O$,
   J. Chem. Phys., 55:1236.

Morse, P.M., 1929, Diatomic molecules according to the wave mechanics. II. Vibrational levels, Phys. Rev., 34:57.

Mulliken, R.S., 1955, Electronic population analysis on LCAO-MO
   molecular wave functions. I., J. Chem. Phys., 23:1833.

Ooi, T., Scott, R.A., Vanderkooi, G., and Scheraga, H.A., 1967,
   Conformational analysis of macromolecules. IV. Helical structure of poly-L-alanine, poly-L-valine, poly- β -methyl-L-aspar-
   tate, poly- γ -methyl-L-glutamate, and poly-L-tyrosine, J.
   Chem. Phys., 46:4410.

Pitzer, K.S., 1951, Potential energies for rotation about single
   bonds, Disc. Faraday Soc., 10:66.

Popov, E.M., Dashevskii, V.G., Lipkind, G.M., and Arkhipova, S.F.,
   1968, Calculating the configuration of peptide chains using
   potential functions, Mol. Biol., 2:491.

Schuster, P., 1976, Energy surface for hydrogen bonded systems,
   in: "The Hydrogen Bond", P. Schuster, G. Zundel, and C. Stan-
   dorfy, eds., Vol. I, North Holland, Amsterdam, p. 25 (see also
   Vol. II and III).

Slater, J.C., and Kirkwood, J.G., 1931, The van der Waals forces in
    gases, Phys. Rev., 37:682.
Umeyama, H., and Morokuma, K., 1977, The origin of hydrogen bonding.
    An energy decomposition study, J. Amer. Chem. Soc., 99:1316.
Urey, H.C., and Bradley, C.A., 1931, The vibrations of pentatomic
    tetrahedral molecules, Phys. Rev., 38:1969.
Warshel, A., Levitt, M., and Lifson, S., 1970, Consistent force
    field for calculation of vibrational spectra and conformations
    of some amides and lactam rings, J. Mol. Spect., 33:84.
Warshel, A., and Lifson, S., 1970, Consistent force field calcula-
    tions. II. Crystal structure, sublimation energies, molecular
    and lattice vibrations, molecular conformations, and enthal-
    pies of alkanes, J. Chem. Phys., 53:582.
Wilson, E.B., Jr., Decius, J.C., and Cross, P.C., 1955, "Molecular
    Vibrations", McGraw-Hill, New York.

THEORETICAL AND EXPERIMENTAL ASPECTS OF PROTEIN FOLDING

Harold A. Scheraga

Baker Laboratory of Chemistry
Cornell University
Ithaca, New York 14853, U.S.A.

INTRODUCTION

This article is based on the second and third of four lectures
that I presented at the International Summer School of Biophysics.
The first lecture, entitled "Recent developments in the theory of
water, aqueous solutions and protein hydration," has already been
summarized in several review articles (Scheraga, 1979; Paterson
et al., 1981; Némethy et al., 1981), and the fourth lecture,
entitled "Enzyme - substrate interactions," has likewise been
summarized in two recent reviews (Pincus and Scheraga, 1981;
Scheraga et al., 1982). In the two lectures constituting this
article, we describe some of our theoretical and experimental work
designed to gain an understanding of the interactions that lead to
the observed conformational behavior of biological macromolecules,
either as individual molecules or as complexes with other molecules.
The theoretical aspects of the problem are dealt with first; this
is then followed by a discussion of experimental studies on protein
conformation.

BASIS OF CALCULATIONS

Several empirical potential functions are used to calculate
the conformational energy of a polypeptide. The one in most
frequent use in our laboratory is ECEPP (Empirical Conformational
Energy Program for Peptides)(Momany et al., 1975; Pottle et al.,
1980). It was parameterized (Momany et al., 1974a) and tested
(Momany et al., 1974b) on the crystals listed in Table 1 of a
paper by Scheraga (1974) and on gas-phase data (Momany et al.,

45

1975). Though designed originally to treat polypeptides, the
potential function is also applicable to polysaccharides (Pincus
et al., 1976). Recently, several revisions have been made
(Némethy et al., 1982) in the parameters of the version that is
currently being distributed by the Quantum Chemistry Program
Exchange, University of Indiana. The overall strategy of the
computations, including the treatment of solvation by a hydration-
shell model, the calculation of the conformational entropy, and
the minimization of the total conformational energy, have been
summarized recently (Scheraga, 1981). In most of our computations,
bond lengths and bond angles are fixed at values characteristic of
each amino acid (Momany et al., 1975), and the variables for
changing conformation are the dihedral angles for rotation about
the single bonds of the backbone and side chain of each residue.
For energetic reasons that are understood (Zimmerman and Scheraga,
1976), the peptide group is maintained in the planar trans
conformation except for peptide groups preceding proline; in such
cases, both trans and cis conformations are allowed.

EARLY CALCULATIONS ON MODEL SYSTEMS

     The simplest peptide considered is the single amino acid with
acetyl and methyl amide blocking groups at the N- and C-termini,
respectively. In the older literature this structure was referred
to as a "dipeptide." It is more correct, however, to call it a
"terminally-blocked amino-acid residue." Initially (Ramachandran
et al., 1963; Némethy and Scheraga, 1965; Scheraga et al., 1965;
Leach et al., 1966a,b), computations were carried out with only a
hard-sphere potential to identify the "allowed" and "disallowed"
regions of conformational space. Subsequently, the computations
were extended to include a more complete potential energy function;
e.g. Fig. 3 of the paper by Scheraga et al. (1967) shows a
superposition of the hard-sphere (ϕ,ψ) map on the full-energy
contours of a terminally-blocked alanine residue, and illustrates
the importance of short-range repulsions in determining the
preferred conformations. As the side chain increases in size and
complexity, the observed region of the backbone diminishes
considerably [see Fig. 2 of the paper by Scheraga et al. (1967)].
Increase of the chain length introduces additional possibilities
for overlap, as illustrated in Fig. 4 of the paper by Scheraga
et al. (1967), in which a hard-sphere (ϕ,ψ) map is superposed on
the full-energy contours for helical structures of poly(L-alanine);
this diagram also illustrates the high stability of the left- and
right-handed α-helical forms with the right-handed one being more
stable than the left-handed one.

     Considerable understanding of the stereochemistry of poly-
peptide chains was gained from theoretical and experimental studies

of helical polyamino acids. The preferred helix senses of many
α-helical homopolyamino acids were computed (Scott and Scheraga,
1966; Ooi et al., 1967; Yan et al., 1968,1970), with good
agreement with experiment. In addition, the influence of side
chain - backbone interactions on the helix sense was elucidated.
For example, the right-handedness of α-helical poly(methyl-L-
glutamate) and the left-handedness of α-helical poly(methyl-L-
aspartate) are attributable to the different interactions in each
polymer between the dipoles of the side-chain ester group and the
nearest backbone amide group (Ooi et al., 1967). These electrostatic interactions outweight the influence of nonbonded interactions
involving the β-carbon, -- the latter leading to a preference for
right-handedness. Other examples are discussed in the paper by
Scheraga (1981).

      In addition to providing explanations for the handedness of
α-helical structures, we have recently accomplished the same thing
by accounting (Chou et al., 1982) for the observed (Chothia, 1973)
right-handed twist of parallel and anti-parallel β-sheets.

      This methodology is also applicable to polymorphic phase
transitions. The stable crystal arrangements of several
homopolymers, including the thermally induced conversion between
α- and ω-helical forms of polyamino acids have been computed
(McGuire et al., 1971; Fu et al., 1974). These involve
intermolecular, crystal-packing degrees of freedom, as well as the
internal ones. For example, whereas poly(p-Cl-benzyl-L-aspartate)
exists as a right-handed α-helix in solution, it is possible to
convert it to an ω-helix in the crystal, where favorable interchain
interactions enable the polymer to adopt this conformation (see
Fig. 1). The entropy for liberation around the helix axis
contributes to the stabilization of the ω over the α form at high
temperature.

HELIX-COIL TRANSITION

      The helix-coil transition has occupied polypeptide chemists
as models for such structural transformations in proteins (Poland
and Scheraga, 1970). Both theoretical and experimental approaches
have been taken in such studies (Scheraga, 1978) in order to obtain
quantitative measures of the tendency of each of the 20 naturally
occurring amino acids to adopt the helix vs. the coil (i.e. all
other, non-helical) conformations. These tendencies are expressed
in terms of the Zimm-Bragg (1959) nucleation and growth parameters,
σ and s, respectively. The values of σ and s may be used directly
to predict the locations of α-helices in proteins. The values of
σ and s have been obtained from experimental studies of thermally-
induced helix-coil transitions in host-guest random copolymers

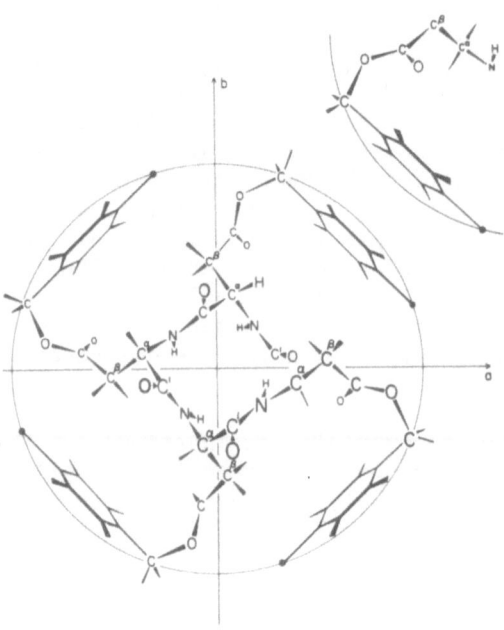

Fig. 1. View of the crystalline ω-helical form of poly(p-Cl-benzyl-
         L-aspartate) (Fu et al., 1974). [Reprinted with permission
         from Macromolecules 7:468 (1974). Copyright 1974 American
         Chemical Society.]

(Scheraga, 1978). For example, Fig. 2 illustrates experimental
curves for the dependence of s on temperature for several of the
amino acid residues, showing that some residues are helix-forming
and some are helix-breaking at a given temperature. Similar data
are obtainable from X-ray data on proteins [see Table 1 of the
paper by Scheraga (1978)]. These parameters have also been
calculated directly (for glycine, alanine and valine) from the
empirical potential functions. Figure 3 provides a comparison of
calculated (Gō et al., 1974) and experimental (Alter et al., 1973;
Chang et al., 1981) curves for the temperature dependence of s for
poly(L-valine) in water. It can be seen that the calculations
match the observed increase in s with increasing temperature,
reflecting primarily the characteristic increase (Némethy and
Scheraga, 1962) in hydrophobic bond strength (involving the valine
side chain) in this temperature range.

MULTIPLE-MINIMA PROBLEM

    While the foregoing calculations could be carried out without
encountering serious problems from the existence of many minima in

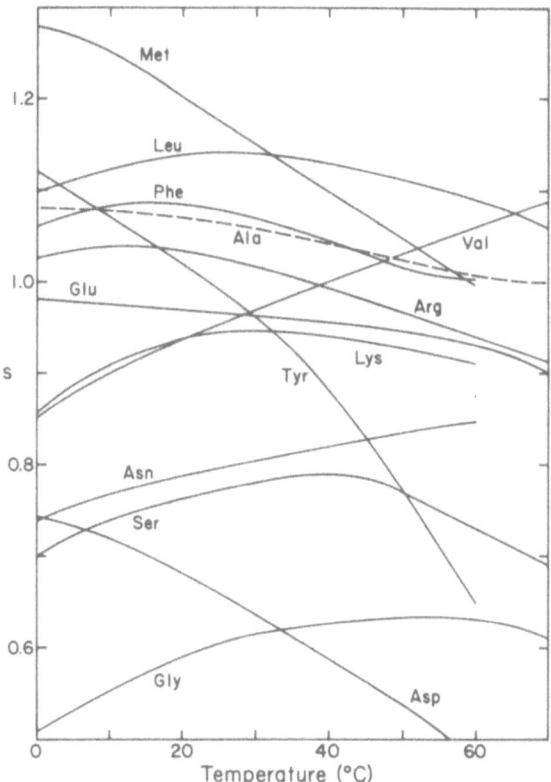

Fig. 2. Experimental s vs. T curves, obtained by the host-guest
         technique (Scheraga, 1978). [Reprinted with permission
         from Pure & Appl. Chem. 50:315 (1978). Copyright
         Pergamon Press.]

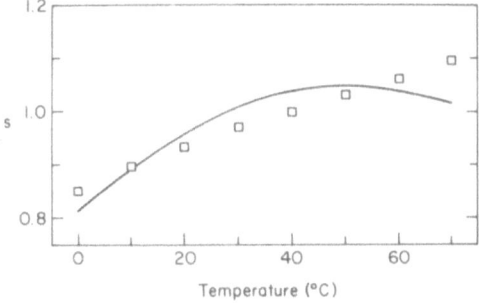

Fig. 3. Comparison of s vs. T curves for poly(L-valine) in water.
         The line is a calculated one (Gō et al., 1974), and the
         squares are the experimental results (Alter et al., 1973).
         [Reprinted with permission from Macromolecules 7:459
         (1974). Copyright 1974, American Chemical Society.]

the conformational energy space, these problems become difficult
when dealing with non-regular structures. In a recent paper
(Scheraga, 1981), I indicated how these problems were surmounted
for small open-chain and cyclic peptides, and for models of fibrous
proteins such as collagen.

These approaches, however, are not suitable for globular
proteins, and different strategies are therefore required.
Essentially, these involve use of approximate procedures to limit
the search of conformation space, and then the abandonment of the
approximations, in carrying out a full-scale energy minimization.
Thus, short-range approximations are used initially to predict the
conformational states of each residue; then medium- and long-range
interactions are introduced to compensate for the shortcomings of
the short-range models.

As indicated above, experimental data on host-guest copolymers,
as well as X-ray data on proteins, are used to predict the conformational state of each residue, taking only short-range interactions
into account. An Ising model treatment has also been used
(Dunfield and Scheraga, 1980) to predict not only the conformational
states but the actual values of the backbone dihedral angles (ϕ,ψ);
this model is presently being improved and extended to long-range
interactions. Medium-range interactions are then introduced to
predict, among other things, the primary nucleation site for folding
(Matheson and Scheraga, 1978), based on maximization of hydrophobic
bonding; for example, the primary nucleation site for the folding
of bovine pancreatic ribonuclease A is the segment comprising
residues 106-118. Finally, by dividing the protein into classes
of different spatial geometric arrangements of their disulfide-
bonded loops, the long-range interactions are introduced. A
constrained optimization is then carried out, taking into account
distance constraints from experimental data, radial distribution of
hydrophobic and hydrophilic residues, and avoidance of interatomic
overlaps. This approach is illustrated in the next section for
bovine pancreatic trypsin inhibitor.

Application to Bovine Pancreatic Trypsin Inhibitor

Use has been made of some of the above constraints in a test
of a protein folding algorithm, as applied to BPTI (bovine
pancreatic trypsin inhibitor).

Even if short-range prediction schemes led to 100% predict-
ability of the conformational states of the residues of a protein,
the structure would not resemble the correct one (Burgess and
Scheraga, 1975a). Long-range interactions are required in order
that the molecule achieve its correct shape. Also, if energy
minimization is started before the molecular shape is approximately
correct, then minimization will be trapped in a non-native local

minimum (Meirovitch and Scheraga, 1981a). One method for
introducing long-range interactions divides the conformational
space of a protein into classes characterized by different spatial
geometric arrangements of the loops formed by disulfide bonds
(illustrated for a hypothetical protein, with two disulfide bonds,
in Fig. 4) and then randomly selects one or more members of each
class for subsequent energy minimization. Because of the limited
number of classes and the availability of an array processor
(Pottle et al., 1980), such an approach is feasible. It is assumed
that one of the conformations selected from the class containing

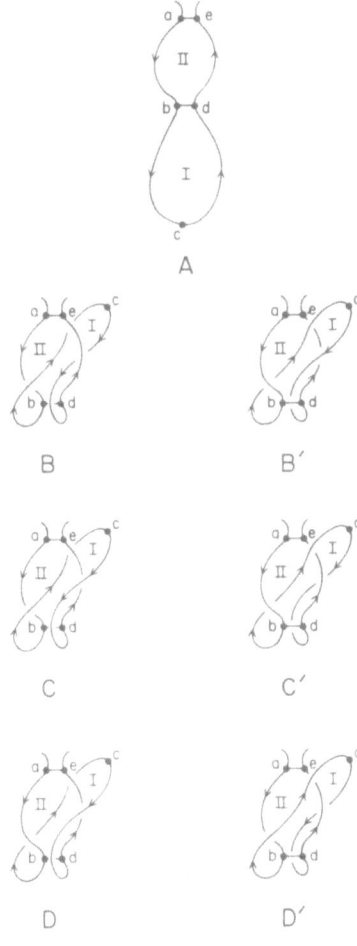

Fig. 4.   Schematic representation of the seven possible geometric
          arrangements of the loops in a hypothetical protein having
          two disulfide bonds (Meirovitch and Scheraga, 1981a).
          [Reprinted with permission from Macromolecules 14:1250
          (1981). Copyright 1981, American Chemical Society.]

the native structure would emerge as the one of lowest-energy.
The selected conformations are defined initially with the help of
a space-filling model, thereby avoiding interatomic overlaps right
from the beginning. This approach was tested on BPTI in a
preliminary way by selecting two starting conformations, -- one
from the class of the native structure (R), and one from a different
class (W), as shown in Fig. 5. After complete (ECEPP) energy
minimization of both, the one obtained from R did have a lower
energy than that obtained from W (94 compared to 168 kcal/mole),
but its root-mean-square deviation from the native structure was
5.9 Å (compared to 8.3 Å for the structure obtained from W).

        Introduction of the correct short-range conformational
information (to the extent of ∿85%), and re-minimization of the
energy reduced the energy from 94 to 10 kcal/mole and the

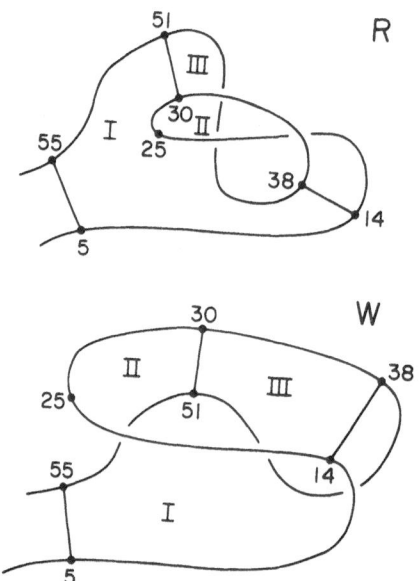

Fig. 5. The three loops of BPTI. R and W are the right and wrong,
         respectively, geometric arrangements of the loops
         (Meirovitch and Scheraga, 1981a). [Reprinted with
         permission from Macromolecules 14:1250 (1981). Copyright
         1981, American Chemical Society.]

root-mean-square deviation from 5.9 to 4.4 Å (Meirovitch and
Scheraga, 1981b).

The computed structure was improved further by introducing
other constraints, and re-optimizing (Wako and Scheraga, 1982b).
This was accomplished by a distance-constraint approach (Goel and
Ycas, 1979) which took into account the hydrophobic and hydrophilic
nature of the amino acid residues (expressed in terms of mean
distances between residues and thereby providing information about
short-, medium-, and long-range distances); this approach also
assigned mean distances in α-helical and extended segments (since
such conformations can be predicted by short-range algorithms),
assigned mean distances in a β-sheet, and incorporated information
about the locations of disulfide bonds between specific half-
cystine residues and the exact distances for five specific pairs
of residues [since experimental data of this type are obtainable
for specific pair interactions (Scheraga, 1967)]. An object
function was defined with which to generate conformations satisfying
the distance constraints; it was minimized for several sets of
distance constraints and starting conformations. The best results
were obtained by starting from the above structure (with a
root-mean-square deviation of 4.4 Å; this was reduced to 2.1 Å).
This optimization (Wako and Scheraga, 1982b) was carried out on a
structure consisting of C^α atoms and virtual bonds. We are now
undertaking the problem of converting from a virtual-bond chain
to a real peptide chain, and then continuing the computation with
a full (ECEPP) energy minimization. It is important to note that
optimization of only the distances, without energy minimization,
is a much less efficient approach, requiring knowledge of a very
large number of distances to obtain a low value of the root-mean-
square deviation (Wako and Scheraga, 1981).

Examination of the folding problem in a two-dimensional space
(Wako and Scheraga, 1982a) indicated that the preference of
hydrophobic and hydrophilic residues for lying inside and outside,
respectively, in a protein molecule appears to be a necessary but
not a sufficient condition for forming the stable native structure
-- because these interactions are non-specific. In addition, as
shown by the computations described above (Wako and Scheraga,
1982b), it is necessary to have information about a few specific
interactions.

Folding Pathways

Based on a variety of calculations, including Monte Carlo
simulations of protein folding (Tanaka and Scheraga, 1975,1977a,b),
a proposed pathway of folding is one that involves nucleation in
several regions that then add additional segments in a subsequent
growth phase, with final coalescence of the larger folded regions.

This is illustrated schematically in Fig. 4 of the paper by Scheraga (1980), in which a discussion is presented as to why this model is preferred over others. This model is also supported by experimental evidence presented below.

EXPERIMENTAL STUDIES OF RIBONUCLEASE

Turning to an experimental approach to the protein folding problem, to see how experimental information complements that obtained in the computations, we consider the stable intermediates in the folding process. As Konishi et al. (1982) have pointed out, in general stable intermediates do not define the folding pathway, although they may in the case of ribonuclease A. The folding/unfolding of this protein may be studied either by keeping the disulfide bonds intact or by first reducing and then oxidizing them. We consider both types of folding processes.

Thermally-Induced Unfolding/Folding

By means of several optical methods, e.g., optical rotation or ultraviolet difference spectrophotometry, ribonuclease can be shown to undergo a reversible, thermally-induced unfolding/folding transition, without rupture of its disulfide bonds (Hermans and Scheraga, 1961). While such optical methods detect average properties of the whole molecule, it is also possible to use methods that detect unfolding/folding in a limited portion of the molecule. For example, proteolytic enzymes, such as trypsin and chymotrypsin, can be used to detect local unfolding, and thereby probe the distribution of conformations at each temperature throughout the transition region (Rupley and Scheraga, 1963; Ooi et al., 1963; Ooi and Scheraga, 1964). From such information about the sequence of unfolding, i.e. about the sequence with which peptide bonds become susceptible to proteolytic attack, it was possible to identify the most stable intermediates at each temperature in the transition region (Burgess and Scheraga, 1975b). Subject to the caveat mentioned above (Konishi et al., 1982), this may be regarded as a pathway for unfolding/folding.

Subsequent experimental evidence supporting this pathway [with slight modifications (Matheson and Scheraga, 1979; Chavez and Scheraga, 1980a)] consisted of: (a) detection of unfolding of the C-terminal residue [by use of carboxypeptidase A (Burgess et al., 1975)], (b) flash photolytic labeling of surface residues (Matheson and Scheraga, 1979), (c) detection of mobility of attached spin labels by e.s.r. spectroscopy (Matheson et al., 1977), (d) examination of tyrosyl residues by Raman spectroscopy (Chen and Lord, 1976), (e) ^{13}C n.m.r. data (Howarth, 1979), and (f) X-ray crystal structure determination of the dominant structures that exist part way into the thermal transition (Petsko, 1980). Thus, the pathway for thermal folding of ribonuclease A appears to proceed through formation of nucleation sites in several regions, followed by folding around these nucleation sites.

Regeneration from Reduced Protein

The folding of ribonuclease has also been studied by oxidizing the reduced protein with a mixture of oxidized and reduced glutathiones, and following the folding process with the help of (fractionated) antibodies raised against the native protein (Chavez and Scheraga, 1980a). These antibodies recognize only the native conformation, and do not cross-react with the unfolded form or with other portions of the molecule, there being four antigenic sites in the native molecule (Chavez and Scheraga, 1979). Kinetic studies indicated that a region of the molecule, containing an antigenic site somewhere in segment 87-104 folded before the other three antigenic sites (Chavez and Scheraga, 1980a) [in segments 1-10, 40-61 and 63-75, respectively (Chavez and Scheraga, 1979)]. Thus, it appears that the formation of the primary nucleation site among residues 106-118 constitutes an interior folded hydrophobic region which induces the folding of a neighboring, surface region, 87-104, which contains one of the antigenic sites (Scheraga, 1980). It appears that folding, with or without disulfide bonds, follows the general folding scheme deduced from the computations (Scheraga, 1980).

These same antibodies were used to determine the equilibrium constant K_{conf} for the interconversion of unfolded and folded forms of ribonuclease and various derivatives thereof (Chavez and Scheraga, 1980b). Among other things, it was found that reduced ribonuclease has a fairly high degree of native structure, with K_{conf} being about 0.06. This value is reduced by one order of magnitude when the sulfhydryl groups are blocked with the bulky negatively-charged carboxymethyl group. Thus, nearest-neighbor interactions essentially determine that the conformation of the reduced protein should be close to native. It then takes only a little help from long-range interactions, and reduction of entropy by disulfide bonds, to stabilize the native structure.

More recently, Konishi et al. (1981,1982) have carried out a very detailed experimental investigation of the equilibrium and kinetics of the regeneration of native ribonuclease from the reduced form, using oxidized and reduced glutathiones. They demonstrated the existence of a pre-equilibrium state and identified the rate-limiting steps for folding. The distributions of intermediates and the folding pathway depend on the concentrations of oxidized and reduced glutathiones. Under some conditions, folding proceeds by a "growth-type" mechanism in which native interactions stabilize nuclei around which folding takes place. Under other conditions, folding proceeds by a "rearrangement type" mechanism in which non-native interactions (including wrong disulfide bonds) are used to start the folding process, but these must then be disrupted to allow folding to proceed properly to completion. These two types of folding mechanisms appear to be applicable to protein folding in general.

CONCLUSIONS

Models of protein folding have emerged from both the theoretical calculations and experimental studies. Not only do the experiments provide a verification of the models deduced from the computations, but they also provide distance constraints that can be incorporated into protein folding algorithms. While proteins of known structure are the vehicles of study during the developmental stage of our ideas about protein folding and stability, the theoretical and experimental methodologies should also be applicable to proteins of unknown three-dimensional structure.

ACKNOWLEDGMENTS

This work was supported by research grants from the National Science Foundation (PCM79-20279) and from the National Institute of General Medical Sciences, National Institutes of Health, U.S. Public Health Service (GM-14312).

REFERENCES

Alter, J.E., Andreatta, R.H., Taylor, G.T., and Scheraga, H.A., 1973, Macromolecules, 6:564.
Burgess, A.W.,and Scheraga, H.A., 1975a, Proc. Natl. Acad. Sci., U.S., 72:1221.
Burgess, A.W., and Scheraga, H.A., 1975b, J. Theor. Biol., 53:403.
Burgess, A.W., Weinstein, L.I., Gabel, D., and Scheraga, H.A., 1975, Biochemistry, 14:197.
Chang, M.C., Fredrickson, R.A., Powers, S.P., and Scheraga, H.A., 1981, Macromolecules, 14:633.
Chavez, L.G., Jr. and Scheraga, H.A., 1979, Biochemistry, 18:4386.
Chavez, L.G., Jr. and Scheraga, H.A., 1980a, Biochemistry, 19:996.
Chavez, L.G., Jr. and Scheraga, H.A., 1980b, Biochemistry, 19:1005.
Chen, M.C.,and Lord, R.C., 1976, Biochemistry, 15:1889.
Chothia, C., 1973, J. Molec. Biol., 75:295.
Chou, K.C., Pottle, M., Némethy, G., Ueda, Y., and Scheraga, H.A., 1982, J. Molec. Biol., submitted.
Dunfield, L.G.,and Scheraga, H.A., 1980, Macromolecules, 13:1415.

Fu, Y.C., McGuire, R.F., and Scheraga, H.A., 1974, Macromolecules, 7:468.

Gō, M., Hesselink, F.T., Gō, N., and Scheraga, H.A., 1974, Macromolecules, 7:459.

Goel, N.S., and Ycas, M., 1979, J. Theor. Biol., 77:253.

Hermans, J., Jr. and Scheraga, H.A., 1961, J. Am. Chem. Soc., 83:3283.

Howarth, O.W., 1979, Biochim. Biophys. Acta, 576:163.

Konishi, Y., Ooi, T., and Scheraga, H.A., 1981, Biochemistry, 20:3945.

Konishi, Y., Ooi, T., and Scheraga, H.A., 1982, Biochemistry, submitted.

Leach, S.J., Némethy, G., and Scheraga, H.A., 1966a, Biopolymers, 4:369.

Leach, S.J., Némethy, G., and Scheraga, H.A., 1966b, Biopolymers, 4:887.

Matheson, R.R., Jr., Dugas, H., and Scheraga, H.A., 1977, Biochem. Biophys. Res. Commun., 74:869.

Matheson, R.R. and Scheraga, H.A., 1978, Macromolecules, 11:819.

Matheson, R.R., Jr. and Scheraga, H.A., 1979, Biochemistry, 18:2437.

McGuire, R.F., Vanderkooi, G., Momany, F.A., Ingwall, R.T., Crippen, G.M., Lotan, N., Tuttle, R.W., Kashuba, K.L., and Scheraga, H.A., 1971, Macromolecules, 4:112.

Meirovitch, H. and Scheraga, H.A., 1981a, Macromolecules, 14:1250.

Meirovitch, H. and Scheraga, H.A., 1981b, Proc. Natl. Acad. Sci., U.S., 78:6584.

Momany, F.A., Carruthers, L.M., McGuire, R.F., and Scheraga, H.A., 1974a, J. Phys. Chem., 78:1595.

Momany, F.A., Carruthers, L.M., and Scheraga, H.A., 1974b, J. Phys. Chem., 78:1621.

Momany, F.A., McGuire, R.F., Burgess, A.W., and Scheraga, H.A., 1975, J. Phys. Chem., 79:2361.

Némethy, G. and Scheraga, H.A., 1962, J. Phys. Chem., 66:1773.

Némethy, G. and Scheraga, H.A., 1965, Biopolymers, 3:155.

Némethy, G., Peer, W.J., and Scheraga, H.A., 1981, Ann. Rev. Biophys. Bioeng., 10:459.

Némethy, G., Pottle, M.S., and Scheraga, H.A., 1982, J. Phys. Chem., in preparation.

Ooi, T., Rupley, J.A., and Scheraga, H.A., 1963, Biochemistry, 2:432.

Ooi, T., Scott, R.A., Vanderkooi, G., and Scheraga, H.A., 1967, J. Chem. Phys., 46:4410.

Ooi, T., and Scheraga, H.A., 1964, Biochemistry, 3:641,648.

Paterson, Y., Nemethy, G., and Scheraga, H.A., 1981, Annals N.Y. Acad. Sci.. 367:132.

Petsko, G., 1980, private communication.

Pincus, M.R., Burgess, A.W., and Scheraga, H.A., 1976, Biopolymers, 15:2485.

Pincus, M.R. and Scheraga, H.A., 1981, Accts. Chem. Res., 14:299.

Poland, D. and Scheraga, H.A., 1970, "Theory of Helix-Coil Transitions in Biopolymers," Academic Press, New York.

Pottle, C., Pottle, M.S., Tuttle, R.W., Kinch, R.J., and Scheraga, H.A., 1980, J. Comput. Chem., 1:46.

Ramachandran, G.N., Ramakrishnan, C., and Sasisekharan, V., 1963,
    J. Molec. Biol., 7:95.
Rupley, J.A.,and Scheraga, H.A., 1963, Biochemistry, 2:421.
Scheraga, H.A., 1967, Fed. Proc., 26:1380.
Scheraga, H.A., 1974, in: "Peptides, Polypeptides and Proteins,"
    E.R. Blout, F.A. Bovey, M. Goodman, and N. Lotan, eds., John
    Wiley & Sons, Inc., New York, p. 49.
Scheraga, H.A., 1978, Pure and Appl. Chem., 50:315.
Scheraga, H.A., 1979, Accts. Chem. Res., 12:7.
Scheraga, H.A., 1980, in: "Protein Folding," R. Jaenicke, ed.,
    Elsevier, Amsterdam.
Scheraga, H.A., 1981, Biopolymers, 20:1877.
Scheraga, H.A., Leach, S.J., Scott, R.A., and Némethy, G., 1965,
    Discuss. Faraday Soc., 40:268.
Scheraga, H.A., Scott, R.A., Vanderkooi, G., Leach, S.J., Gibson,
    K.D., Ooi, T., and Némethy, G., 1967, in: "Conformation of
    Biopolymers," G.N. Ramachandran, ed., Academic Press, London.
Scheraga, H.A., Pincus, M.R., and Burke, K.E., 1982, in: "Structure
    of Complexes Between Biopolymers and Low-Molecular-Weight
    Molecules," W. Bartmann, ed., Heyden & Sons Ltd., London,
    in press.
Scott, R.A.,and Scheraga, H.A., 1966, J. Chem. Phys., 45:2091.
Tanaka, S.,and Scheraga, H.A., 1975, Proc. Natl. Acad. Sci., U.S.,
    72:3802.
Tanaka, S.,and Scheraga, H.A., 1977a, Proc. Natl. Acad. Sci., U.S.,
    74:1320.
Tanaka, S.,and Scheraga, H.A., 1977b, Macromolecules, 10:291.
Wako, H.,and Scheraga, H.A., 1981, Macromolecules, 14:961.
Wako, H.,and Scheraga, H.A., 1982a, Biopolymers, in press.
Wako, H.,and Scheraga, H.A., 1982b, J. Protein Chem., in press.
Yan, J.F., Vanderkooi, G., and Scheraga, H.A., 1968, J. Chem. Phys.,
    49:2713.
Yan, J.F., Momany, F.A., and Scheraga, H.A., 1970, J. Am. Chem. Soc.,
    92:1109.
Zimm, B.H.,and Bragg, J.K., 1959, J. Chem. Phys., 31:526.
Zimmerman, S.S. and Scheraga, H.A., 1976, Macromolecules, 9:408.

SOME ASPECTS OF THE MACROMOLECULAR CHEMISTRY

OF CARBOHYDRATE POLYMERS

Vittorio Crescenzi

Institute of Physical Chemistry
University of Rome
Rome, Italy

INTRODUCTION

Saccharides occur everywhere as simple sugars involved in
energy metabolism, as polymers which perform specialized tasks for
our life or which contribute to extracellular support in plants,
microorganisms and animals, as well as components associated with
species of great biological importance, as for example, nucleic
acids, many enzymes, antibodies, hormones, membrane proteins and
lipids. As a logical consequence, research on carbohydrates is particularly active. This is true not only from a purely fundamental
standpoint but, especially considering carbohydrate polymers, also
from an applied point of view in as much as many natural polysaccharides (and their derivatives) find a number of technological applications, not to mention biomass utilization.

In what follows, limiting attention to some aspects of the
macromolecular chemistry of carbohydrate polymers, a schematic description of the more common chain structures and functions of different classes of natural polysaccharides will be given. Finally
some typical features of the behavior of polysaccharides in solution, going from dilute aqueous solutions of ionic species to more
concentrated solutions giving rise to mesophases or gels, will be
qualitatively reviewed.

1. POLYSACCHARIDE STRUCTURE AND FUNCTIONS

The Monomers

    Discussion on any kind of polymers is logically preceded by
at least a quick survey of properties of parent monomers. This is
particularly relevant in the case of polysaccharides because monosaccharides are rather special monomers, with peculiar configura-
tional-conformational properties and reactivity. Monosaccharides
are multifunctional compounds, usually with three or four OH groups
of approximately equal chemical reactivity; reducing sugars have,
of course, additional reaction capabilities.

    Derivatization of a single, selected OH group thus normally
requires the careful execution of a series of reactions. Synthesis
of a given disaccharide or trisaccharide may be therefore a considerable task, taking into account that in addition there is the
problem of anomeric (α or β glycosydic links), and this is in
marked contrast to the situation in peptide chemistry. Among other
things, there is in fact a very large difference in the number of
isomeric oligopeptides and oligosaccharides that, at least in principle, can be obtained from the same number of corresponding monomers. For instance, given three appropriate monomers X Y Z one can
obtain only 6 isomeric tripeptides but about a thousand isomeric
trisaccharides.

    With more than three monomeric sugars, branching may also
occur and the number of possible isomeric oligosaccharides further
increases.

    In other words, sugars display an enormous potential for
structural diversity of the oligomers or polymers of which they may
constitute the building blocks. This is particularly important from
the biological point of view since it makes sugars excellent candi-
dates as carriers of biological "information". In fact, unlike peptides and oligonucleotides, in which the information content is
based only on the number and sequence of different monomeric units,
heterosaccharides "information" resides also in the position and
anomeric configuration of the glycosydic units and in the occurrence
of branch points. Such "information" capabilities were not, however, evident in facts untill the 60´ s when the unexpected disco-
very was made that the glycans in glyconjugates, e.g. glycoproteins,

which had been thought to have neither a role nor any biological significance, are on the contrary extremely important.

It is now realized that carbohydrate moieties serve as important recognition markers on glycoproteins in solution as well as on cell surfaces. Moreover, there are examples of the role of cell surface sugars in intracellular adhesion, a key step in processes such as fertilization and cellular differentiation.

To appreciate points very qualitatively enumerated above, in particular concerning glycoproteins which have opened a very fascinating field of research, the recent literature offers authoritative presentations (Montreuil, 1980; Sharon et al., 1981). However this presentation will be limited essentially to non-informational biopolymers.

For the configuration-conformational properties of simple carbohydrates, as well as their reactivities, comprehensive treatises should be consulted (Pigman and Horton, 1972). It has to be mentioned on closing, that recent achievements in macromolecular science have established the ways to the synthesis of linear homopolysaccharides and copolysaccharides (normally by cationic polymerization) of different suitably protected anhydrosugars (Schuerch, 1981). These syntheses open the route to the availability for our studies of a variety of tailor made carbohydrate polymers. A few synthetic polysaccharides have in fact already been used to investigate immunological and allergic reactions in experimental animals and the interaction of plant lectins with carbohydrates (Schuerch, 1981).

The Polymers

For polysaccharides the number of different chains found in nature is very large. There is, on the other hand, an obvious preference in nature for certain sugar monomers (out of the nearly 200 presently known) as well as for certain covalent chain structures.

The monosaccharides more commonly found in linear polysaccharides are shown in Fig. 1.1. With polysaccharides a classification of practically encountered covalent structures may be made using the following three main groups (Rees, 1977): a) periodic sequences, b) interrupted sequences, c) aperiodic sequences.

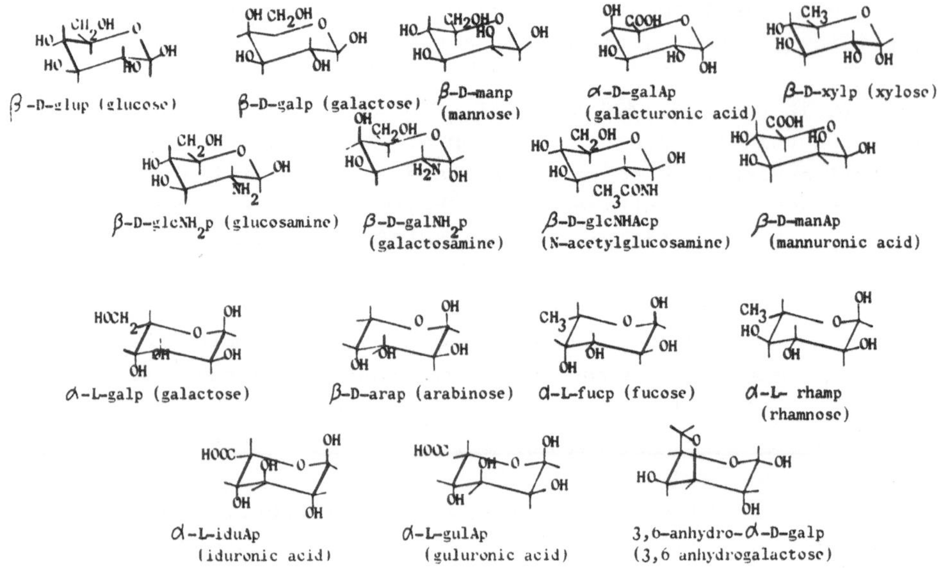

Fig. 1.1. Monosaccharides more commonly found in polysaccharides. The ring conformations (only pyranose (p) rings) are either $^4C_1$ (e.g. α-D-glucopyranose) or $^1C_4$ (e.g. β-L-galactopyranose) in order to minimize the number of axial OH groups (or bulkier groups). Other conformations of higher energy are of course possible (Rees, 1977).

Typical examples are schematically given in Table 1. In Table 2 is given a simplified summary of possible chain conformations (helical) of polysaccharides of the periodic or interrupted-periodic type. It can be shown that the primary influence that determines the type of overall chain shape is the geometry within each repeating unit rather than the interactions between them, once these are minimized pairwise in each distinct case by appropriate Ψ and Φ values (see Fig. 1.2). This is at least as far as chain geometries in the solid state are concerned. (This is just an extension of the "conformational equivalence" principle set forth years ago for any type of linear, regular chain). In solution the situation may become quite more complicated, in particular because of often strong solvent-chain interactions. If, however, the so-called "unperturbed-state" of the chains may be experimentally achieved, average chain dimensions become amenable to comparison with theoretical prediction (Flory, 1969).

Table 1. Sequence of Sugar Units in Carbohydrate Chains[a]

Name of Polymer	Sequence
"Periodic Type"	
Amylose	
Cellulose	
Chitin	
Hyaluronate	
O-antigen	
"Interrupted Type"	
Alginate	
Carrageenan	
"Aperiodic Type"	
Carbohydrate chains of glycoproteins	

[a] The different symbols (⊗, ⊘, etc.) represent different sugar residues. For instance in "alginate", ⊘ may be β–D-mannuronate, and ⊗   α–L-guluronate, respectively.

Conformational states that can be populated by each repeating saccharide units play, in fact, a major role in the solution behavior of the macromolecule in the aforementioned conditions. In any case, however, to fully understand the conformational behavior of biopolymers of our concern it is necessary to know not only the free energy difference between the ring conformations in the "monomers" but also how this difference is affected by neighboring units in the chain. Considering the structures given in Table 1, to which in reality others might be added of even more complicated nature, it should appear evident that their elucidation must have been in most cases a difficult achievement. In fact, for a simple oligosac-

Table 2. Possible Helical Chain-Conformations of Linear, Re-
        gular Carbohydrate Polymers[a]

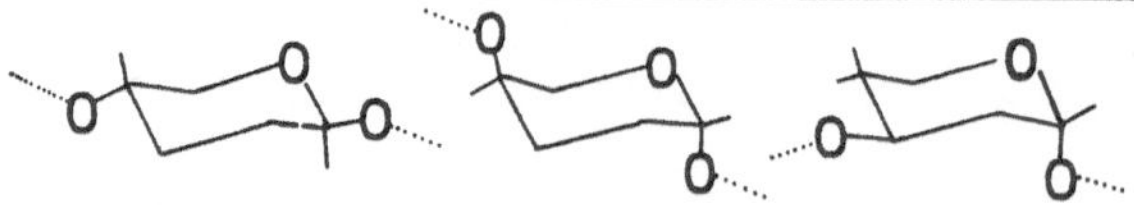

β -1,4-glcp   α -1,4-galp   α-1,3-glcp

ribbons (2 hold helices)
n = 2 ⟶ ±4;
h ∼ length of a unit

examples: cellulose, chitin, mannan..

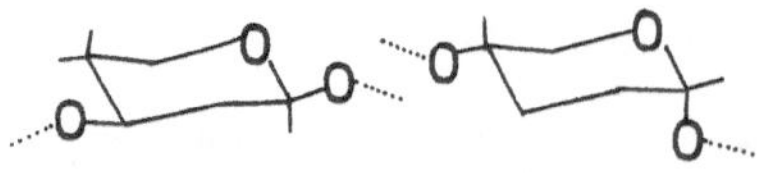

β-1,3-glcp   α -1,4-glcp

hollow helices
n ⟶ ±10;
h ≪ length of a unit

examples: amylose, curdlan, and many polys. with interrupted-
        periodic chains (e.g. agarose, carrageenans, glyco-
        saminoglycans)

---

[a] n is the number of units per turn of the helix, and h is the
projected length of each unit along the helix axis. It is interesting to mention that it has been found very recently,
that the linear polymer of α -(1,6)-glucp (linear dextran),
normally classified as a "loosely jointed polymer", can give
crystals with the polysaccharide chains in a highly stretched
conformation (of the cellulose type) (Guizard, 1981). The same
might be true also for pustulan (poly- β -(1,6)-glucp)). Finally, the rare types of 1,2-linked chains (e.g. poly-β -1,2-
glcp, a polysaccharide from Agrobacter Tumefacens) have
"crumpled chains" for which ordered conformations appear difficult.

charide (or for the repeating unit of a regular polysaccharide) in
order to find out the structure it is normally necessary to determine points a) through e), as schematically indicated in Table 3.
One particular example of the usefulness of glycosydase enzymes
and methylation analysis to determine sequences and linkage positions of sugars in branched oligosaccharides is afforded in a recent
review paper by Sharon et al. (1981).

   Additional, relatively simple chemical procedures which result
often very helpfully in elucidation of basic structural features

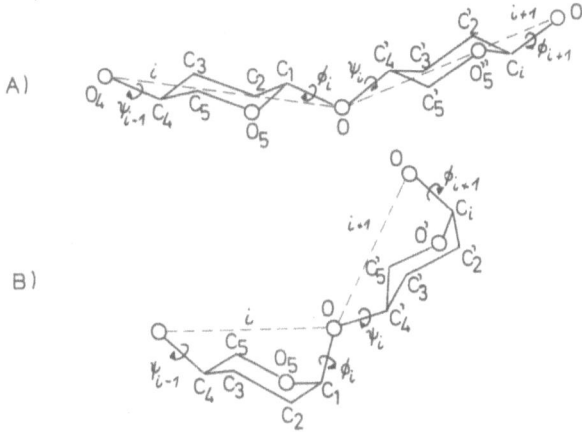

Fig. 1.2. Representative conformations of disaccharide repeating
units along: A) cellulose, and B) amylose chains. Φ and
Ψ are the rotation angles around the glycosidic bond
joining anomeric carbon C(1) to carbon C(4). Assuming
the glcp conformation (^{4}C1) to be rigid, the chains in
solution may be thought to be built up by virtual segments (dotted lines).

of carbohydrate polymers are based on glycol-cleavage oxidation
(Perlin, 1980). A schematic example is presented in Fig. 1.3, in
which species obtainable (and easily identifiable with standard
analytical methods) from amylose upon oxidation by periodate, followed by $NaBH_4$ reduction and final acid hydrolysis, are shown. In
another particular example, selective oxidation of terminal units
of hexose chains, followed by chemical and enzymatic quantitative
essays of resulting products, may be used to determine number-
average chain lengths (Sturgeon, 1980).

Quite naturally, in addition to the few chemical approaches
simply enumerated above, investigations with ^{1}H and ^{13}C NMR techniques (as well as IR and chiroptical techniques) prove extremely
useful (Gorin, 1981). Moreover whenever the specimen under study
can be suitably crystalized, X-ray techniques may accurately solve
structural problems (Atkins et al., 1974; Sundarajan and Marchessault, 1979; Marchessault and Deslandes, 1980).

Table 3. Simplified Scheme of Main Chemical Steps Normally
         Necessary to Elucidate the Structure of an Oligosac-
         charide

---

a) <u>Composition of monosacharides</u>: acid hydrolysis followed
                                        by chromatographic analysis
b) <u>Molecular weight</u>:   possibly by means of
                                        "vapour pressure osmometry"
                                        or other techniques
c) <u>Ring size of the units</u> and   methylation, hydrolysis,
d) <u>Position of glycosidic links</u>: and appropriate derivati-
                                        zation for gas chromato-
                                        graphic-mass spectral ana-
                                        lysis
e) <u>Sequence of the units and</u>   partial hydrolysis followed
   <u>their anomeric configuration</u>: by high-performance liquid-
                                        chromatographic analysis

For step e), sequential use of appropriate glycosydases (exo-
and/or endo-), followed by procedures similar to those outlined for steps c) and d), may be extremely useful (Sharon et
al., 1981).

    As a final comment, let us recall that a further complication
in the quantitative, structural analyis of certain carbohydrate
polymers may result from the fact that oligo- and polysaccharides,
whether free or attached to proteins, are not primary gene products
(in contrast to polypeptide chains which are under direct control
of the genetic code). They are, in fact synthesized by enzymes
(glycosyltransferases) in the absence of any template. This synthesis may not be very accurate, and can result in microhetero-
geneity in heterosaccharide chains.

    Keeping in mind the classification of carbohydrate polymers
schematized in the Tables 1 and 2, we may now inquire where such
biopolymers are found and what their functions are.

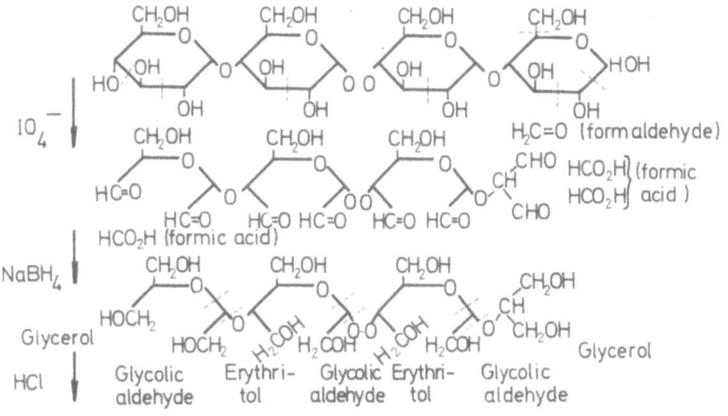

Fig. 1.3. Elementary reaction scheme of the Smith degradation of amylose.

The majority of carbohydrate chains have functions in a) forming structure, b) storage energy, c) binding of water and ions, and in holding species in the neighborhood of cells. Some polysaccharides (d) perform very specialized functions through highly specific interactions with other biologically important molecules or biological macromolecules (e.g. the glycans of cell membrane glycoproteins and heparin, respectively), and some (e) are particular products of various microbial species.

Structural Polysaccharides

Among structural polysaccharides one must recall cellulose and chitin (Fig. 1.4), the two most abundant compounds in the world, being synthesized at an estimated rate of 10^{10} - 10^{11} tons per year by plants and crustaceans, insects, spiders and fungi, respectively. Their molecular weights are in the range 10^5 - 10^6 and they are considered to be polydisperse materials, as it often appers to be the case also for different natural carbohydrate polymers. This, among other things, introduces some additional difficulties in their structural and physico-chemical characterization.

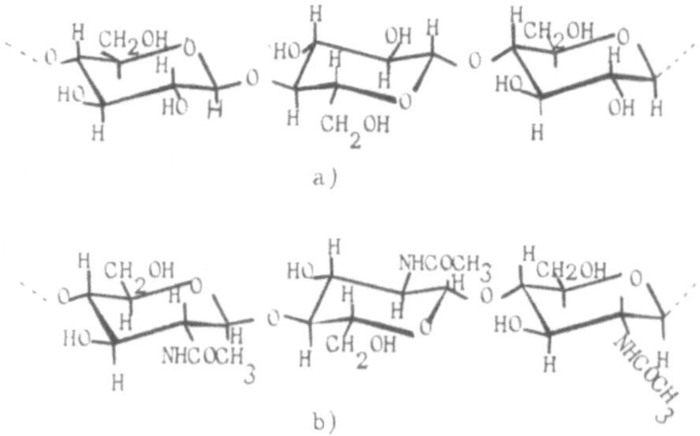

Fig. 1.4. Schematic drawings of cellulose (a) and chitin (b). In
both cases the monomers are linked by 1,4-β -linkages.
Chitosan, another interesting carbohydrate polymer is a
polymer of (1,4- β -linked) glucosamine.

       Both cellulose and chitin may be found in highly crystalline
forms and, unfortunately for certain chemical applications, are
almost invariably very difficult to dissolve without some chain degradation. Very crystalline cellulose samples have been obtained,
for instance, from cotton, from an algae (Valonia), and from cul-
tures of Acetobacter Xylinium.

       Recent evidence indicates that the microfibrils in native cel-
lulose range from 18 to 200 Å, depending on the origin (see Table
4).

       It is also interesting to notice that, structurally, native
cellulose is almost always in form I (Blackwell et al., 1978), with
the notable exception of cellulose from the cell walls of Halicys-
tis algae which is in form II. In form I the chains would have a
parallel alignment while in cellulose II they would be antiparallel. Relying upon this hypothesis, the proposed structures for the
two forms (Blackwell et al., 1978) would be those schematically
depicted in Fig. 1.5.

Table 4. Average Size of Cellulose Microfibrils from Different
Sources and Possible Interconversions of the Cellulose Polymorphs.

Source of cellulose microfibrils	Average lateral dimensions/$\overset{o}{A}$
Valonia	200
Tunicata	90–100
Cotton	50
Ramie	40
Wood	30
Quince Slime	25–30
Primary Cell Walls	18–20

Possible interconversions of
cellulose polymorphs

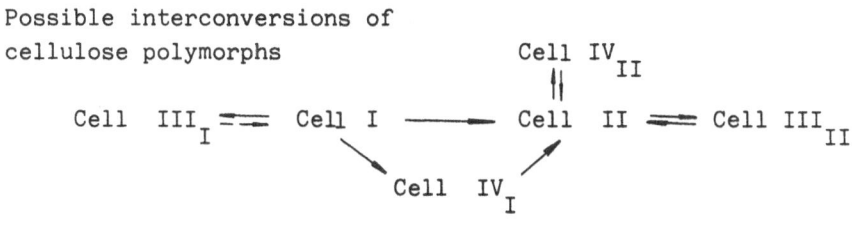

Deeper insights into cellulose structural arrangement are presently being obtained by examination of highly crystalline samples
also by means of cross-polarization magic angle spinning ^{13}C NMR
spectroscopy. It would be outside the scope of this elementary presentation to deal with the results of such experiments. The reader
is referred directly to the original literature in order to appreciate the power of the above mentioned, relatively new NMR technique which is proving valuable in the structural investigation
of biopolymers, e.g. cellulose (Earl and Vanderhart, 1981) and
chitin (Vincedon, unpublished results) as well as of synthetic
macromolecules in the solid state (McBrierty and Douglass, 1981).

Finally, other interesting linear, homopolymeric, "structural"
polysaccharides to be found in algae are Xylan (1,3-β -D-xylopyra-
noside) and Mannan (1,4-β -D-mannopyranoside) serving as the fibrillar substances instead of cellulose, while in some fungi and
yeasts such a function would be performed by 1,3-α -D-glucan.

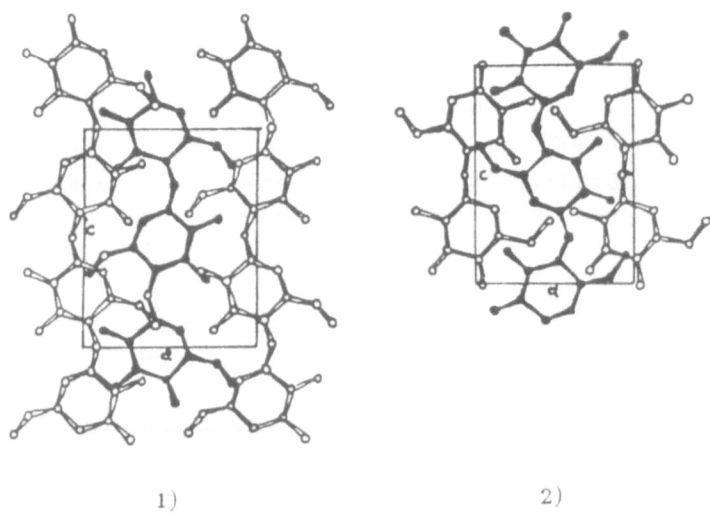

1)   2)

Fig. 1.5. 1) Projection of the parallel chain model for cellulose
         I. Projection perpendicular to the ac plane. The center
         chain (black) is staggered with respect to that at the
         origin. 2) Projection of the antiparallel chain model
         for cellulose II. Projection perpendicular to the ac
         plane (from Blackwell et al., 1978).

     The ability of these support substances to form regular se-
cundary, tertiary and quaternary structures play an essential role
in their biological functions. Existing knowledge on this point
shows examples of convergent evolutionary development at the molecular level, leading to similar architecture of different structural materials in different groups of organisms. For instance, chemically different structural polysaccharides like cellulose and
chitin have a corresponding conformation and packing (see Table 2).
On the other hand, 1,3-β-D-xylan forms wide triple helices (see
Table 2) packed in an hexagonal array (Atkins et al., 1969) giving
rise to strongly adherent microfibrils.

     "Energy-Storage" Polysaccharides. Among the "energy reserve" po-
lysaccharides one must recall that in nature starch is distributed
almost as widely as cellulose. It is a mixture of two polysaccharides,
amylose and amylopectin, both of which have covalent structures
based on chains of 1,4-linked-α-D-glucopyranose units (Fig. 1.6).

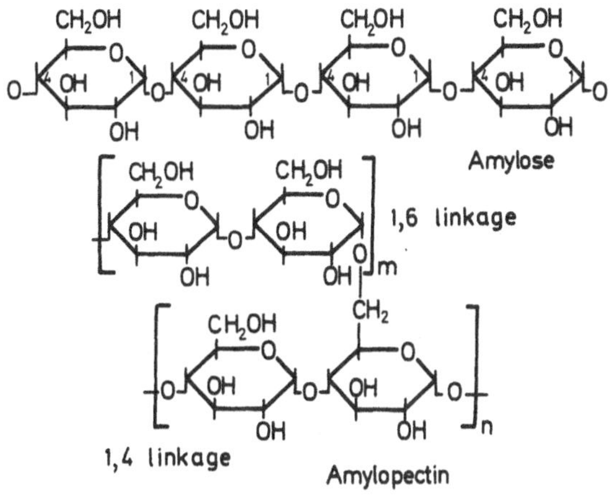

Fig. 1.6. Chemical structures of amylose and amylopectin.

Amylose is essentially linear, whereas in amylopectin chain segments of about 20 or more units are joined together by α-1,6-linkages to form a branched structure. Contrary to cellulose, amylose (and amylopectin) can be quite soluble in water: this is one demonstration of the dramatic influence that the nature of the glycosydic linkage (i.e. either β or α) may have on the physico-chemical properties of otherwise identical polysaccharidic chains. In further contrast with cellulose, whose conformational formula (see Fig. 1.2 and Table 2) reveals the necessity of stiff, straight chains (serving excellently their purpose in nature), the conformational formula of amylose suggests how the α-glucosydic linkages may yield a helical arrangement of the chains.

Amylose was the first natural polymer shown to be composed of helical chains. In the well known complex with iodine, the iodine molecules are lined up inside the helix, which would be stable in dilute aqueous solution. A possible double helical arrangement of amylose (French and Murphy, 1977) in the solid state is schematically given in Fig. 1.7.

The double helical molecular structures, determined by X-ray analysis of crystalline A and B forms of amylose are reported in the literature (Wu and Sarko, 1978a; 1978b). Other reserve polysaccharides of importance are xylans (1,4-β-D-xylopyranoside)found in

Fig. 1.7. Native starch double helix (from French and Murphy, 1977).

Watsonia (bulbs of a South African plant) and the granular mannans
(1,4-β-D-mannopyranoside) from ivory nuts.

Support and Binding Polysaccharides. There exists a large family of carbohydrate polymers from surface plants, seeds, and algae,
which perform a variety of functions in the living organisms as
supporting materials, to hold water, to hold ions, ets.

An incomplete list of the more common polymers belonging to this
family (and collectively named "polysaccharide gums"), having also widespread practical applications (Rees and Welsch, 1977) is given in Table 5.

The more interesting and practically used, water soluble polysaccharides listed in Table 5 are: a) guaran (guar gum), b) the
alginates, c) the carrageenans. The idealized repeating unit of
guaran is reported in Fig. 1.8. The polymer is composed of chains
of 1,4-linked β-D-mannopyranosyl units, with every second unit
bearing (on the average) on carbon C6 a single unit side chain that
consists of an α-D-galactopyranosyl group. One of the important

Table 5. Some Natural Polysaccharide-Gums.

Origin	Traditional Name	Gelling Propts.
Plant exudates	Arabic	X
	Tragacanth	
	Karaya	
	Ghatti	
Seaweed extracts	Agar	X
	Alginates	X
	Carrageenans	X
	Furcellaran	X
Plant seeds	Guar	
	Locust bean	
	Psyllium	
	Quince	
	Tamarind	
Plant extracts	Pectin	X
	Arabinogalactan	

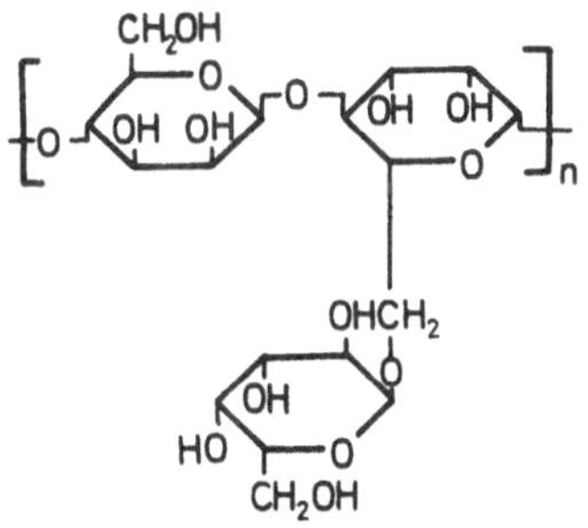

GUARAN

Fig. 1.8. An average repeating unit of guaran (purified guar-gum). Chains are built up by (β-(1,4)-D-manp) residues with a side (6,1)-linked β-D-galp residue.

properties of guar-gum is its ability to hydrate rapidly in cold
water to produce highly viscous solutions.

With the term "alginates" one normally refers to the major
structural polysaccharides of marine brown algae. Chemically they
are 1,4-linked block copolymers of β-D-mannuronate and its C_5 epi-
mer α-L-guluronate, with residues arranged in homopolymeric sequences of both types, and in mixed sequences that contain some alternating character (Fig. 1.9).

Being anionic polyelectrolytes, the alginates can bind metal
ions, in particular multivalent counterions, quite extensively and
can form gels in a manner strongly dependent upon the mannuronic/
guluronic acid content. Finally, the carrageenans (Fig. 1.10),
which occur in certain genera of red seaweeds, are sulfated galac-

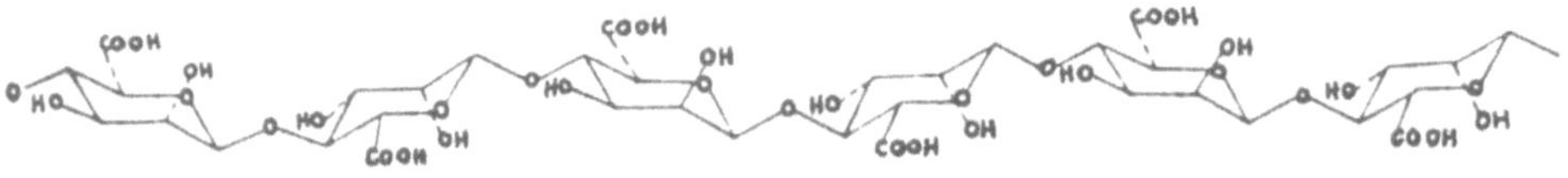

Poly(M)

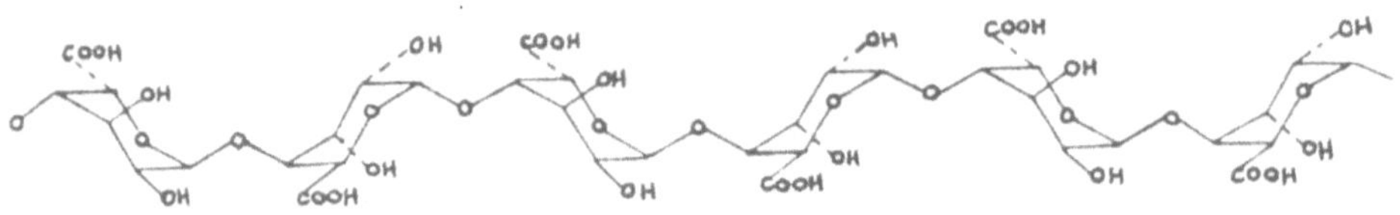

Poly(G)

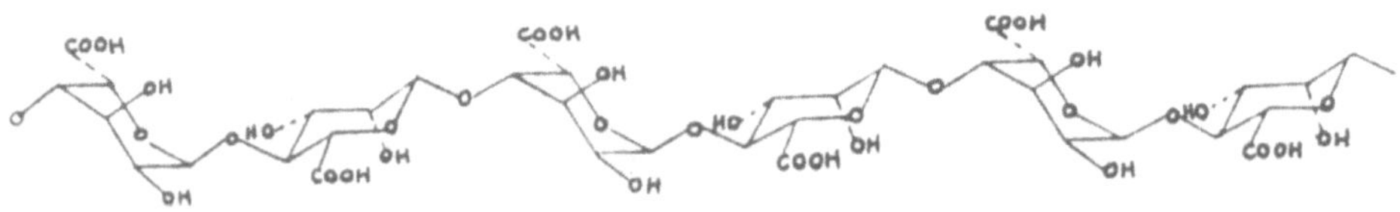

Poly(GM)

Fig. 1.9. Monomers enchainment in poly-α-D-mannuronic acid (Poly
(M)), in poly-α-L-guluronic acid (Poly(G)), and in their
alternating copolymers (Poly(GM): an idealized alginate
structure). The M residues are drawn in the $^4C_1$ conformation, while the G ones are in the more stable $^1C_4$ form.

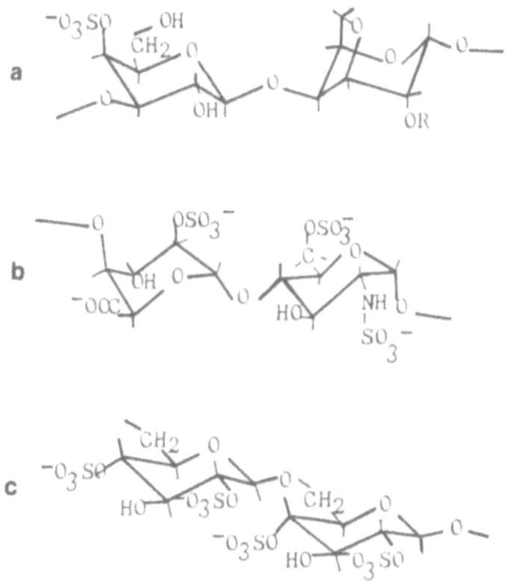

Fig. 1.10. Idealized repeating units of: (a) carrageenans segments
(R = H : K-carrageenan; R = SO$_3^-$: i-carrageenan);
(b) heparin; (c) dextran sulfate.

tans based on a disaccharide repeating structure with residues
linked alternatively α-1,3 and β-1,4. Double helical structures
for carrageenans in the solid state have been proposed (Rees and
Welsh, 1977). In reality, both iota and kappa carrageenan (Fig.
1.10) contain a small portion of 1,4-linked residues in the form
of galactose-6-sulfate: they are thus susceptible to periodate oxy-
dation, permitting selective chain cleavage at these residues. The
resulting carrageenan "segments" can assume ordered conformations
(likely double helical ones) in dilute aqueous solution. In contrast, intact carrageenan chains can proceed further in the process
of intermolecular aggregation giving rise to gelation, a phenomenon
strikingly dependent on the nature of added salts (see chapter 2).

The Glycosaminoglycan Polysaccharides. An important class of
carbohydrate polymers, the glycosaminoglycans, are widely distributed in animal tissues (Sharon, 1975). Probably the most thoroughly
studied polymer of this class is heparin, having special anticoagulant and antilipemic properties. Like other aminoglycans, heparin
is made up of uronic acid and amino-sugar building blocks in ap-

proximately equimolar proportion. More than 70% of the structure of
"conventional" heparins can be accounted for by the repeating unit
drawn in Fig. 1.10, that is from 1,4-linked L-iduronic acid (O-sulfated at position 2) and N-sulfated glucosamine (O-sulfated at position 6). This important polysaccharide is heterogeneous in regard to the degree of sulfation, and is polydisperse, with a degree
of polymerization usually ranging from 10 to 50[*]. Bearing up to
four anionic fixed charges per repeating unit, heparin is a strong
polyelectrolyte and can strongly interact with basic species and
with basic sites on proteins.

As a matter of fact, the association of heparin with plasma
protein is currently thought to be the basis for its biological activity. The anticoagulant activity of heparin is largely due to
formation of a very stable, specific complex with Antithrombin III,
and its antilipemic activity comes from the complex formation with
lipoproteinlipase. Such interactions would be regulated by divalent
inorganic cations, in particular Ca^{2+} ions. An elegant experiment
has shown that only segments of certain heparin chains contain the
"active site" for Antithrombin. In this experiment, heparin bound
on an antithrombin affinity column was treated with the enzyme
heparinase which cleaved the bound heparin chains only at sites not
involved in binding with the protein (Casu, 1979). The "active site"
turned out to be located in a dodecasaccharide. Very recent reports
indicate that the real active site may be even shorter, and constituted by only four saccharide units which, interestingly enough,
would be less sulfated than expected.

Microbial and Fungal Polysaccharides. Last but not least, we
may consider (very superficially) the family of the microbial polysaccharides, in particular the extracellular microbial carbohydrate
polymers (Sandford, 1979). The number of microbial species which
can produce substantial amounts of different polysaccharides is
large. Moreover the degree of polymer production by a given orga-

---

[*]The question remains if these are intrinsic properties of heparin
in the living tissues or they may be brought about to some extent
by its isolation procedure. Heparin chains, in fact, are present
in tissues as proteoglycans, i.e. bound to a protein, from which
they are released by combined action of proteolytic enzymes and
a base.

nisms usually can be improved by variation in culture conditions. Extracellular microbial polysaccharides are interesting polymers because of their often very regular, peculiar chain structure and also because some of them are finding widespread commercial use.

Similar considerations apply to the so called Fungal Polysaccharides (see Marchessault and Deslandes, 1980). An incomplete list of such polysaccharides is given in Table 6. Among extracellular gums which have met the greatest commercial success one should recall xanthan, the repeating unit of which is depicted in Fig. 1.11. Xanthan, normally with a molecular weight of the order of 10^6, dissolves in water giving very viscous solutions which are pseudo plastic, i.e. the measured viscosity decreases rapidly as the shear rate is increased. Its physico-chemical equilibrium properties in aqueous media have recently been the subject of rather detailed studies (Brant, 1981). Other structurally and biologically quite interesting microbial polysaccharides composed of regular repeating units are those which can be obtained from the seventy-seven serologically different strains of <u>Klebsiella</u> known so far.

Table 6. Examples of Polysaccharides from Microbial (Fungi)
and Bacterial Sources.

Name	Organisms	Chemical repeating unit
Dextran	Leuconostoc mesenteroides	$-((1,6)-\alpha-D-glcp)-$, branched
Xanthan[x]	Xanthomonas campestris	$-((1,4)-\beta-D-glcp-(1,4)-\beta-D-glcp)-$ (3,1) $\beta-D-manp-(1,4)-\beta-D-glcAp-(1,2)-\alpha-D-manp-6-OAc$ / \ 4 6 \ / C / \ CH_3 .COOH
Scleroglucan	Sclerotium glucanicum (fungi)	$-((1,3)-\beta-D-glcp-(1,3)-\beta-D-glcp-(1,3)-\beta-D-glcp)-$ (6,1) $\beta-D-glcp$
Alginates	Azotobacter vinelandii	$-((1,4)-\beta-D-manAp)_m-((1,4)-\alpha-L-gulAp)_n$
Curdlan	Alcaligenes faecalis	$-((1,3)-\beta-D-glcp)-$
Pullulan	Anereobasidium pullulans (fungi)	$-((1,4)-\alpha-D-glcp)_4-(1,6))-$
Cellulose	Acetobacter xylinium	$-((1,4)-\beta-D-glcp)-$

[x]One pyruvate group occurs, on the average, on every two repeating units.

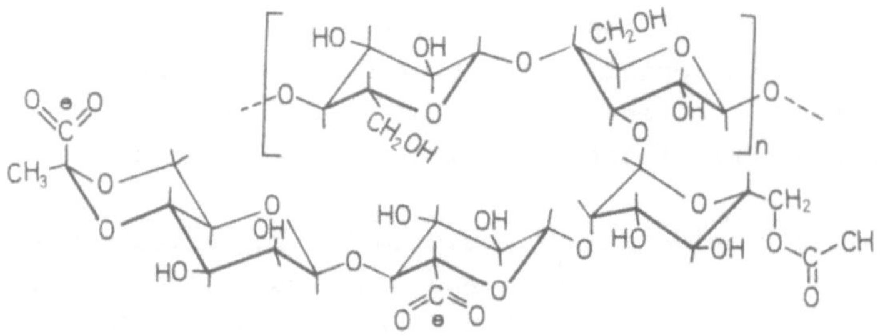

Fig. 1.11. Repeating unit of the exocellular microbial polysac-
          charide Xanthan (see Table 6). In reality, pyruvate
          group is attached to approximately one-half of the side-
          chain terminal mannose residues.

Likewise, particularly important and structurally intriguing are
the capsular polysaccharides from Pneumococcus.

2. POLYSACCHARIDE SOLUTIONS

"Polysaccharide Solutions" is a very wide and rather complicated topic. It encompasses, in fact, aqueous solutions and non-
aqueous ones of both charged and/or non-charged saccharide polymers
which, superimposed on the behavior to be expected for macromolecules with relatively stiff backbones, exhibit conformation dependent properties strongly influenced by the nature of the solvent
medium. In what follows, discussion will be limited to a few interesting examples of the special solution behavior often displayed
by polysaccharides. Attention will be focussed on certain equilibrium properties, without pretending to put the matter into any rigorous physico-chemical or biological framework. The presentation,
a simple survey of experimental data and phenomenological observations pertaining mainly to aqueous solutions, can be divided into
two parts:
A. Dilute solution properties of ionic-polysaccharides, in particular: a thermodynamic approach to counterions binding in i-carrageenan (segments) aqueous solutions.
B. More concentrated solutions of both charged and non-charged
   carbohydrate polymers giving rise to mesophase formation or to
   gel formation.

A. Counterions Binding in Sulfated Polysaccharide Solutions

One of the most typical features of polyelectrolyte solutions which has since many years attracted the interest of many workers, certainly is "counterions-binding". The phenomenon has been studied employing a number of experimental approaches, and has stimulated the minds of many theoreticians.

One of the salient properties of polysaccharide sulfates in aqueous media is the strong binding of counterions, in particular of divalent metal ions, eventually leading to gelation depending upon polymer and counterions concentration, and upon temperature. Qualitatively speaking this is what can be expected considering that the chains of such polyelectrolytes have a relatively high fixed-charge density. On the other hand, steric regularity of natural polysaccharide sulfate chains and chemical constitution of their repeating units can add to their "polyelectrolytic" behavior certain very interesting features, including counterions-binding ones, specifically dependent on structure and conformation of the macroions. In that context, and to afford just one example, we may consider a few microcalorimetric data (Crescenzi et al., 1981a; Crescenzi and Rizzo, unpublished) on the enthalpy of Cu^{2+}, Mg^{2+}, and Ca^{2+} ions binding by $\underline{i}$-carrageenan (segments), heparin, and dextran sulfate (Fig. 1.10) (a sample with two sulfate groups per glucose residue), respectively, in dilute aqueous solution. The latter two polymers have similar anionic charge densities, approximately twice that of $\underline{i}$-carrageenan (a sample having 1.7 sulfate groups per repeating unit, on the average) but, of course, quite different chain architectures (see Fig. 1.10).

Heparin and $\underline{i}$-carrageenan (segments) should have relatively stiff backbones with an essentially regular structure, and therefore liable (in principle) to assume ordered chain conformations in solution[*].

---

[*]In heparin, which bears both $-SO_3^-$ and $-COO^-$ groups, there is in reality a non-strictly regular substitution of ionic residues along the chains. The heparin sample mentioned here, had a carboxyl to sulfate groups ratio equal 1/2.3.

   In contrast, in dextran sulfate, the product of sulfation of
dextran, a natural polymer with an inherently rather flexible backbone, both chain branching (a modest branching though) and a somewhat irregular distribution of $-SO_3^-$ groups tend to confine the po-
lyanions in a randomly coiled state.

   The results of the calorimetric experiments are reported in
Figs. 2.1 and 2.2. Fig. 2.1 shows that binding of divalent metal
ions by dextran sulfate and by heparin is an endothermic process.
Consequently, it must be driven by the entropy. This may be safe-
ly ascribed to the release of water molecules from the electrostric-
ted hydration sheaths of interacting species (in particular Cu^{2+}
ions), as suggested by the rather large volume increases upon binding, experimentally observed for different polyelectrolyte-coun-
terions systems (Paoletti et al., 1981).

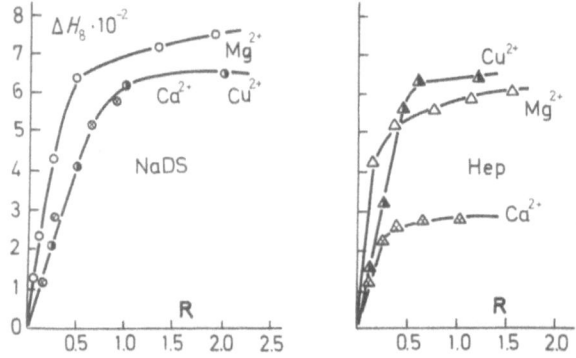

Fig. 2.1. Microcalorimetric data on the heat of interaction (ΔH_B)
        of heparin and dextran sulfate with divalent counterions
        in dilute aqueous solution at $25^{\circ}C$. Polymer concentration:
        $5 \cdot 10^{-3}$ equiv./l. ΔH_B is the observed heat, corrected
        for dilution effects, and normalized per equivalent of
        polysaccharide in solution. R is the stoichiometric ratio
        of added equivalent M^{2+} concentration to polysaccharide
        equivalent (constant) concentration (from Crescenzi et
        al., 1981a).

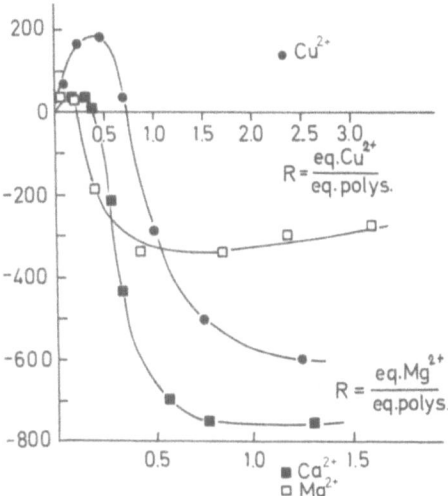

Fig. 2.2. Microcalorimetric data for binding of different M^{2+} ions
        by i-carrageenan segments in dilute aqueous solution at
        $25°C$ (see legend to Fig. 2.1) (from Crescenzi et al.,
        1981a; Crescenzi and Rizzo, unpublished).

        The results for i-carrageenan (Fig. 2.2) do follow a strikingly
different pattern. Copious evidences in favor of the occurrence of
a salt induced disorder $\rightleftharpoons$ order conformational change of i-carrageenan chains come from a number of different, independent experiments (Crescenzi et al., 1981a; Crescenzi and Rizzo, unpublished). The change in sign of ΔH_B upon increasing R (Fig. 2.2)
can then be traced back to such a conformational change promoted
by added divalent counterions via an efficient screening of electrostatic repulsions among macroions fixed charges. The conformational change would be an exothermic process, with an onset at R
values depending on the counterion (Fig. 2.2), superimposed to and
more than compensating the typical endothermicity of the M^{2+}/SO_3^-
interaction (see Fig. 2.1). The ability of counterions in trig-
gering the i-carrageenan conformational change, as monitored by

the change in sign of ΔH_B, would then follow the sequence: $Mg^{2+} > Ca^{2+} > Cu^{2+}$ (see Fig. 2.2).

That such a sequence reflects a specific counterions effects and not just simply a decreasing extent of binding is strongly suggested by potentiometric data (collected using ion-specific electrodes (Crescenzi and Rizzo, 1982), showing that Cu^{2+} ions are, in fact, bound by i-carrageenan to a larger extent than Ca^{2+} ions. It is also interesting to report that, elaboration of data of Fig. 2.2 (Crescenzi et al., 1981b) leads to the estimate that at $25^{\circ}C$ the enthalpy change associated with the isothermal disorder → order conformational transition of i-carrageenan segments with either Ca^{2+} or Cu^{2+} ions is about -1 kcal per mole of disaccharide repeating unit. In the case of Mg^{2+} ions, similar calculations lead to the value of approximately -0.5 kcal/mole for the transition enthalpy. This might indicate that the conformational change of i-carrageenan segments, whether it does or does not involve formation of double helices (a matter still subject to some controversy and about which we cannot enter into details here), is a process whose extent, if not its final ordered state too, depends on the nature of counterions. This section must close here, however, as it was aimed at providing some original evidence on this point which on the other hand, finds analogies in a variety of different ionic-polysaccharide/counterions systems.

Ample discussions on the interesting matter concerning possible ordered conformations of ionic-polysaccharides in solution and the influence thereon of the nature of added salts, may be found in recent literature (Morris et al., 1980; Crescenzi et al., 1981b; Smidsrod, 1980).

It has to be mentioned that different non-ionic polysaccharides whose chains belong to the periodic-sequence group (Table 2) can also assume ordered conformations in dilute aqueous solution, whose stability is governed by temperature and which moreover, can collapse into random coiled conformations in organic solvents. This is, for example, the case of Schizophyllum commune polysaccharide (Schizophyllan) whose chains consist of linearly linked β-1,3-D-glucose residues with one β-1,6-D-glucose side chain for every three main chain residues.

In dilute aqueous solutions, schizophyllan would assume a triple-helical conformation (which is completely dispersed in di-

methylsulfoxide or in concentrated NaOH (Kashiwagi et al., 1981)
with geometrical features similar to those of other polysaccharides, e.g. β -1,3-xylan (Atkins and Parker, 1969) and β -1,3-glucan
(Deslandes et al., 1980) in the solid state.

In conclusion, the polysaccharide chains in dilute aqueous
solutions can feature such diverse ordered conformational states
as single, double or triple-helical ones, beyond the trivial random
coil state; they do therefore provide a fertile material for investigations.

B. Concentrated Solutions

Mesophases Formation. For some substances whose molecules are
markedly asymmetric in shape, the tendency toward an ordered arrangement is so great that the crystalline form does not melt directly to a liquid phase but first passes through an intermediate
stage (mesophase), which at a higher temperature undergoes a transition to the liquid state. The intermediate states are also called
"liquid crystals" since they display some of the properties of
each of the adjacent states. Two main types of anisotropic melt
may be obtained, namely: smectic, in which the molecules are oriented in well defined planes (with defined orientation but without
periodicity), and nematic in which the planar structure is lost
but the orientation is preserved. With highly anisometric species
bearing centers of asymmetry, one can also obtain cholesteric mesophases; in this case each molecule is slightly displaced in relation to the next giving rise to a helical arrangement (see Fig.
2.3). What said above can be qualitatively extended to (concentrated) solutions of different polymers in certain solvents. The
situation may be summarized with the words of P.J. Flory (1980):
.....''The random disorder characteristic of the more commonplace
synthetic polymers in the amorphous state and in solution requires
a minimum degree of chain flexibility. For chain molecules failing
to meet this requirement, theory predicts the occurrence of an ordered phase or of liquid-crystalline domains. Examples confirming
this prediction include α -helical polypeptides, polyisocyanates,
p-phenylene polyesters and polyamides and, belatedly recognized,
cellulose and cellulose derivatives. The fact that polysaccharide
chains are often sufficiently rigid to be endowed with the capacity
to yield mesomorphic phases may provide the clue to the explanation
of their special properties. Theory is most explicit for rigid rod-

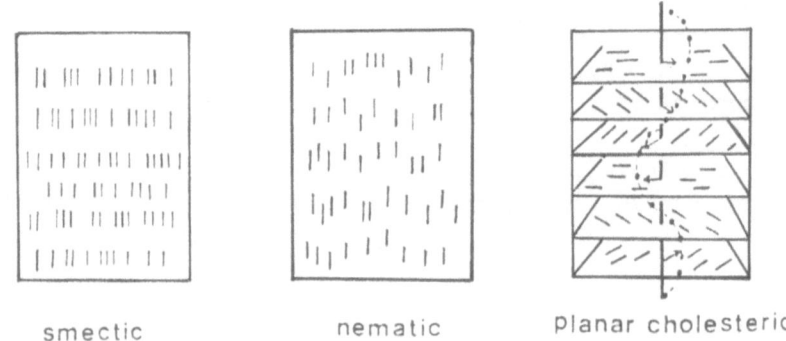

smectic   nematic   Planar cholesteric

Fig. 2.3. Schematic drawing of the arrangements of rod-like mole-
          cules in smectic, nematic and cholesteric mesophases,
          respectively.

like particles. Separation of their mixtures with a diluent into
two phases, one isotropic and the other nematic, is unambiguously
predicted on the basis of particle asymmetry alone. Partial flexibility of a semi-rigid polymer chain reduces its tendency to form
a nematic phase: nevertheless an appreciable degree of flexibility
may be tolerated."....

     As said above, the fact that cellulose, cellulose derivatives
as well as other polysaccharides can give rise, in concentrated
solutions, to the formation of mesophases has been only belatedly
recognized. At present, however, the research on these systems is
active and quite a few interesting cases have been finally disclosed in the scientific literature. For instance, Chanzy et al.
(1980) have recently shown that cellulose solutions (30-35% by
weight) in N-methyl morpholine N-oxide (with small amounts of water)
display optical birefringence indicative of liquid-crystalline
ordering of cellulose chains. Simple smearing of such solutions
followed by "regeneration" in water and methanol leads to highly
oriented, crystalline cellulose films and fibres. More recently
the formation of mesophases in solutions of a variety of natural
polysaccharides always in the N-methyl morpholine N-oxide/water
solvent system has been reported (Chanzy et al., 1982). Interestingly, also in the case of relatively concentrated aqueous solutions of the exocellular microbial polysaccharide xanthan it has
been shown that cholesteric mesophase formation can take place

(Maret et al., 1981). Besides the scientific interest in all of these findings[*] their practical relevance should also be stressed in that they may lead the route to the preparation of high modulus, high tenacity fibers.

In order to briefly illustrate some salient features of meso-phase formation in polysaccharide solutions, it is expedient to consider a few data for hydroxypropylcellulose (HPC).

HPC is a polymer (Fig. 2.4) soluble in a surprisingly large number of solvents, from water to piperidine, having widely different hydrogen bonding capabilities and polarities. Moreover, liquid-crystalline phases are easily discernible with concentrated HPC solutions in a variety of solvents. For example, in a 30% (w/w) HPC solution in acetic acid, the periodicity lines characteristic of large-pitch cholesteric structures are visible (Werbowyj and Gray, 1981). It is also found that at a given temperature the concentration required to form an ordered phase depends on the solvent but only slightly on HPC average molecular weight. In this connection one may recall that the critical volume fractions, Φ_2, of rods required for phase separation (isotropic solution + anisotropic phase) is approximately (Flory, 1956):

$$\Phi_2 \sim (8/y)(1 - 2/y), \tag{1}$$

where y is the axial ratio of the rods. Viewing the HPC chains as a set of freely jointed rods, each with an axial ratio y, one may then qualitatively understand that the phase separation is mainly controlled by the y value and not by the number of such rod-like segments per chain (that is, by the molecular weight). Incidentally, for HPC the y value should be about 13; assuming a diameter of 12 Å, this would correspond to rod-like stretches of about 170 Å, comprising about 30 monomeric units. The complete phase diagram for the system HPC-H_2O is reported in Fig. 2.5. It contains a number of interesting features discussion of which, however, is

---

[*]The paracrystalline state seems one of the most suited to biological functions, as it combines the fluidity and diffusibility of liquids while preserving the possibilities of "internal structure" characteristic of crystalline solids. As a matter of fact, this is indeed the case in a number of instances.

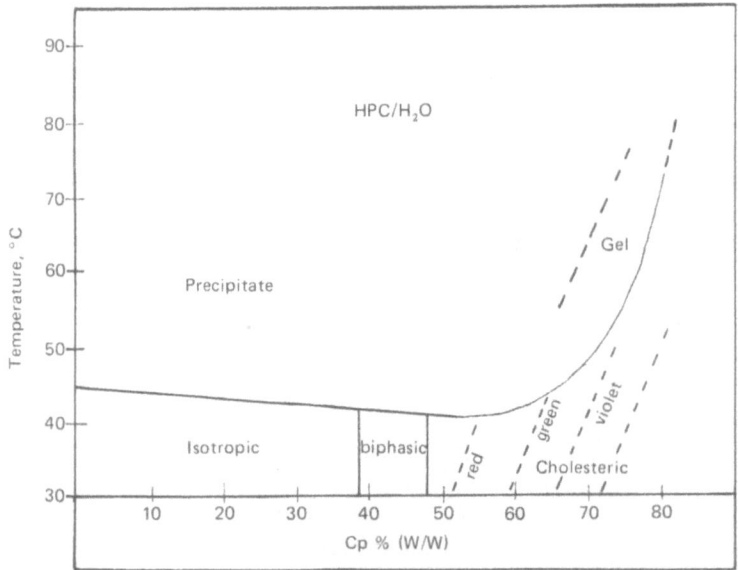

Fig. 2.4. Chemical structure of hydroxypropylcellulose, with an
          average of three hydroxypropyl groups per anhydroglucose
          residue.

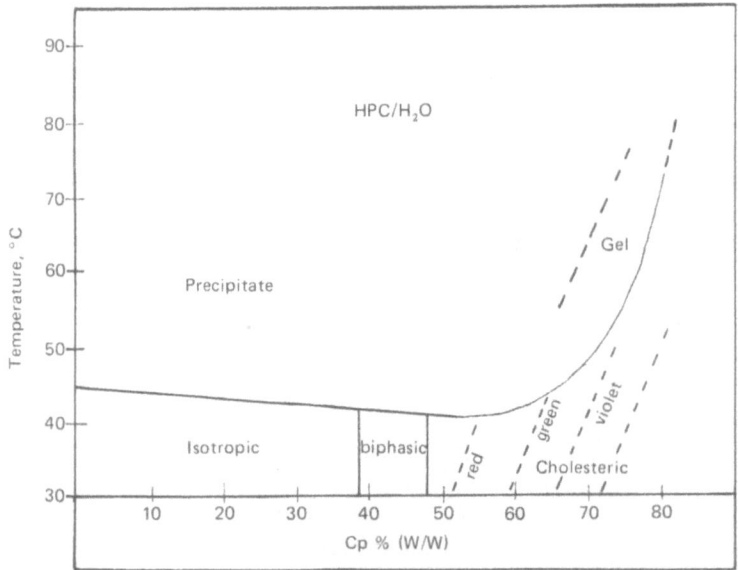

Fig. 2.5. Phase diagram for the hydroxypropylcellulose-water system.
          The different colors that may be seen on increasing the
          polymer concentration in the cholesteric mesophase-region
          are due to the varying pitch of the helical overall ar-
          rangement, acting as a diffraction grating for sun light.

beyond   the scope of this brief presentation (Bianchi and Valente, 1982).

In concluding this section, it may be useful to point out that the fact that different polysaccharidic chains of sufficient chain stiffness (a necessary but not sufficient property for ordered-phase separation) in concentrated solutions do not crystallize or do not become locked into an unoriented gel may be due to a number of favourable reasons. Among these, and disregarding possible kinetic effects, probably important is the presence of appropriate substituent groups (e.g. the hydroxypropyl groups of HPC) or of "bound" solvent species (e.g. N-methyl morpholine N-oxide in the case of cellulose) along the chains which at high volume fractions might allow the chains to "slip"to an optimal arrangement eventually yielding liquid-crystals. The latter arrangement, incidentally, may be solvent dependent and has not to be necessarily reminiscent of that prevailing in the pure polymer crystalline state, whenever this can in practice exist (Chanzy et al., 1982).

Gel Formation. In "dilute" solutions of random coil macromolecules intermolecular interactions are, by definition, negligible and the chains can be fully solvated. The other extreme is represented by the solid state in which structurally regular, linear macromolecules, such as cellulose, can pack orderly in a sterically regular chain conformation. The gel state has features in common with the two extreme states mentioned above, with interchain junctions comprising the association of long but structurally and conformationally regular chain segments (as in the solid state) and with non-associated regions which are conformationally irregular and extensively hydrated, serving to "solubilize" (or, better, to swell) the gel network. A critical requirement for a gelling macromolecule may therefore be sufficient flexibility and structural irregularity to prevent complete association and precipitation, as well as the need for sufficient regularity to allow stable, interchain junctions to form in a cooperative manner. The one feature identified almost universally as an essential characteristic of a gel is its solid-like mechanical behavior. When deformed, its response is that of an elastic body with a low modulus of elasticity. If plastic flow occurs, it will so respond only above a finite yield stress: deformation at lower stresses is recoverable, and hence elastic.

One of best known gel forming polysaccharides is <u>agarose</u>
(Fig. 2.6).

For agarose the formation of left-handed double helices is
proposed to play an important role in the formation of junction-
zones. It is further proposed that the double helices associate
to form aggregates to act as "super junctions" (Rees, 1981), causing high concentration agarose gels to be opaque. The α-D-galac-
tose containing sequences would be helix-breaking and would connect
in a disordered fashion the junction-zones.

Polysaccharides with diastereoisomeric sequences with respect
to agarose are the carrageenans (Fig. 2.6). In this case, the block
sequences p would interact in aqueous solution yielding double-
helices while the helix-breaking sequences q would be the spacers
causing each chain to associate with more than one partner as is
necessary to arrive at a three-dimensional, irregular network (gel).[*]
In reality , the mechanism of <u>i</u>-carrageenan gelation, which markedly depends on the nature of simple added salts, might occur in
two steps, as schematically represented in Fig. 2.7 (Rees, 1981).

A further interesting case of gelation, brought about specifically by divalent metal ions, is that of alginates (see Fig.
1.9). Studies of alginate gelation have shown that the primary process of interchain association is by dimerization of the L-guluronate sequences in a regular 2-fold conformation (analogous to that
of the free acid in the solid state) with specific interchain
chelation of calcium or of other divalent cations of appropriate
size. A schematic representation of the "egg-box" model for alginate gelation is given in Fig. 2.8: the random coils would convert
to buckled ribbons which pack to contain arrays of Ca^{2+} ions (Rees,
1981).

Finally, it is interesting to recall one case of mixed gel
formation, e.g. the phenomenon taking place when aqueous solution

---

[*]As already mentioned, these sequences may be chopped out of native <u>i</u>-carrageenan chain by means of the Smith-degradation. The
resulting <u>i</u>-carrageenan segments (Fig. 1.10), essentially devoid
of structural defects (and with practically no gelling tendency)
have been dealt with in chapter 1.

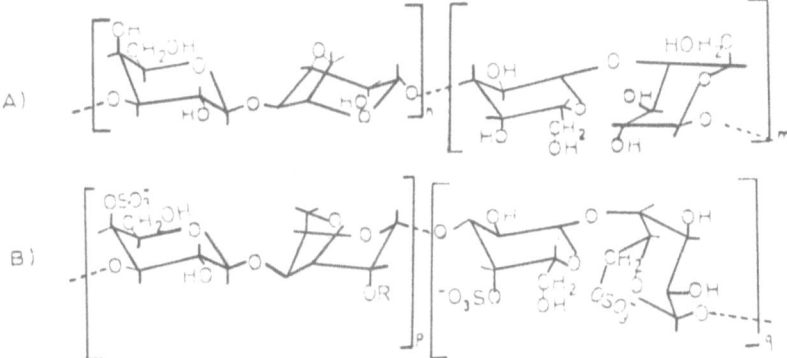

Fig. 2.6. (A) agarose, (B) native carrageenan chains.

of xanthan (see Fig. 1.11) and of certain plant galactomannans are mixed together. While xanthan alone would not gel, mixed gels are in fact easily formed at low concentrations (ca. 0.5%) with locust-bean gum, a polysaccharide with relatively long unsubstituted (by galactose branches) mannan sequences. In contrast, with guaran (see Fig. 1.8) in which the net ratio of substituted to unsubstituted mannose residues is close to unity, no gel is formed despite visco-simetric evidence of "interactions" with xanthan (Dea et al., 1977). These may be considered to be typical examples of specific polysac-

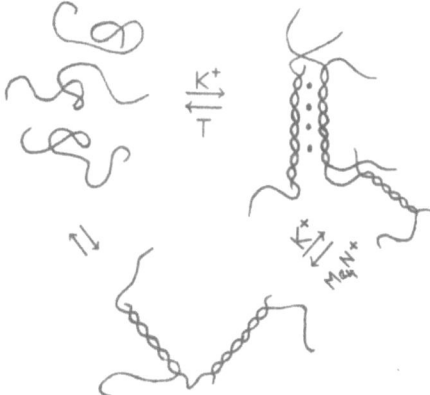

Fig. 2.7. Simplified scheme of possible sequence of events leading to gelation of intact carrageenans in aqueous media (adopted from Rees, 1981).

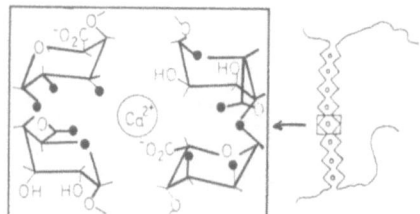

Fig. 2.8. The "egg-box" model for alginate gelation induced by Ca^{2+} ions. Oxygen atoms involved in interactions with a single Ca^{2+} are indicated in the enlarged section as filled circles (adapted from Rees, 1981).

charide-polysaccharide interactions eventually leading to supramolecular structures.

CONCLUDING REMARKS

In concluding this review quite obviously an incomplete one, of certain aspects of the macromolecular chemistry of carbohydrate polymers, I hope that in recalling well known facts and in presenting a few original pieces of evidence I may have given the impression that the world of polysaccharides has been developing at the fascinating rate, deserving attention and keen interest. The interdisciplinarity of the field is most wide, since it includes the problems of molecular biology, organic chemistry, macromolecular chemistry, as well as technological problems. Since these notes may not at all make justice of all these facts, the reader is strongly recommended to take advantage of references given herein.

REFERENCES

Atkins, E.D.T., Isaac, D.H., Nieduszynski, J.A., Phelps, C.F., and Sheehan, J.K., 1974, Polyuronides: their molecular architecture, Polymer, 15:263.
Atkins, E.D.T., and Parker, K.D., 1969, The helical structure of β-D-(1,3)-xylan, J. Polymer Sci. C, 28:69.
Atkins, E.D.T., Parker, K.D., and Preston, R.D., 1969, Helical

structure of the β -1,3-linked xylan in some siphoneous green
algae, Proc. Roy. Soc. B, 173:209.

Bianchi, E., and Valente, A., 1982, to be published.

Blackwell, J., Kolpak, F.J., and Gardner, K.H., 1978, The structure
of cellulose I and II, Tappi, 61:71.

Brant, D.A., 1976, Conformational theory applied to polysaccharide
structure, Quart. Rev. Biophys., 9:527.

Brant, D.A., (ed.), 1981, "Solution Properties of Polysaccharides",
ACS Symp. ser. 150, American Chemical Society, Washington, D.C.

Casu, B., 1979, Fractionation and characterization of glycosaminoglycans of mammalian origin, Pharm. Res. Comm., 11:1.

Chanzy, H., Chumpitazi, B., and Peguy, A., 1982, Solutions of polysaccharides in N-methyl morpholine N-oxide, Carbohydr. Polymers, in press.

Chanzy, H., Peguy, A., Chaunis, S., and Monzie, P., 1980, Oriented
cellulose films and fibres from a mesophase system, J. Polymer
Sci., 18:1137.

Crescenzi, V., Airoldi, C., Dentini, M., Pietrelli, L., and Rizzo
R., 1981b, Calorimetric data on salt-induced conformational
transitions of ionic-polysaccharides in aqueous solution,
Makrom. Chem., 182:219.

Crescenzi, V., Dentini, M., and Rizzo, R., 1981a, Polyelectrolytic
behavior of ionic-polysaccharides, in: "Solution Properties
of Polysaccharides", D.A. Brant, ed., ACS symp. ser. 150,
American Chemical Society, Washington, D.C.

Crescenzi, V., and Rizzo, R., 1982, unpublished results.

Dea, I.C.M., Morris, E.R., Rees, D.A., Welsh, E.J., Barnes, H.A.,
and Price, J., 1977, Association of like and unlike polysaccharides: mechanism and specificity in galactomannans, interacting bacterial polysaccharides, and related systems, Carbohydr. Res., 57:249.

Deslandes, Y., Marchessault, R.H., and Sarko, A., 1980, Triple
helical structure of (1,3)- β -D-glucan, Macromol., 13:1466.

Earl, W.L., and Vanderhart, D.L., 1981, Observation by high-re-
solution carbon-13 NMR of cellulose I related to morphology
and crystal structure, Macromol., 14:570.

Flory, P.J., 1956, Phase equilibria in solutions of rod-like particles, Proc. Roy. Soc., London, A, 234:73.

Flory, P.J., 1969, "Statistical Mechanics of Chain Molecules",
Interscience, New York.

Flory, P.J., 1980, 179th ACS National meeting: Cellulose, paper and
textiles division, Houston, Texas.

French, A.D., and Murphy, V.G., 1977, Computer modeling in the study
     of starch, Cereal Foods World, 22:61.
Gorin, P.A.J., 1981, Carbon-13 NMR spectroscopy of polysaccharides,
     Adv. Carbohydr. Chem. Biochem., 38:13.
Guizard, C., 1981, Ph.D. Thesis, University of Grenoble, France.
Kashiwagi, Y., Norisuye, T., and Fujita, H., 1981, Triple helix of
     Schizophyllum Commune polysaccharide in dilute solution. 4.
     Light scattering and viscosity in dilute aqueous sodium hydro-
     xide, Macromol., 14:1220.
Marchessault, R.H., Deslandes, Y., 1980, Texture and crystal struc-
     ture of fungal polysaccharides, in: "Fungal Polysaccharides",
     P.A. Sandford and K. Matsuda, eds., ACS symp. ser. 126, Ame-
     rican Chemical Society, Washington, D.C.
Maret, G., Milas, M., and Rinaudo, M., 1981, Cholesteric order in
     aqueous solutions of the polysaccharide xanthan, Polymer
     Bull., 4:291.
McBrierty, V.C., and Douglass, D.C., 1981, Recent advances in the
     NMR of solid polymers, J. Macromol. Sci., Macromolecular Rev.,
     16:295.
Montreuil, J., 1980, Primary structure of glycoprotein glycans.
     Basis for the molecular biology of glycoproteins, Adv. Car-
     bohydr. Chem. Biochem., 37:157.
Morris, E.R., Rees, D.A., and Robinson, G., 1980, Cationic-specific
     aggregation of carrageenans helixes: Domain model of polymer
     gel structure, J. Mol. Biol., 138:349.
Paoletti, S., Cesaro, A., Ciana, A., Delben, F., Manzini, G.,
     and Crescenzi, V., 1981, Thermodynamics of protonation and
     of copper(II) binding in aqueous alginates solutions: Alginic
     acid from Azotobacter Vinelandii, in: "Solution Properties
     of Polysaccharides", D.A. Brant, ed., ACS symp. ser. 150, Ame-
     rican Chemical Society, Washington, D.C.
Perlin, A.S., 1980, Glycol-cleavage oxidation, in: "The Carbohy-
     drates Chemistry and Biochemistry", W. Pigman and D. Horton,
     eds., Vol. I B, 2nd ed., Academic Press, New York.
Pigman, W., and Horton, D., (eds.), 1972, "The Carbohydrates ,
     Chemistry and Biochemistry", Vol. I A, Academic Press, New
     York.
Rees, D.A., 1977, Polysaccharide shapes, in: "Outline Studies in
     Biology", Chapman and Hall, London.
Rees, D.A., 1981, Polysaccharide shapes and their interactions.
     Some recent advances, Pure Appl. Chem., 53:1.
Rees, D.A., and Welsh, J.E., 1977, Secondary and tertiary structure of

polysaccharides in solutions and in gels, Angew. Chem. Int. Ed. Engl., 16:214.

Sandford, P., 1979, Extracellular microbial polysaccharides, Adv. Carbohydr. Chem. Biochem., 36:265.

Schuerch, C., 1981, Synthesis and polymerization of anhydro sugars, Adv. Carbohydr. Chem. Biochem., 39:157.

Sharon, N., 1975, "Complex Carbohydrates", Addison-Wesley, Reading, Mass.

Sharon, N., and Lis, H., 1981, Glycoproteins: Research booming on long-ignored, ubiquitous compounds, Chem. Eng. News., 59:21.

Smidsrod, O., 1980, "Structure and Properties of Charged Polysaccharides", IUPAC 27th Symp., Pergamon Press, London.

Sturgeon, R.J., 1980, Enzymic method for determination of glycerol: Chain length of hexans, in: "Methods in Carbohydrate Chemistry", R.L. Whistler, and J.N. Miller, eds., Vol. VIII, Academic Press, New York.

Sundarajan, P.R., and Marchessault, R.H., 1979, Bibliography of crystal structures of polysaccharides, Adv. Carbohydr. Chem. Biochem., 3:315.

Vincedon, M., 1982, to be published.

Werbowyj, R.S. and Gray, D. G., 1981, Ordered phase formation in concentrated hydroxypropylcellulose solutions, Macromol., 13:69.

Wu, H-C.H., and Sarko, A., 1978a, Packing analysis of carbohydrates and polysaccharides. VIII. The double helical molecular structures of crystalline B-amylose, Carbohydr. Res., 61:7.

Wu, H-C.H., and Sarko, A., 1978b, Packing analysis of carbohydrates and polysaccharides. IX. The double helical molecular structure of crystalline A-amylose, Carbohydr. Res., 61:27.

HYDRATION AND INTERACTIONS IN AQUEOUS

SOLUTIONS OF IONS AND MOLECULES

Felix Franks

Department of Botany
University of Cambridge
Cambridge, United Kingdom

WATER AND LIFE

I need hardly convince you that water is central to biology in controlling and modifying the reaction of life. The importance of water in exerting a controlling, or modifying influence on protein function was really only recognized some ten years ago. The whole problem of solvation only appeared in enzyme studies in 1970 and we have not advanced very far since then.

Life in an aqueous environment can be treated at three levels. We first of all have the so-called water structure which arises from the peculiar intermolecular nature of water and which is perturbed by pressure, temperature and solutes. At the molecular level, we must examine discrete water molecules in protein crystals or in polysaccharides to see how they act in stabilizing or de-stabilizing certain structures. Then we have, on a more macroscopic level, water as a transport medium that distributes nutrients and carries waste products away; this is a fascinating study in itself. Then, finally, there is water as the physical environment for many living organisms in oceans, rivers and lakes. Many organisms live all their lives in water, and the physical properties of water have really shaped the development of such organisms in no small amount.

Since this is the problem of the interdisciplinary nature, let me give you one example of how various disciplines look at water, and at mild perturbations. Let us look at the effect of heavy water, D_2O (Franks, 1982). To the physicist D_2O substitution is no problem; he can easily account for the properties of D_2O by allowing for the difference in zero-point energies between H_2O and D_2O. The physicists tell us that the ratio p_{D_2O}/p_{H_2O} of the vapor pressure is 1.05. Sure enough, that is exactly what you find near the critical point of water. The problem is that at room temperature this ratio is 0.87. We have the situation that at low temperature H_2O is the more volatile of the two and somewhere at above 200° the vapor pressure curves cross so that at high temperatures D_2O is more volatile. No zero-point energy correction can explain that. To the physicist, this is a trivial problem, but the chemist realizes that the complications arise from the intermolecular, librational modes in water which are very important: they are the hydrogen bonding, or lattice modes. They have no zero-point energy and it is the trade-off of intermolecular against intramolecular degrees of freedom that is responsible for the change-over of the vapor pressure curves. This is also responsible for most of the remarkable properties of water. For instance, the maximum density in water is at 4°, but in heavy water it is at 11°. We also know that D_2O is lees ionized than H_2O and one result that has not yet been explained at all is the ratio of the self diffusion coefficients, 1.3. It cannot be explained in terms of the masses of the two nuclei. The chemist, therefore, is well aware that there are considerable problems with D_2O substitution.

The physiologist realizes that D_2O does not support life. It is very toxic. Only the very simplest forms of life, some of the protozoa, can be acclimatized to D_2O gradually. To give some examples: in excitable systems, a 40 percent substitution of D_2O leads to a very marked reduction in ATP synthesis and a loss of excitability. In some plants a 50 percent substitution of D_2O inhibits protein synthesis completely for hours after which it picks up gradually, but the rate of protein degradation increases by 100 percent. In tobacco, a 60 percent substitution of D_2O reduces the plant size, the rate of flowering and also rather interestingly, robs the plant of the discrimination in favor of ^{12}C over ^{13}C in its metabolic reactions, so that the plant becomes non-discriminating between the isotopes. The physiologist realizes that D_2O is a disaster, the chemist has a vague feeling that problems exist and the physicist can explain it all in terms of zero-point energy.

In biological systems hydration is important. It helps to maintain the right cell environment. Figure 1 shows the intracellular
compositions of two plants that resemble each other quite closely.
One is a freshwater plant, Nitella, and the other is a seawater
plant, Valonia. Looking at them in terms of their intracellular
ionic compositions, they are not too dissimilar; although the environment is utterly different in pond water and in seawater.

PHYSICAL PROPERTIES OF WATER

Let us now consider those physical properties of aquoeus solutions that impinge on the biological scene. I shall probably deal
mainly with simple molecules and then extend the treatment into the
context of large molecules.

We need first of all to explain the properties of water, for
instance the low heat of fusion, which is only 15 percent of the

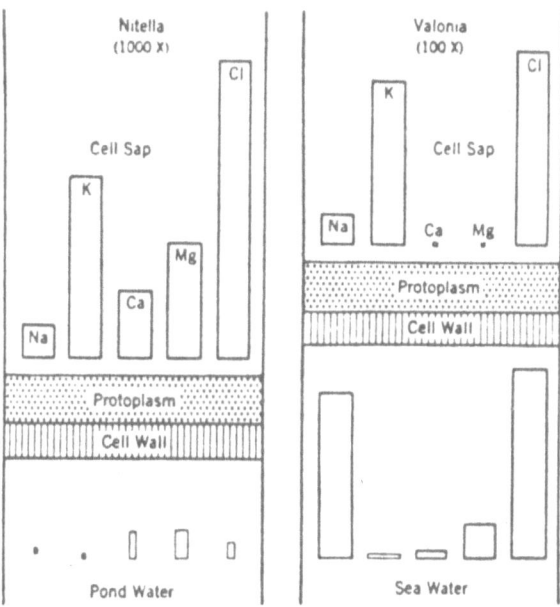

Fig. 1. Intracellular and environmental ionic composition of Nitella
and Valonia.

heat of sublimation. This is quite atypical for a liquid and means
that in the liquid state there is residual structure. We also have
to explain the doubling of the specific heat when ice melts which
is again rather unusual. We have to explain the very large liquid
range - 100 degrees - and the very high critical temperature, 374°C.
We have to explain the maximum density effect, and the vapor pressure isotope effect. Recently, much work has been done on undercool
ed water, and this is particularly fascinating. Any self-consistent
theory will have to account for the properties of the liquid from
the undercooled state right up to the critical point and beyond.

Much can be explained by reference to the radial distribution
function of liquid water (Fig. 2). The broken line is the radial
distribution function of liquid water and the drawn out line is the
radial distribution function of liquid argon. Both are normalized
to the respective molecular diameters so that at $R^{*} = 1$ we find the
nearest neighbor molecule, and so on.

The feature which is of the greatest importance in water is the
second peak which does not appear at $R^{*} = 2$, as it would be in a
close-packed substance. The other feature of interest is the area
under the first peak. The peak corresponds to the nearest neighbours,
that is, the hydrogen bond distance, and by integration we obtain
the number of nearest neighbours which in water is 4.4, and in

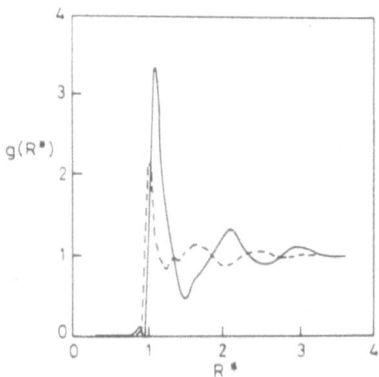

Fig. 2. Radial distribution functions $g(R^{*})$ of liquid argon (——)
and liquid water (---); $R^{*} = r/\tau$, where τ is the van der
Waals molecular diameter.

liquid argon is 11. This indicates that liquid water resembles ice
and maintains some of the geometry of ice. The second peak occurs
at a distance of 1.5 R* because of the tetrahedral geometry of water.
We therefore realize that water has a coordination number of just
over four and that the tetrahedral structure is well established.
Structure making and structure breaking in water must be defined in
terms of this second peak. Any process that will sharpen this second
peak I will call "structure making". Anything that destroys this
second peak I will call "structure breaking". The first peak is trivial because it only describes the hydrogen bond distance between
two water molecules.

The other factor we must realize in water is the complexity of
the coupling between various vibrational modes, arising from this
peculiar structure. Figure 3 gives a summary of the Raman frequencies
associated with the symmetric OH (OD) stretch valency in water and
in D_2O. For an isolated molecule in the gas state this is the frequency of the OH (OD) vibration. In liquid water there is a shift
to a lower frequency and in ice another shift to a yet lower frequency. If coupling between the two protons within the molecule is
allowed, then we observe line splitting. If we also allow coupling
between the OH groups in different molecules then each of these lines
will be further split into two lines, and this makes for an immensely
complex Raman spectrum, because Fig. 3 describes only one of the
three degrees of freedom. There is also the angular bending mode and
the asymmetric -OH stretch. Hence, in water this OH coupling is ex-

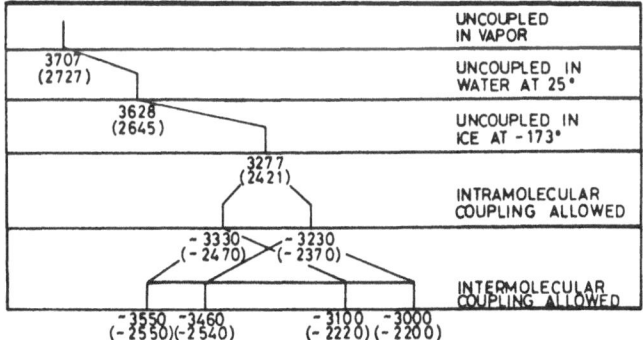

Fig. 3. The effects of coupling on the O-H (O-D) valence stretching
        band in ice.

tremely important, so that the substance really exhibits some solid-
like features.

Let me digress for a moment to mention undercooled water. The
density of water is a complex function of temperature (Fig. 4). When
liquid water is cooled down below its normal freezing point there is
a rather rapid decrease in density but φ(T) is by no means symmetrical about the temperature of 4°C. Unfortunately, nucleation
takes place at $\sim -40^{\circ}$ but if one were to fit a function to the
density curve, then the density would diverge at -45°. It would approach to ice value. What would this mean in terms of the Clausius-
Clapeyron equation? We thus have evidence of an instability, either
a critical type phenomenon or at least the spinodal-type instability.
This is shown by all the physical properties of undercooled water.

Next, let us examine the specific heat of undercooled water
(Fig. 5). One rather striking feature of liquid water is the temperature independence of the specific heat between 0° and 100°C. How-

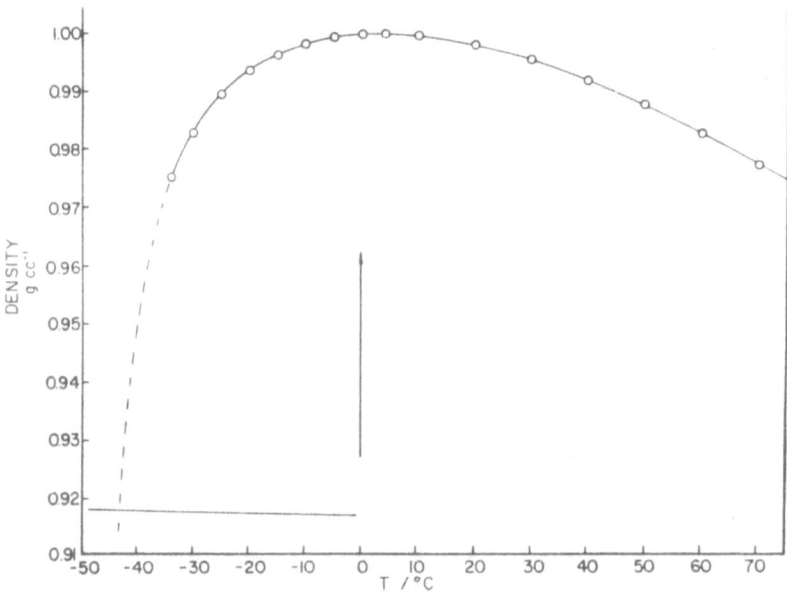

Fig. 4. Density of liquid water as a function of temperature; the
        density of ice is also shown.

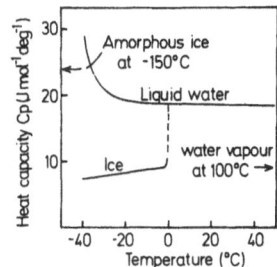

Fig. 5. Heat capacity of undercooled water and ice.

ever, C_p also appears to diverge at $-45°$. Some recent low-angle X-ray scattering studies on undercooled water at $-20°C$ have suggested correlation lengths of the order of 8 to 10 nm. This corresponds to enormous domains of structure in the liquid, and it shows up the co-operativity in water (Bosio et al., 1981).

These then are the properties of undercooled water which are so important in the subject of low temperature resistence of living organisms. Plants, insects and microorganisms seem to know all about these things and can turn them into effective mechanisms of survival.

ION HYDRATION

Let us now consider the perturbation of water structure by solutes; there are three distinct types of solutes. First of all, let us discuss ions: $M^+X^-\cdot H_2O$. We would like to study the environment of an ion in solution, the hydration shell. We also want to study the interaction between ions in solutions, including their hydration shells. The classical theories, based on the Debye-Hückel formalism, only deal in electrostatics, and therefore the solvent continuum only has one property, a dielectric permittivity.

The later developments of the Debye-Hückel theory contain various corrections designed to make the experiment agree with the theory. These approaches had led to a point of diminishing return in the 1960's and there is now a new generation of theories which allow for the fact that the solvent is composed of molecules. Probably the

work that has had the most impact on ionic solvation is that of
Enderby and his group at Bristol who apply neutron diffraction to
solutions of electrolytes (Enderby and Nielson, 1979). Even for the
simplest electrolyte. $M^+X^- \cdot H_2O$, with only four atomic species, the
scattering intensity curve contains ten radial distribution functions which can hardly be resolved.

Enderby et al. (1979) circumvented this problem by performing
different experiments using two different isotopes of one of the
atomic species, say M^+. Six of these radial distribution curves then
cancel, leaving four to be resolved: MX, MM, MH, MO. Since solvation
only accounts for about 10 percent of the observed scattering intensity, good data are needed. Most of the scattering is in fact due
to the solvent.

The results for a 1.5 molal solution of $NiCl_2$ in D_2O are shown
in Fig. 6. The first peak corresponds to the Ni-O distance and the
second peak to the Ni-D distance; the ratio of the areas under these
peaks is of course 1:2, and integration of the peaks yields a hydration number of 5.8±0.2 which is invariant of concentration. This
fact, taken together with the sharpness of the peaks reflects the
stability of the first hydration shell of the Ni^{2+} ion. From the
position of the two peaks and a knowledge of the geometry of the D_2O
molecule it becomes apparent that the D_2O molecules are tilted at an
angle to the Ni-O axis; this angle increases with increasing concen-

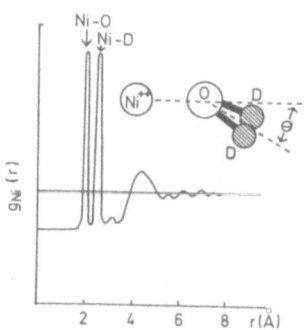

Fig. 6. Neutron scattering radial distribution function of Ni in
         aqueous solutions of $NiCl_2$, showing also the hydration shell
         geometry.

tration, so that the hydration sphere is sensitive to distortion as the packing fraction of hydrated ions increases. In very dilute solutions this angle tends to zero. The broad peak, centered on 4.5 Å, could be due either to Ni-Cl or to the second hydration shell.

Similar experiments have been performed in which isotope substitution has been applied to the anion, in this case Cl^-. The corresponding picture that emerges for the chloride ion hydration shell is that, once again, the hydration number is approx. 6, and that the D_2O molecules are disposed around the anion as shown in Fig. 7. In no case can any structure be detected beyond 5 Å.

Experiments on other electrolyte solutions have so far revealed that the Ca^{2+} and Na^+ ions are surrounded by hydration shells closely resembling that of Ni^{2+}, the main difference being that the peaks are not as sharp. A surprising result, and one that conflicts with other data, is that the Cl^- hydration shell is independent of the nature of the cation: Ca^{2+}, Na^+ or Rb^+. Not much can be said yet about cation-cation distributions, but it is claimed that the experimental results at reasonable concentrations clearly show that the Cl^- does not penetrate the cation primary hydration shell.

While there are still several doubtful aspects concerning the interpretation of neutron scattering data, it is certain that this technique will contribute very significantly to the eventual elucidation of the behaviour of aqueous electrolyte solutions. There is, however, still a long way to go. In fact, there are still major problems with quite simple concepts, e.g. the correct assignment of

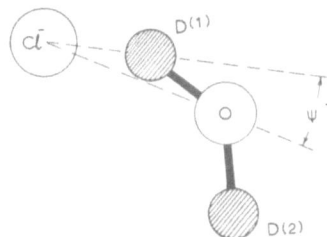

Fig. 7. Hydration shell geometry of the aqueous chloride ion, as
        obtained from neutron scattering data.

ionic radii. This is of importance not only in studies of the
transport properties of electrolytes in solution but in biochemistry and physiology, where calculations involving the diffusion
of ions, active or passive, and permeation of ions through barriers, such as membranes, always require knowledge about the
radius of the diffusing species.

The next problem then is: how do ions interact in solution?
Friedman and Krishnan (1973) differentiate between three types of
models. There is brass-balls-in-the-bathtub-model, where the ions
are taken to be hard spheres in the bathtub (solvent); then there
are the chemical and the Hamiltonian models, which start from
fairly rigorous statistical mechanics and calculate the interaction of ions in solution.

Different types of problems can arise. The cation and anion
are not hydrated in the same way, i.e. the orientations and spacings of water molecules are different. This leads to complex hydration shell interactions. It is also simple-minded to assume
that the perturbation of water is necessarily of a short-range
nature; it may decay exponentially with correlation lengths.
Unfortunately the second shell water molecules cannot be defined
from the neutron diffraction data, and neither can ion-ion interactions, which is remarkable. There are different ways in which
the hydrated ions can interact; there may be net repulsions when
they interact, but the Debye-Hückel theory does not account for
the orientation of solvent molecules, and whether there might be
additional repulsions or attractions due to hydration. The more
recent theories do include such effects and they can account quite
well for the properties of electrolytes in solution.

We know that there are specific ion effects in biology, and
in chemistry too. These find expression in the lyotropic series.
The lyotropic series was discovered by Hofmeister (1888) who found
that certain ions would precipitate proteins and others would keep
them in solution. He produced this series of ionic effects and
speculated that it originates from the different affinities that
ions have for water. Since then we have rediscovered the lyotropic
series about every five years but we still do not understand its
origin. An example of the series is the contribution of the various
ions to the destabilisation energy of proteins (Franks and Eagland,
1975). So-called "salting-in" ions (I^-, CNS^-) promote instability;

while "salting-out" ions (PO_4^{---}, SO_4^{--}) promote conformational stability. The ions operate in exactly the same series by affecting
the solubility of argon in water: sulphates and phosphates reduce
the solubility of argon, and perchlorate, iodide and thiocyanate
increase the solubility of argon in water. All the phenomena are
present in very simple systems. We have no idea how this comes
about. While electrostatics can account for salting-out, it can
not explain salting-in.

SOLUTIONS OF HYDROPHOBIC MOLECULES

     Now let us consider the non-electrolytes. Here we have two
very distinct types of behaviour. There are the so-called hydrophobic, and the hydrophilic effects. The hydrophobic effect can
be shown schematically from a consideration of the thermodynamics
of hydrocarbon solutions. Usually a non-ideal solution arises
because the two components either strongly attract each other or
strongly repel each other: the effects are shown in the enthalpy.
Figure 8 shows various types of behaviour, as reflected in the
excess thermodynamic functions (Rowlinson, 1969). The drawn out
lines are free energies, the broken lines are enthalpies and the
dotted lines are the entropy curves. A positive free energy means
a positive deviation from ideal behaviour. In "normal" systems
ΔG^E follows the ΔH^E curve fairly well, e.g. benzene-MeOH. In
water-dioxan, again with positive ΔG^E deviations, the entropy
term is negative and large. The same is true for the water-ethanol

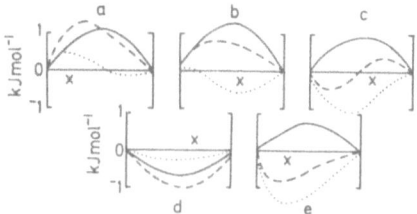

Fig. 8. ΔG^E (——), ΔH^E (----) and $T \Delta S^E$ (....) of binary sys-
         tems: a) benzene/methanol, b) benzene/ethanol, c) water/
         dioxan, d) water/hydrogen peroxide, e) water/ethanol;
         x is the mole fraction of the second component.

system, again with positive deviations from ideal behaviour; but
again the reason is that $T \Delta S^E$ is large and negative. Finally in a
well-behaved system, water-hydrogen peroxide, ΔG^E is negative because of a large negative excess enthalpy.

We have two types of behaviour here: there is the normal behaviour where the sign of ΔG^E is determined by the sign of ΔH^E,
and the other type of behaviour where the sign of ΔG^E is determined by $T \Delta S^E$. This latter is one of the symptoms of "hydrophobia".
Notice, by the way, that $\Delta G^E(x)$ looks most uninteresting in all
cases. It seems to be always symmetrical about x = 0.5.

How does one explain the hydrophobic effect? The current idea
is that water tries to maintain its hydrogen bonding at all costs.
Since the water structure is so open and has so much empty space,
then by bending the tetrahedral angle slightly and by very slightly lengthening hydrogen bonds one can produce many different topo-
logical situations that will accommodate non-polar, non-interac-
ting residues (Stillinger, 1973). Figure 9 shows one such example
of a hard sphere inserted into water. The water molecules can respond by changing slightly the tetrahedral angles to form a cage

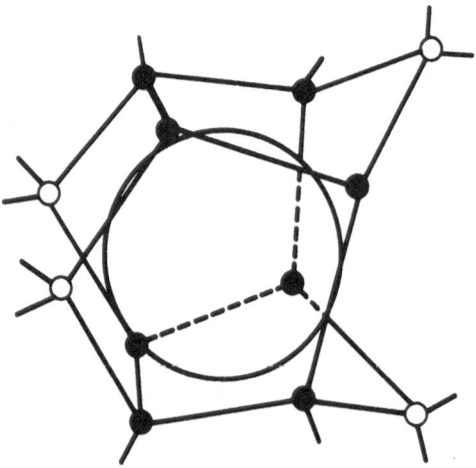

Fig. 9. Hydrophobic hydration of a nonpolar solute molecule,
        showing the hydrogen bonded water cavity (after Stillinger,
        1973).

around the sphere. The water maintains its hydrogen bonding, but the structure differs from that of ice. Each water molecule is still tetrahedrally bonded to four others, as in ice, but beyond that the geometry differs. This structure is stabilized by the polarizability of the guest molecule or atom, which more than cancels the 2 kJ mole^{-1} of unfavourable free energy of this new lattice. In terms of entropy, the water has lost some of its configurational degrees of freedom, because the OH vectors are not allowed to point to the centre of the cavity; the hydrogen bonding is confined to the edges of the polyhedron or away from the polyhedron. This leads to a reduction in the entropy. The process has been examined by simulations of the neon atom in water. In fact, most of the simulation work has been performed on "neon water". The water molecule is taken to be a neon atom, because water and neon are isoelectronic. Four charges are then placed into the neon atoms tetrahedrally to simulate water. The simulation uses a Lennard-Jones potential for neon and the electrostatic interactions due to the four charges. One can now remove the charges from one of the water molecules and replace it into the computer so that it reverts to a real neon atom and then one can study what happens to the water around this neon atom. Figure 10 shows the (simulated) OH pair correlation function which can be obtained experimentally from neutron scattering (Geiger et al., 1979). I said earlier that if the second peak shows sharpening, this is indicative of structure-making; here the second peak is sharpened for the water which is the nearest to the neon atom. Thus, the water which makes up the hydration shell of a non-polar molecule has a narrower second peak, also the minimum between the peaks is deeper. The hydration shell is therefore better defined structurally than the water which is further away.

The simulation can now be extended to two neon atoms in water. The hydrophobic interaction predicts a partial cancellation of the rather unfavourable effect of pushing non-polar groups into water, by allowing two of them to approach each other and releasing water. One can place these two neon atoms close to each other and then start the molecular dynamics simulation. If the prediction is correct, these atoms or molecules will stay together. In practice, the neon atoms drift apart and come to rest at some distance which is not the nearest neighbour distance. At equilibrium, water has pushed in between these inert atoms and forced them apart some distance.

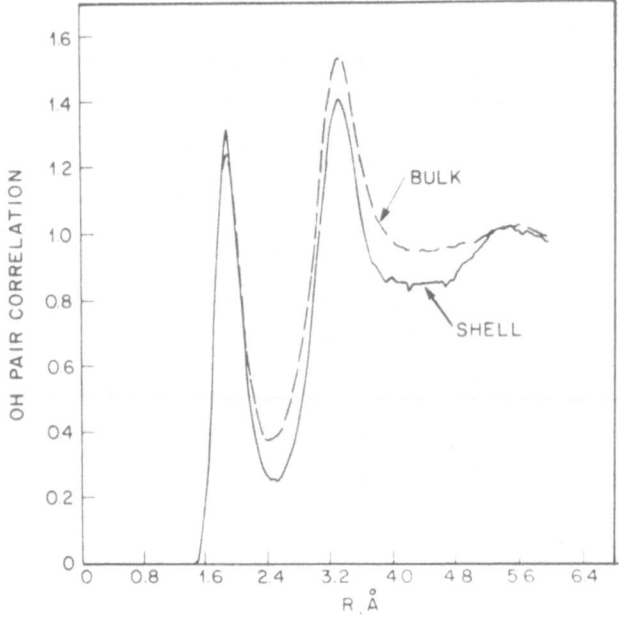

Fig. 10. O-H pair correlation function for bulk water and the
         water molecules in the hydration shell of a nonpolar
         solute molecule (reproduced, with permission, from Geiger
         et al., 1979).

This should not be too surprising: in the crystalline ana-
logues of these systems, the clathrate hydrates, one never finds
two gas atoms occupying the same cavity. They always occupy their
own cavities although there is plenty of room in each cavity. This
seems to be the equilibrium separation and has important impli-
cations in the treatment of the hydrophobic interaction in terms
of thermodynamics. The hydrophobic interaction is thus not a complete reversal of the introduction of a neon atom into water; it
is rather more subtle than that.

Let us consider one consequence of the hydrophobic interaction. Any molecule with an alkyl group and only one functional
group behaves qualitatively like a neon atom in solution. Alcohols,
amines, ethers, ketones and amides all behave this way in solution.
The free energy sacrificed per CH_2 group of alkyl chains is almost

constant, but it is an entropy dominated process and therefore all
the free energy calculations that involve exposed surface of
proteins, i.e. the accessible surface area, are meaningless and
predict the incorrect sign for the temperature derivative of the
free energy. As I mentioned (Shrake and Rupley, 1973) before, the
free energies are most uninteresting - they are easily fitted by
one parameter, for instance the surface tension. However, a fit at
only one temperature can be quite meaningless. The hydrophobic
interaction is not an attraction: one should not describe it by a
potential of mean force (Pratt and Chandler, 1980), like a Lennard-
Jones function, because it derives from the structural entropy
of the water.

The hydrophobic interaction is very marked in the tetraalkyl-
ammonium salts. The larger the alkyl group, the more these salts
behave like hydrophobic non-electrolytes. One of the more sophisti-
cated treatments of calculating electrolyte properties nowadays
can estimate the pair distribution function of the ions: there
are three such distribution functions: cation-cation, cation-anion
and anion-anion. On the basis of electrostatics, two of these seem
without interest, because like charges repel each other. The main
peak would correspond to the interaction between a positive and a
negative ion. Figure 11 shows this for tetrapropylammonium bromide.
Note the cation-anion radial distribution curve showing electrostatic attraction. But notice also the peak in $g_{++}(r)$. This indicates cation pairing, because the ions behave like hydrophobic molecules and there is an entropic cation-cation interaction. Of

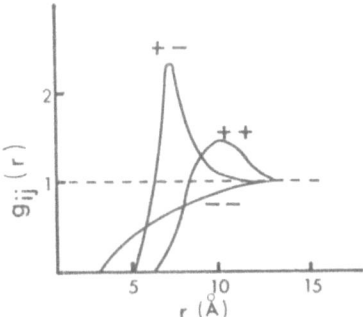

Fig. 11. Ionic pair correlation functions for aqueous tetra-n-pro-
         pylammonium bromide, calculated from activity data.

course there are no features of any interest in $g_{--}(r)$; chloride
ions do not attract each other. Here then we have some rather
strange behaviour and this is of course reflected in the thermodynamic properties of such salts. The Debye-Hückel theory cannot account for that because it is only concerned with electrostatic interactions. I have laid such great emphasis on hydrophobic effects
because of their universality in maintaining biological structures,
from the integrity of single polypeptide chains in globular proteins to membrane structures, fibrillar assemblies (tubulin, actin)
and complex multisubunit assemblies, such as enzymes and viruses.
Historically, recognition of the importance of hydrophobic effects
derives from 1959 (Kauzmann, 1959), and since then its involvement
in many different biochemical processes has been recognized. However, its origin and exact molecular description are still something of a mystery. Even the ways in which biochemists estimate
the hydrophobic contributions (the "hydrophobicity") to the stabi-
lities of biological structures (Bigelow, 1967) are fundamentally
quite incorrect, because the hydrophobic interaction is treated,
either explicitly or implicitly, like any other attractive interaction. I hope that the above discussion has made clear that apolar
residues in water do not attract each other but are driven together
by solute-water repulsion.

HYDROPHILIC, POLAR SOLUTES

Let us now consider the other types of molecules, those that
interact with water through hydrogen bonding, especially those
that have exchangeable protons: amines, alcohols and the polyalco-
hols. To emphasize the special effects of hydrogen bonding we can
study the polyhydroxy compounds; the carbohydrates are a coherent
group that shows up all the simplicity and all the complexity of
such interactions (Franks, 1979).

The whole gamut of theoretical treatments becomes infinitely
more difficult with carbohydrates, even simple sugars. Atom-atom
potential functions are fairly crude because there are motions in
molecules which affect the stereochemistry to a very marked extent
but which cannot really be included rigorously in any calculations.
Figure 12 shows a five-membered sugar ring, the furanose type ring.
The two configurations show the effects of an in-plane flexibility
in the five-membered ring. These rings are not rigid, they do show

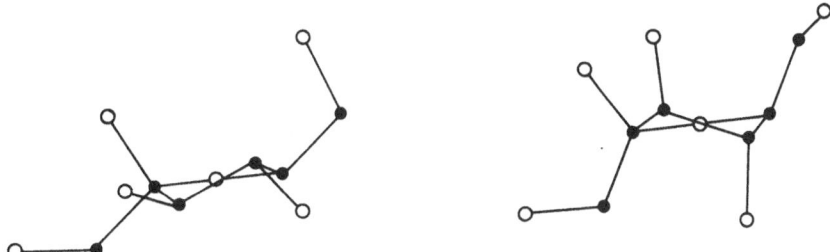

Fig. 12. Effects of ring flexibility of furanose rings on the
         peripheral hydroxy group orientations (reproduced, with
         permission, from Franks, 1979).

librational degrees of freedom. Very slightly changing the plane
of the ring produces a completely different stereochemistry at the
periphery. This very markedly affects the way that such molecules
can interact with the solvent via the -OH groups, but such effects
cannot really be incorporated into <u>ab initio</u> calculations.

     Nevertheless, the theoreticians do calculate, for instance,
anomeric equilibria and the proportions of α and β -sugars in
anomeric mixtures. This is done with atom-atom potential functions
and energy minimization   but there is one very great weakness:
many of these potential functions have been derived from measurements made in aqueous solutions (Angyal, 1969), because that is
how all the early measurements were made.

     Let us consider a sugar, glucose, that has a simple anomeric
equilibrium with only two coexisting species: the pyranose α and
the pyranose β -anomers. The α -sugar has the C1-OH group in the
axial position; in the β -sugar it is in the equatorial position.
The sugar, reaches equilibrium in solution, with a constant α / β
ratio. From calculations it has been deduced that glucose will
exist as mixture containing 44% α-anomer (Dunfield and Whittington,
1977). The energy barriers between α and β-anomers are very low
so that there is a ready conversion between them. Experimental
measurements showed that in dimethyl sulphoxide and dioxan this
agrees very well with the predicted value, 44-45%. However, in
water the percentage of α-sugar is 27%. The theoreticians ratio-
nalize such discrepancies by including 1.5 kJ mole^{-1} hydration
energy, but only for equatorial -OH groups. One may wonder why,

but the estimated α/β ratio then corresponds with the experimental value, except that in D_2O the proportion of α -anomer is only
34%. There is really no satisfactory way of building hydration effects into such calculations. One cannot just place water molecules in a few specific positions; the water-water interactions
are also perturbed, but are not taken into account in such calculations. The solvent water quite clearly prefers the β -sugar.
There are speculations that the orientation of the equatorial
hydroxyl groups on the sugars match the orientations and spacings
of -OH groups in liquid water. This is shown, perhaps not too suc-
cessfully, in Fig. 13 (Suggett, 1976).

It seems that water will always prefer that configuration
which has the largest number of equatorial OH groups. Let us therefore examine a sugar that is rather more complex - ribose. With
glucose there are only the two species: pyranose α and pyranose
β-forms. With ribose there are six species, as shown in Fig. 14.
The energy differences between them are so small that they all
actually coexist in solution. Five of these can be identified by
various NMR techniques. Two have the same spin coupling and cannot
be differentiated. Figure 15 shows the equilibrium mixture composition in D_2O as a function of temperature (Franks et al., 1982).
The β -pyranose (C-1) is the dominant configuration in this mixture, and it is also the one that most resembles β -glucose in
that it has the most equatorial OH groups. The other forms are
present to varying extents. As the temperature increases, so the
preponderance of the β -pyranose decreases. In DMSO the equilibria
look completely different; the β -pyranose is not the dominant
species to the same extent; the composition agrees more closely
with the calculated values. Water will therefore affect the equilibria to give the highest concentration of that sugar which has
the largest number of equatorial OH groups.

The OH groups on sugars are not at all equivalent; they are
very sensitive to solvation. Some years ago it became possible by
supercooling techniques to make visible the sugar OH group
protons with ^{1}H NMR. Normally this cannot be done, because the OH
signals are hidden under the solvent peak, but at subzero temperatures these peaks appear downfield from the water signal (Harvey
et al., 1978). Invariably the largest downfield shift is due to
the β -anomeric OH. The spectra are extremely specific to the sugar,
they are not equivalent for the whole series of hexoses. Each OH

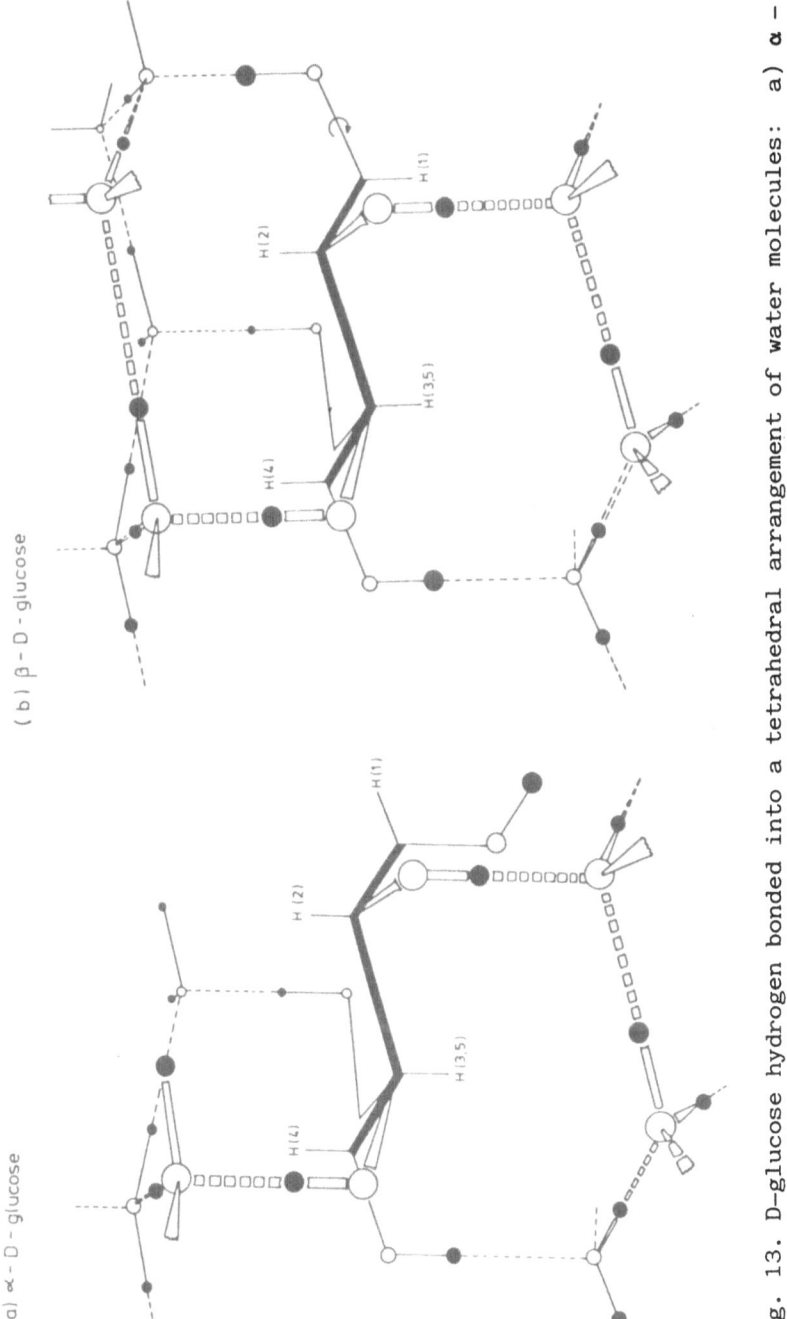

Fig. 13. D-glucose hydrogen bonded into a tetrahedral arrangement of water molecules: a) α – glucose, b) β –glucose. The pyranose ring is indicated in side view by the prominent line. O and H atoms are represented by open and solid circles, respectively, and co- valent and hydrogen bonds by solid and broken lines. The C6 protons have been omitted for the sake of clarity (reproduced, with permission, from Suggett, 1976).

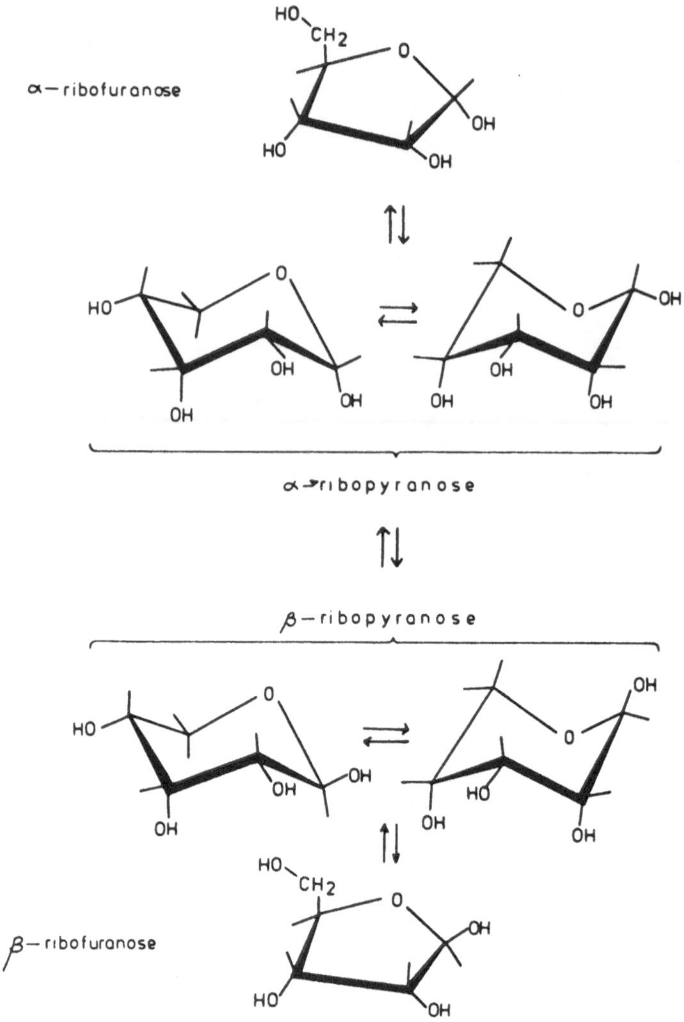

Fig. 14. Conformational equilibria in D-ribose.

group is sensitive to the solvent – or the solvent can discrimi-
nate between different OH groups. From the line shapes of 300 MHz
spectra one can measure the proton exchange rates with the solvent.
It is found that β -glucose exchanges protons much more rapidly
than does α -glucose (Bociek and Franks, 1979). Again this may be
because of the rather facilitated orientation of the β -anomeric
OH group.

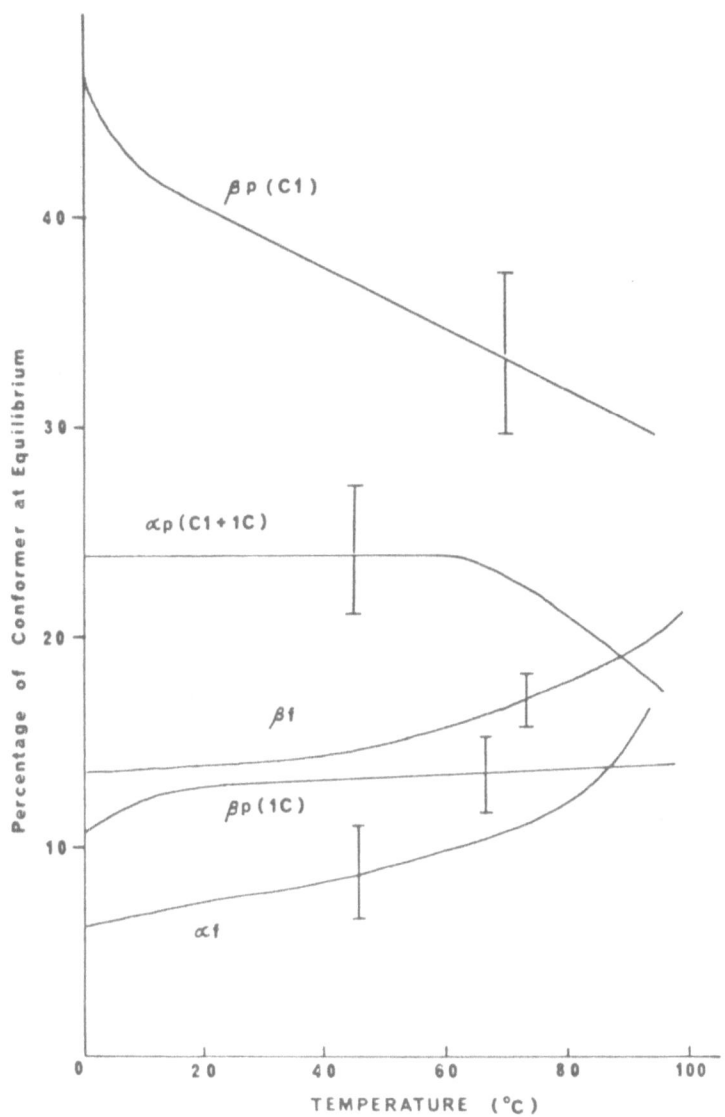

Fig. 15. Temperature dependence of the equilibrium composition of
D-ribose in aqueous (D_2O) solution; p and f signify pyranose and furanose, and C1 and 1C refer to pyranose
ring conformations; see Fig. 14.

At the next level of complexity are the disaccharides, where
we encounter additional degrees of freedom. The glycosidic linkage
is described by two rotational angles which will provide further
complications with hydration effects, because now the solvent can
also affect the values of these angles. The solvent tries to enforce a certain molecular configuration and can do so by altering
these angles, and thus the conformation of the whole molecule.
This can be analysed by optical rotation, with some rather interesting results. The sugars that were studied a few years ago were
α -methyl-maltoside, cellobiose and trehalose. They have one factor
in common, they are composed of two glucose units, but with different linkages, maltose: α -1-4, cellobiose:β- 1-4, and trehalose:
α -1-1. There are thus different conformations of this molecule,
which consists of the same two sugar units.

It is possible to estimate how much of the measured optical
rotation is in fact associated with the glycosidic linkage (Rees
and Thom, 1977). Knowing the optical rotations of the two monomers, and of the molecule in the crystal state, then measurement
of the optical rotation in solution makes it possible to calculate
how much of that is due to the two angles and what is likely to be
the configuration of the molecule in solution. The results are
shown in Table 1. A Monte Carlo simulation in vacuo predicts a
value of -28° for maltose and α-methyl-maltoside. This is very
close to the experimental values in dimethyl sulphoxide and dioxan.
For aqueous solutions there is absolutely no correspondence with
the calculated value. The rotation is even of the opposite sign.
The solvent is again able to force into the carbohydrate molecule
some configuration which is unique to water. Let us progress
further in the degree of complexity, to the interaction between
sugars in solution. This requires information obtained from osmotic
second virial coefficients; and very little work has been done in
this field. This always strikes me as remarkable, because it is
the sugars which are responsible for most biological recognition
processes.

As a first approximation the second virial coefficient
measures the size of the sugar molecule. However, there
are two sugars which have the same molecular weight, and
are stereoisomers: xylose and ribose. It turns out that a
1 M solution of ribose is as near as can be ideal, it
has a zero second virial coefficient, whereas xylose has quite

Table 1. The Contribution of the Glycosidic Linkage to the
Optical Rotation of Disaccharides (after Rees and
Thom, 1977).

| Molecule in solution | Solvent | | | Crystal | Monte Carlo solution |
	DMSO	Dioxan	Water		
Trehalose	-18°		-19°		
α –methyl maltoside	-19°	-24°	-45°	-110°	-13°

a sizeable second virial coefficient. Xylose molecules seem to
have a way of interacting, of "seeing" one another, while ribose
molecules do not, at the same concentration. Such sugar interactions
are usually small compared to the hydrophobic effects. They are
very stereospecific and that is probably why they have been neglected for so long. Yet, for a polysaccharide, a small effect for
each pair of sugar residues becomes a very large effect on the
interactions of the molecule as a whole.

SOLUTIONS OF MACROMOLECULES

     Now let me leave the sugars and turn to polymers – proteins
and carbohydrates. Polymer hydration can be studied in four different ways (Franks, 1979). The solid state, especially the crystal,
is of course strong in experimental information, providing distances and angles, which define structure. Unfortunately, in a
crystal the energy minimization techniques depend very much on
molecular packing. Packing economy dictates the configuration of
molecules, and it is uncertain how far one should extrapolate to
the solution state. Then there is the isolated molecule, e.g. the
protein in vacuo, much used by theoreticians for quantum mechanical
calculations and computer molecular dynamics simulations. This
state has no physical existence. Third, we have the isolated molecule, but in the solvent medium, in what thermodynamics refers
to as infinite dilution. Here we are very rich in experimental
methods, but woefully short on theory. In fact, there is no theory
worth speaking of. Finally we have the state which is of real importance: the in vivo state, or the functional state. This is more
often than not a concentrated system. It is usually turbid, and
it might be a gel. It is hard to study experimentally and there

are no adequate theories either. One can draw certain analogies
between the isolated molecule in vacuo and the isolated molecule
in solution, and between the solid state and the in vivo state.

Let us first examine water in protein crystals. This can be
done by diffraction, but we are not now dealing with electron densities but with probability electron densities, because the water
molecules can exchange, even in a crystal. However, there are
several rules about where water molecules are to be found relative
to the peptide chain. One very common place is near a metal atom,
zinc, for instance. H_2O takes one of the coordination positions
of zinc and it disappears when the substrate approaches. Another
place where water molecules appear is between charged residues
inside a folded protein, to attenuate the charge repulsion. Thus
in papain there are two lysine residues in close juxtaposition
with a glutamate and a water molecule in between them. Finally,
water molecules hold different parts of the same peptide chain
together, or they link two different peptide chains by hydrogen
bonds. Sometimes one also finds water molecules linking side chains
in the amino acids, although this is rarer. One of the great un-
solved questions is: what is the involvement of water in maintaining the native state of a protein. In this state a certain fraction
of the apolar residues are found to be in the centre of the protein,
so when the folding takes place, many non-polar groups are with-
drawn from contact with water and placed in the interior. A quantitative estimate of the free energy has historically been obtained
from free energy of transfer experiments of amino acids from water
to oil (von Hippel and Wong, 1965; von Hippel et al., 1973). For
ribonuclease this "hydrophobic" contribution to protein stability
is 150 kJ $mole^{-1}$, as shown in Table 2. In addition there is a small
destabilizing effect, -20 kJ $mole^{-1}$, due to the ionic residues
which are to be found on the inside of the protein. There is also
a large destabilizing effect due to the configurational entropy,
and finally, the favourable contribution from intramolecular hydrogen bonds and packing, dispersion energies, dipole-dipole interactions and core repulsion. This includes local effects of the
intramolecular hydrogen bonds in a β-sheet structure or in an α -
helix, and distant effects resulting from the folding process.
The sum total of all these contributions is a very large negative
contribution almost balanced by a very large positive contribution,
producing a very marginal stability (Pain, 1979). The solvent is
implicated in all of these contributions, via electrostatics, in
the hydrogen bond, and through the hydrophobic interaction.

Table 2. Estimated Contributions to the Stabilization Free
         Energy of Ribonuclease(kJ/mole) (after Pain, 1979)

Conformational entropy	-450
Hydrophobic stabilization	150
Electrostatic free energy of buried charged groups	- 20
Peptide hydrogen bonding and van der Waals interactions	370
Free energy of stabilization (from denaturation experiments)	50

Let us discuss in detail just one example of the very complex
stability-instability relationship of a protein/solvent system,
that of ribonuclease in a mixed solvent. Mixed solvents, such as
aqueous guanidinium chloride are used to unfold proteins, as is
also ethanol. But ethanol/water mixtures themselves show some
rather complex properties in aqueous solution, even in the absence
of a protein. Figure 16 shows the standard free energy of unfolding
in the mixed solvent relative to water (Brandts and Hunt, 1967).
$\Delta(\Delta G^{\theta}) > 0$ means favouring the native state, and $\Delta(\Delta G^{\theta}) < 0$
means favouring the unfolded state. $\Delta(\Delta G^{\theta})$ is shown as a function of ethanol concentration. At 25° or $30^{\circ}C$ ethanol acts to
unfold the protein, but at 10° the first additions of ethanol stabilize the native state. It is easy to fit the free energy at any
one temperature by all kinds of models, but it is of course the
enthalpy and specific heat changes which are responsible for this
rather remarkable behaviour, that at low temperatures ethanol
favours the native state, and at high temperatures if favours unfolding.

One other way of changing the stability of a protein via the
solvent is, of course, a change in the temperature. Figure 17 shows
this temperature effect in a generalized manner (Franks, 1982).
Once again the free energy of unfolding is considered. The reason
why $\Delta G^{\theta}(T)$ is so very curved is that it contains a large specific

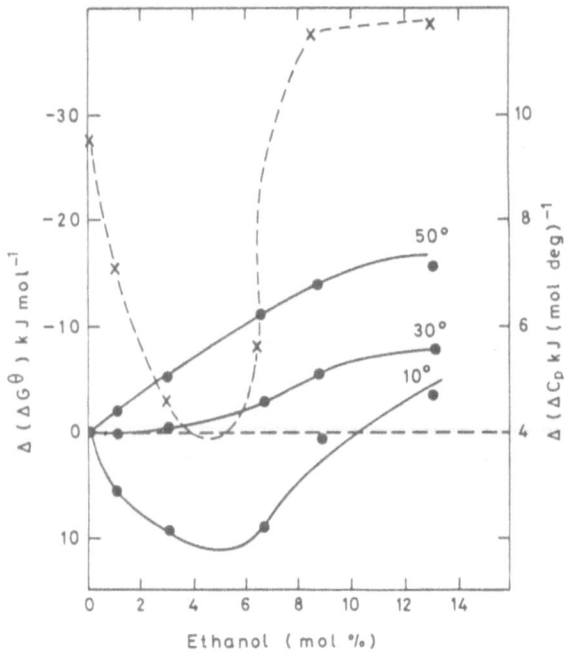

Fig. 16. Free energy of denaturation of ribonuclease (relative to
         aqueous solution) in water/ethanol mixtures. The correspon-
         ding heat capacity change is indicated by the broken
         line (after Brandts and Hunt, 1967).

heat contribution. Nevertheless, in terms of the protein _in vivo_
there is a temperature of maximum stability, and it is interesting
that for a given group of proteins, this seems to be a constant
temperature; for instance, for the tryptic enzymes it is 12°C. The
curve can move about and this is one way in which nature can produce thermophilic and psychrophilic proteins. At ΔG^{θ} = 0 the
protein will unfold both at the high and low temperature limits.
Unfortunately the solution freezes near 0°C , but if it does not
freeze, then the protein can be expected to unfold in the undercooled state, and will be cold labile. The temperature range over
which a protein is viable can be increased in one of three ways.
The whole curve can be lowered, making ΔG^{θ} more negative, or the
whole curve can be shifted along the temperature axis. This is what
is usually found in thermophilic mutants (Jaenicke, 1981): the

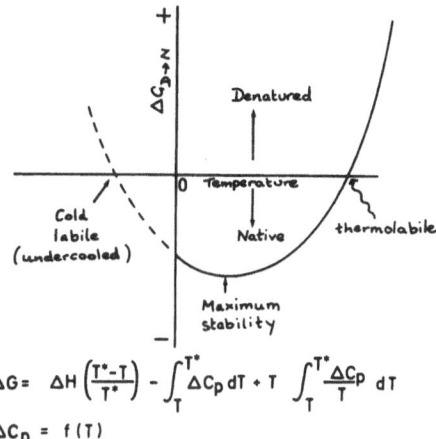

$$\Delta G = \Delta H \left(\frac{T^* - T}{T^*}\right) - \int_T^{T^*} \Delta C_p \, dT + T \int_T^{T^*} \frac{\Delta C_p}{T} \, dT$$

$$\Delta C_p = f(T)$$

Fig. 17. Denaturation of a protein as a function of temperature;
  T^* corresponds to the transition temperature(s), where
  $\Delta G = 0$ (reproduced from Franks, 1982, with permission).

curve has the same shape but has been shifted to higher temperatures. The third way is to make the curve shallower. In that way one
would sacrifice stability but the native protein would be stable
over a larger temperature range. There are thus three ways in
which the curve can be manipulated to enhance the stability in the
cold or the warm region. They all occur in nature.

METASTABLE, BOUND WATER

     Finally, mention must be made to the so-called unfreezable or
bound water which exercises the minds of many people. This is a
difficult topic because we leave the realm of equilibrium thermodynamics and enter the fuzzy world of metastability and kinetic
barriers. It is found that most complex molecules, especially
macromolecules and capillary systems immobilize a fraction of the
water molecules so that this water does not freeze, even in the
presence of ice. This we relate to hindered diffusion rather than
to binding (Packer, 1977). Figure 18 shows schematically the problems with the diffusion of water. The two capillary systems might
be phospholipid bilayers, or they might be pores or vascular systems. We have a region of thickness x and another region of thick-

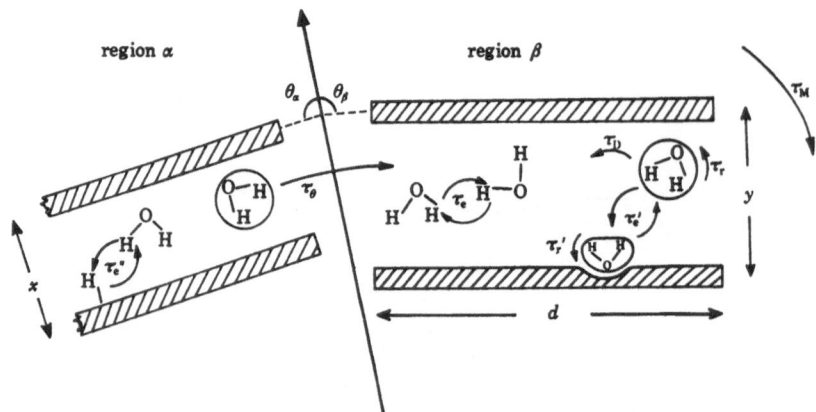

Fig. 18. Diffusion and exchange behaviour of water in heterogeneous
systems, as measured by NMR relaxation, showing the time
constants for the various motions. The two regions shown
are oriented at angles θ_α and θ_β to some external reference axis and perform rotations with a time constant τ_M
(reproduced, with permission, from Packer, 1977).

ness y and they are oriented at different angles to some reference
axis which in practice is very often the axis of the magnetic field
in an NMR spectrometer. These thicknesses are not large compared
to the molecular dimensions; they might be something of the order
of 2 nm. NMR relaxation measures the decay of magnetization, and
from there to a credible interpretation of diffusional properties
of molecules involves several assumptions and some model building.
There are essentially two types of processes: exchange and diffusion. For instance, there is exchange between a proton from water
and a proton that sits on the substrate. This might be a carbohydrate substrate, or silica. There is also exchange of protons between water molecules and actual molecular exchange between the
so-called bulk and some site of adsorption; the adsorbed molecule
has a finite residence time after which it exchanges.

In addition there is normal self-diffusion, characterized by
τ_d and the rotation of the water molecule, with a time constant
τ_r. Furthermore, diffusion from one such region into another
region takes place. Finally, we have rotation of the whole region

about some axis, which is essentially a low frequency motion, but there will be certain water molecules whose motions will be governed by the slow motions of this whole region. All these effects are included in the measured spin-lattice and spin-spin relaxation rates. It takes some very refined measurements with ^1H, ^2H and ^{17}O to disentangle these effects, even then it is still not totally agreed what is really being measured.

Some of the water is severely perturbed and will not freeze. It is certain by now that the diffusion of water in confined spaces is anisotropic, giving rise to spectral line splitting. There are preferred directions of diffusion and preferred axes of rotation, which make it harder for this water to freeze, but it is a dynamic problem and not a problem of water binding. Water cannot bind, because its hydrogen bonds are always weak. At temperatures approaching -60° and -80° the driving force for freezing is very strong and even then a proportion of the water does not freeze. In biology this perturbed water is of immense importance because it seems that the limiting water content which is necessary for maintaining structures inviolate against denaturation or desiccation is identifiable with this unfreezable water (Franks, 1982).

ACKNOWLEDGEMENTS

I am indebted to the Leverhulme Trust, The Royal Society and the Agricultural Research Council for grants in support of our work on the biophysics and biochemistry of low temperatures.

The following figures have been reproduced from Water - A Comprehensive Treatise, F. Franks, ed., Plenum Press, New York, with the permission of the authors: Fig. 2 (A. Ben-Naim, Vol. 1), Fig. 3 (F. Franks, Vol. 1), Fig. 4 (C.A. Angell, Vol. 7), Figs. 6 and 7 (J.E. Enderby and G.W. Nielson, Vol. 6).

REFERENCES

Angyal, S.J., 1969, The composition and conformation of sugars in solutions, Angew. Chem. 8:157.

Bigelow, C.C., 1967, On the average hydrophobicity of proteins and the relation between it and protein structure, J. Theoret. Biol., 16:187.

Bociek, S., and Franks, F., 1979, Proton exchange in aqueous solutions of glucose, J. Chem. Soc. Faraday Trans. I, 75:262.

Bosio, L., Teixeira, J., and Stanley, H.E., 1981, Enhanced density fluctuations in supercooled H_2O, D_2O, and ethanol-water solutions: evidence from small angle X-ray scattering, Phys. Rev. Lett., 46:597.

Brandts, J.F. and Hunt, L., 1967, The denaturation of ribonuclease in water and in aqueous urea and aqueous ethanol mixtures, J. Amer. Chem. Soc., 89:4826.

Dunfield, L.G. and Whittington, S.G., 1977, A Monte Carlo investigation of the conformational free energies of the aldohexopyranoses, J. Chem. Soc. Perkin II, 654.

Enderby, J.E. and Nielson, G.W., 1979, X-ray and neutron scattering by aqueous solutions of electrolytes, in "Water - A Comprehensive Treatise", Vol. 6, F. Franks, ed., Plenum Press, New York, p. 1

Franks, F., 1979, Solvent interactions and the solution behaviour of carbohydrates, in "Polysaccharides in Foods", J.M.V. Blanshard and J.R. Mitchell, eds., Butterworths, London.

Franks, F., 1982, Physiological water stress, in "Water Biophysics", F. Franks and S.F. Mathias, eds., John Wiley and Sons, Chichester.

Franks, F. and Eagland, D., 1975, The role of solvent interactions in protein conformation, Crit. Rev. Biochem., 3:165.

Franks, F., Robinson, G., and Lillford, P.J., 1982, to be published.

Friedman, H.L., and Krishnan, C.V., 1973, Thermodynamics of ion hydration, in "Water - A Comprehensive Treatise", Vol. 3., F. Franks, ed., Plenum Press, New York, p. 1.

Geiger, A., Rahman, A., and Stillinger, F.H., 1979, Molecular dynamics study of the hydration of Lennard-Jones solutes, J. Chem. Phys., 70:263.

Harvey, J.M., and Symons, M.C.R., 1978, The hydration of monosaccharides, and NMR study, J. Solution Chem., 7:571.

von Hippel, P.H., Peticolas, V., Schack, L., and Karlson, L., 1973, Ion binding to polyacrylamide and polystyrene columns, Biochem., 12:1256.

von Hippel, P.H. and Wong, K.Y., 1965, The effects of various non-electrolytes on the thermal ribonuclease transition, J. Biol. Chem., 240:3909.

Hofmeister, F., 1888, Zur Lehre von der Wirkung der Salze, Arch. Exp. Pathol. Pharmakol., 24:247.

Jaenicke, R., 1981, Enzymes under extremes of physical conditions, Ann. Rev. Biophys. Bioeng., 10:1.

Kauzmann, W., 1959, Some factors in the interpretation of protein denaturation, Adv. Protein Chem., 14:1.

Packer, K.J., 1977, The dynamics of water in heterogeneous systems, Phil. Trans. Roy. Soc., B278:59.

Pain, R.H., 1979, The conformation and stability of folded proteins, in "Characterization of Protein Conformation and Function", F. Franks,ed., Symposium Press, London.

Pratt, L.R.,and Chandler, D., 1980, Hydrophobic interactions and osmotic second virial coefficient for methanol in water, J. Solution Chem., 9:1.

Rees, D.A. and Thom. D., 1977, Polysaccharide conformation. 10. Solvent and temperature effects on optical rotation and con- formation of model carbohydrates, J. Chem. Soc. Perkin II, 191.

Rowlinson, J.S., 1969, in "Liquids and Liquid Mixtures", J.S. Rowlinson,ed., Plenum Press, New York.

Shrake, A.,and Rupley, J.A., 1973, Environment and exposure to solvent of protein atoms in lysozyme and insulin, J. Mol. Biol., 79:351.

Stillinger, F.H., 1973, Structure in aqueous solutions of nonpolar solutes from the standpoint of scaled-particle theory, J. Solution Chem., 2:141.

Suggett, A., 1976, Motions and interactions in aqueous carbohydrate solutions, III., J. Solution Chem., 5:33.

Many of the subjects discussed in this lecture are treated in detail in

Water - A Comprehensive Treatise, F. Franks,ed., Vol. 1-7, Plenum Press, New York, 197-1982. The following chapters are of particular relevance to this article:

Angell, C.A., 1982, Supercooled water, Vol. 7, p. 1.

Davidson, D.W., 1973, Clathrate hydrates, Vol. 2, p. 115.

Derbyshire, W., 1982, The dynamics of water in heterogeneous sys- tems with emphasis on subzero temperatures, Vol. 7, p. 339.

Enderby, J.E. and Nielson, G.W., 1979, X-ray and neutron scattering by aqueous solutions of electrolytes, Vol. 6, p. 1.

Finney, J.L., 1979, The organization and function of water in protein crystals, Vol. 6, p. 47.

Friedman, H.L. and Krishnan, C.V., 1973, Thermodynamics of ion hydration, Vol. 3, p. 1.

Franks, F., 1975, The hydrophobic interaction, Vol. 4, p. 1.

Franks, F., 1982, The properties of aqueous solutions at subzero
    temperatures, Vol. 7, p. 215.

Lilley, T.H., 1973, Raman spectroscopy of aqueous electrolyte
    solutions, Vol. 3, p. 265.

Ninham, B.W., Chan, D.Y.C., Mitchell, D.J., and Pailthorpe, B.A.,
    Solvent structure and hydrophobic solutions, Vol. 6, p. 239.

Suggett, A., 1975, Aqueous solutions of polysaccharides, Vol. 4,
    p. 519.

BIOMEMBRANES

Derek Marsh

Max-Planck-Institut   für biophysikalische Chemie
Göttingen, Fed. Rep. Germany

1. BIOLOGICAL MEMBRANE STRUCTURE

Introduction

Biological membranes are the supramolecular structures which
define the boundary of the cell and its organelles, and which control communication between the exterior and interior. As such they
constitute one of the basic organizing principles of biological organisms. In this chapter a review will be given of the overall composi-
tional and structural features of biomembranes, which lead to their
dynamic and functional properties. The structure of phospholipids and
of integral membrane proteins will be described. Results on the
rotational and translational mobility of membrane components will be
discussed to indicate the dynamic nature of membrane structure and
the possible functional role of membrane fluidity.

Overall Membrane Composition and Structure

Membrane Composition. Biological membranes consist mostly
of protein and lipid in roughly comparable quantities by weight.
The overall composition of a variety of different membranes is
given in Table 1.1. Values range from the relatively protein-rich
inner mitochondrial and bacterial membranes which contain 70-80%

Table 1.1. Percentage Composition by Weight of Plasma and
Intracellular Membranes (from Marsh, 1975).

Membrane	Protein	Lipid	Carbohydrate
Myelin (human CNS)	18	79	3
Erythrocyte (human)	49	43	8
Liver (rat, plasma)	58	42	(5-10)
Mitochondria (guinea pig liver)			
inner membrane	76	24	(1-2)
outer membrane	55	45	
Microsomes (bovine liver)			
rough ER	55	45	
smooth ER	47	53	
B. subtilis	80	20	
E. coli (plasma)	68	32	

protein in line with their complex enzymatic and transport functions, to the relatively inert nerve myelin membrane which contains
only 18% protein. Plasma membranes, and membranes derived topolo-
gically from them, contain modest amounts of carbohydrate (up to
10% by weight) which is entirely covalently linked, either in the
form of glycoproteins or glycolipids, and which is exclusively
located at the outer surface of the plasma membrane.

Membrane Lipids. The major lipid components of cell membranes
are the phospholipids and the neutral lipid, cholesterol. Representative chemical structures of these lipids are given in Fig. 1.1,
and the lipid composition of various membranes is given in Table
1.2. The major membrane phospholipids are the zwitterionic lipids
phosphatidylcholine (PC), phosphatidyl ethanolamine (PE) and sphingomyelin (Sph). Most mammalian membranes also contain approximately
10-15% of anionic phospholipids: phosphatidyl serine (PS), phosphatidyl glycerol (PG), phosphatidyl inositol and cardiolipin. Bacterial membranes often contain higher proportions of negatively
charged lipids - up to 100% of the total phospholipids in the case
of halophilic organisms. The cholesterol composition is strongly
dependent on the type of membrane. The cholesterol content of most
intracellular organelle membranes and of bacterial membranes is
very low, whereas eukaryotic plasma membranes contain approximately 25% by weight ($\sim$ 50 mole %) of cholesterol.

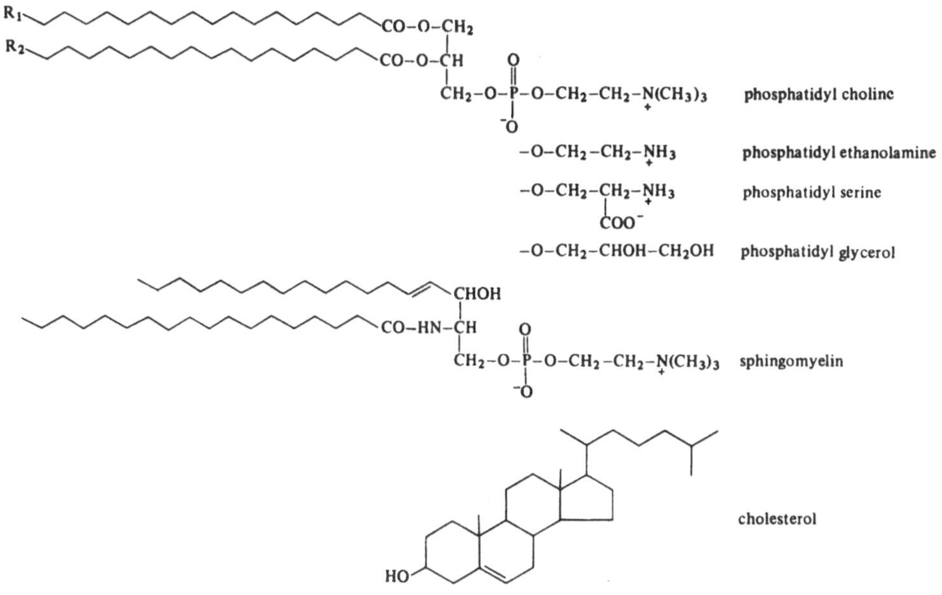

Fig. 1.1. Chemical structure of lipids commonly found in biological membranes.

The most important structural feature of phospholipids is the
amphiphilic nature of the molecule. As is seen in Fig. 1.1, the two
hydrophobic fatty acid chains, R_1 and R_2 are separated from the
charged polar headgroup by the glycerol backbone. This segregation
of hydrophobic and polar parts of the molecule means that most
phospholipids spontaneously form bilayers when dispersed in water.
These bilayers are membranous structures with the phospholipid
polar headgroups pointing out into the aqueous phase and the hydrocarbon chains directed in towards the centre of the bilayer, thus
forming a hydrophobic core which is largely impermeable to water
and other hydrophilic solutes.   .

The fatty acid composition of the phospholipid chains, R_1 and
R_2, influences the fluidity of the hydrocarbon core: shorter chainlength and greater unsaturation favouring higher fluidity. The
presence of cholesterol tends to decrease fluidity except of very
rigid chains, and it serves to stabilize the lipid bilayer against

Table 1.2. Percentage Lipid Composition by Weight of some
Typical Cell Membranes (from Korn, 1966).

	Myelin	Erythro-cyte	Mito-chondria	Micro-somes	E. coli	B. mega-terium
Cholesterol	25	25	5	6	0	0
PC	11	23	48	55	0	0
PE	14	20	28	18	82	45
Sph	6	18	0	0	0	0
PS	7	11	0	9	0	0
PG	0	0	1	0	7	45
Remainder[*]	37	3	18	13	11	10

[*]Mostly cerebroside in the case of myelin, and cardiolipin
in the case of mitochondria and E. coli.

large changes in fluidity. The phospholipid chains in biological
membranes normally vary from 16 to 20 carbon atoms in length and
most frequently contain none or just one cis-double bond. The unsaturated chains are usually found at the sn-2 position of the glycerol backbone and the saturated chains at the sn-1 position. Polyunsaturated chains are also found in some cases with up to 4
double bonds per chain. The lipids of rod outer segment disc membranes, for instance, are highly unsaturated. The chain composition
of most membranes normally ensures that the lipids are in a more-or-
less fluid state at physiological temperatures.

    A notable feature of the lipid regions of biological mem-
branes is that the different phospholipid types may be asymmetrically distributed across the bilayer. For the erythrocyte membrane
for example, it has been demonstrated by surface labelling and
phospholipase digestion that the sphingomyelin and phosphatidylcholine are located in the outer half of the bilayer, whereas the
phosphatidylethanolamine and phosphatidylserine are localized to
the inner half (Zwaal et al., 1973).

Membrane Proteins. The membrane proteins are divided into two fundamental classes: the extrinsic or peripheral proteins, and the intrinsic or integral proteins. The extrinsic proteins are bound superficially to the membrane by predominantly electrostatic forces and can normally be removed by aqueous extraction procedures: changing the pH and ionic strength, and adding chelating agents. Among the extrinsic proteins are typical soluble globular proteins such as cytochrome c and also more extended molecules such as spectrin which forms a cytoskeletal lattice on the inner surface of the erythrocyte membrane. The integral proteins penetrate and mostly, if not exclusively, span the hydrophobic region of the membrane. They have mostly a more apolar amino acid composition than soluble proteins and can only be extracted from the membrane with the use of detergents. Molecular weights and subunit structures of some typical integral membrane proteins are given in Table 1.3. Many are rather large proteins and certainly have sufficient molecular mass to contribute substantially to the hydrophobic interior of the membrane. The integral proteins are bound within the membrane by hydrophobic forces but since they are exposed to the aqueous phase they also contain polar segments in addition to the apolar intramembranous section. The amphiphilic polarity distribution across the membrane can be matched either by the distribution of polar and apolar residues along the amino acid sequence or by a differential polarity of the protein subunits as explained later.

Overall Membrane Structure. The diagram in Fig. 1.2 depicts the familiar fluid mosaic model of membrane structure in which integral membrane proteins are embedded in a phospholipid bilayer, and extrinsic proteins are associated either with the phospholipid headgroups or the polar sections of the integral proteins. Evidence for a bilayer arrangement of lipids in biological membranes comes from both X-ray diffraction and spectroscopic studies. The diffraction experiments reveal the static features of the bilayer structure whereas the spectroscopic studies reveal the dynamic features characteristic of molecular organization in bilayers. The evidence for the disposition of the integral membrane proteins comes from surface labelling and proteolytic digestion experiments, and also from the fact that they can only be extracted with detergents. Additional information on the location of the integral proteins also comes from freeze-fracture electron microscopy of membranes and reconstituted systems which reveals the presence of the protein-associated, intramembranous particles with the bilayer.

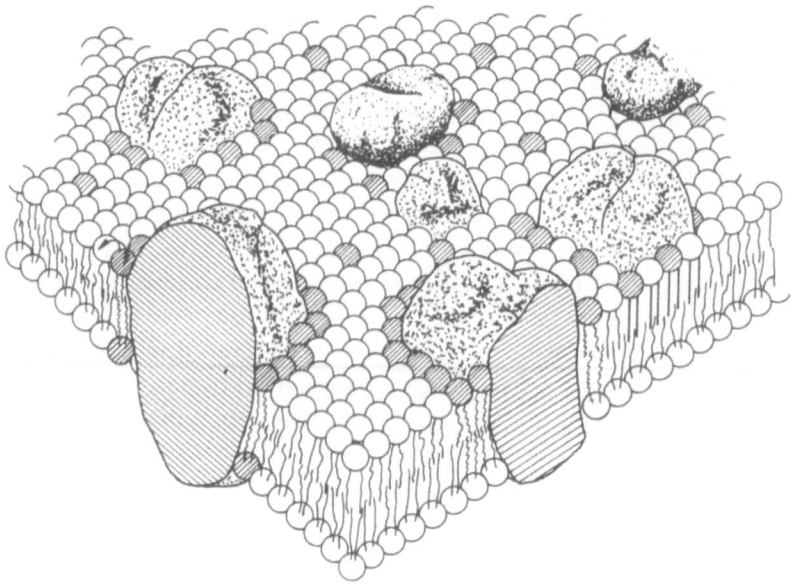

Fig. 1.2. Disposition of lipids and proteins in the fluid mosaic
         model of membrane structure proposed by Singer and
         Nicholson (1972).

Crystal Structure of Phospholipids

Relatively few single crystal structures have been obtained
of phospholipids, but those which are known reveal the same major
structural features. The results are of direct relevance to bio-
membrane structure since phospholipids crystallize in bilayers and
the characteristic features of the molecular structure are preserved
in fully hydrated, fluid phospholipid bilayers.

The single crystal structure of dilauroyl phosphatidylethanolamine is given in Fig. 1.3. The two most notable features of the
structure are the "bent-down" conformation of the headgroup and
the non-equivalence of the acyl chains, with a bend in the sn-2
(or β -) chain at the C-2 atom. Apart from the bend in the sn-2
chain, both chains are in the all-trans conformation. The bend in
the sn-2 chain ensures that the chains lie parallel and close-
packed to one another in the bilayer configuration. In fluid,

Table 1.3. Molecular Weights and Subunit Compositions of
        Various Integral Membrane Proteins (from Marsh and
        Watts, 1982).

Protein	Molecular Weights (daltons)	Number of Subunits
(Na^+, K^+)-ATPase (renal)	280,000	2 x 2
Acetylcholine receptor (Torpedo)	250,000	4
Gap junction "connexon"	150,000– 180,000	6 x 1
Cytochrome c oxidase	200,000	7 – 10
Ca^{2+}-ATPase	115,000$\pm$7,000	1
Band III	95,000	1
Rhodopsin (bovine)	37,000	1
Bacteriorhodopsin	26,000	1
Cytochrome b_5	16,000 (5,000)[a]	1
Glycophorin A	30,000 (3,700)[b]	1

[a]Hydrophobic segment   [b]Monomer

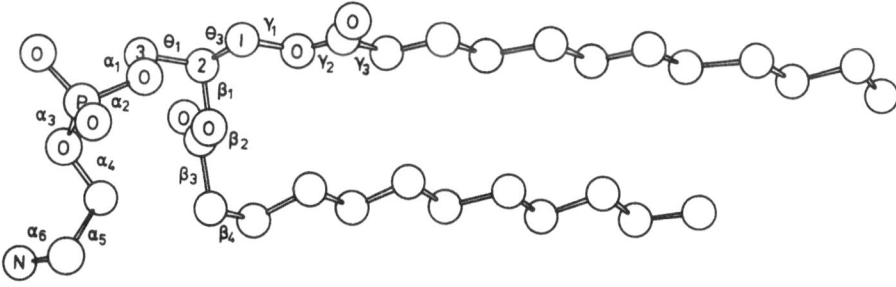

Fig. 1.3. Crystal structure of 1,2-dilauroyl-rac-glycero-3-phospho-
        ethanolamine. H-atoms are not shown (Hitchcock et al.,
        1974).

hydrated bilayers the close-packing of the chains is relaxed by
the creation of a limited, fluctuating population of gauche rotational isomers about the C-C bonds of the chains. The bend in the
sn-2 chain, which arises from the planar configuration of the glycerol backbone, is preserved in the fluid bilayers and gives rise
to a motional inequivalence of the two chains.

Similar overall results are observed in the crystal structure
of dimyristoyl phosphatidylcholine. The conformational torsion
angles of the various lipids are given in Table 1.4. There are two
inequivalent structures for the dimyristoyl phosphatidylcholine
molecule which differ mostly in the orientation of the headgroup
relative to the chains. These arise from the packing restrictions
imposed by the bulkier phosphocholine group. Nevertheless the overall headgroup structures are similar for all three lipids, all
having the "bent-down" structure which allows closest approach of
the positively-charged nitrogen to the adjacent negatively-charged
phosphates. In the case of phosphatidylethanolamine this also allows the possibility for intermolecular hydrogen bonding between
amine and phosphate.

Amino Acid Sequence of Integral Proteins

The full primary sequence is known for fewer integral proteins
than for soluble proteins. Glycophorin A, the major glycoprotein
of the human erythrocyte membrane, was the first to be sequenced
and illustrates one important structural aspect concerning integral
proteins. The complete sequence of the molecule, which has a single,
highly glycosylated, polypeptide chain, is given in Fig. 1.4. The
most striking feature of the primary structure is the non-uniform
distribution of polar and apolar amino acid residues, and of sugar
moieties, throughout the sequence. There is a hydrophobic stretch
of 23 completely non-polar residues (enclosed by the box) in the
centre of the sequence that is identified as the intramembranous
section of the protein. As seen in Fig. 1.4, this non-polar section
is sufficient to span the 35 $\overset{\circ}{A}$ hydrophobic region of the lipid
bilayer when the residues are arranged in an α-helical conformation. Evidence for an α-helical structure comes from circular
dichroism measurements on the tryptically cleaved hydrophobic
peptide in non-polar solvents.

Table 1.4. Torsion Angles in the Crystal Conformations of
           Phospholipids.

	1,2-dilauroyl-rac- -glycero-3-phos- phorylethanolamine $CH_3COOH(DLPE)$[a]	1,2-dimyristoyl-sn- -glycero-3-phospho- rylcholine $2H_2O$ (DMPC)[b]		1-lauroyl- -propandiol- -3-phospho- rylcholine $H_2O(LPPC)$[c]
		A	B	
α_1	−154	162	170	162
α_2	58	68	−76	86
α_3	66	63	−46	45
α_4	106	139	−161	129
α_5	67	−51	64	84
θ_1	−52	58	169	28
θ_3	−172	−176	175	78
β_1	97	76	124	
β_2	179	180	169	
β_3	−119	−87	−126	
β_4	65	63	63	
γ_1	−178	−173	103	156

Remaining β and γ torsion angles are approximately all-trans,
antiplanar (180°).

[a] Elder et al. (1977); [b] Pearson and Pascher (1979);
[c] Hauser et al. (1980).

The glycophorin sequence is thus characterized by an amphipathic distribution of polar and apolar moieties which matches the
polarity profile across the membrane. The N-terminal section which
is exposed to the exterior of the red cell is hydrophilic and contains many charged residues and all the carbohydrate moieties, in
particular sialic acid. This then passes through the central hydrophobic region to the hydrophilic C-terminal which again contains
several charged residues and is exposed to the red cell interior.
The polypeptide chain passes only once through the membrane although several molecules may be associated in an oligomeric structure or with other integral proteins.

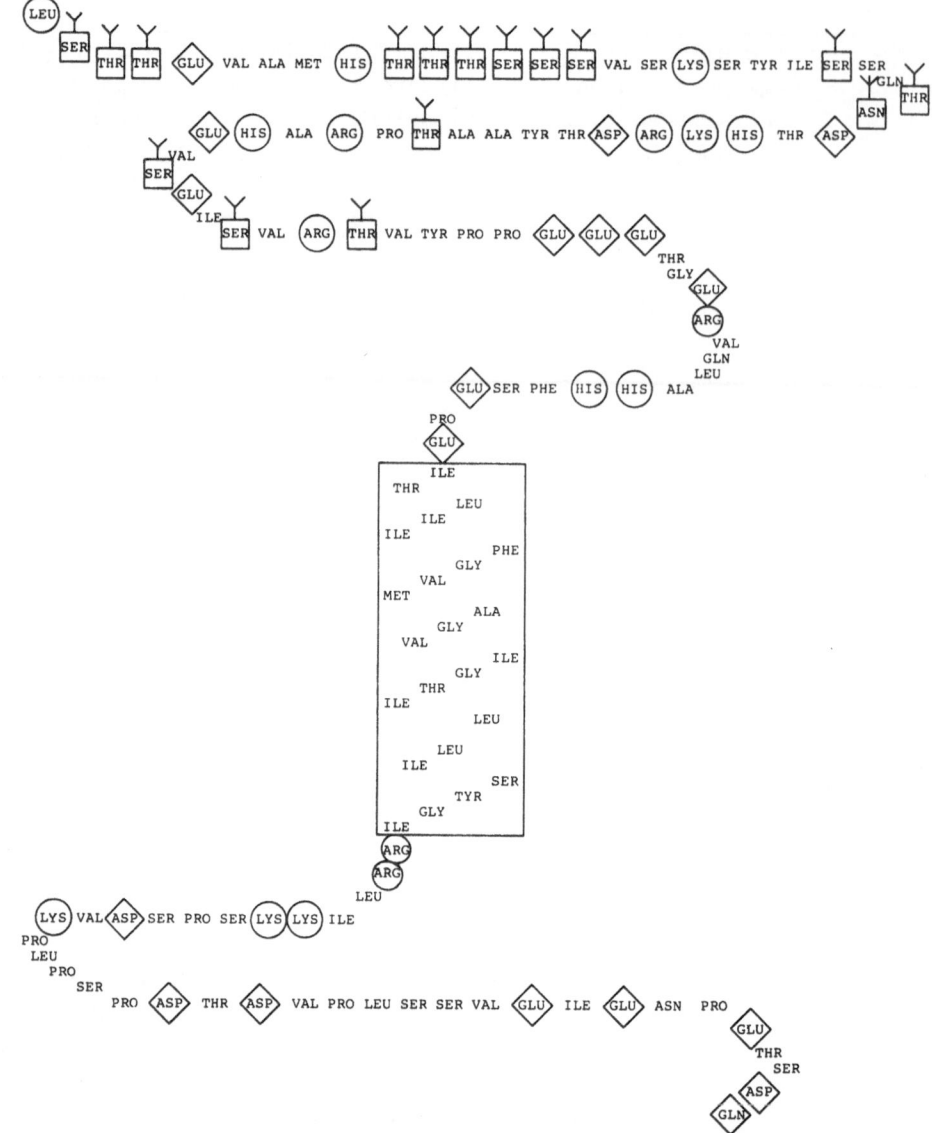

Fig. 1.4. Amino acid sequence of human erythrocyte glycophorin A
          (Tomita et al., 1978) with presumed intramembranous α-
          helical segment. O and ◇ indicate positively and ne-
          gatively charged residues, respectively, and □ in-
          dicates glycosylated residues. The N-terminus is Leu
          and the C-terminus Gln.

Three-Dimensional Structure of Integral Proteins

    The best characterized structure of an integral membrane
protein is that of bacteriorhodopsin, the light-driven proton pump
from the purple membrane of H. halobium. The structure as determined by electron diffraction to a resolution of 7 Å is given in
Fig. 1.5. The protein consists of 7 helices, most probably α -he-
lices, of 35-40 Å in length, which span the membrane and are oriented essentially perpendicular to the membrane surface. Since the
helices must account for 70-80% of the total molecular weight only
a small part of the protein will project from the surface of the
membrane, in agreement with its predominantly apolar amino acid
composition. The complete primary sequence is known and has given
rise to a molecular model which places all the charged residues in
the interior of the molecule and in which the faces of the helices
which are directed outward towards the lipid phase consist solely
of hydrophobic, uncharged residues (Engelman et al., 1980).

    The polypeptide structure of bacteriorhodopsin thus appears
to be exactly the reverse of that found for soluble proteins; the
surface of the protein is covered with hydrophobic or apolar residues and all polar residues are buried in the interior. It is quite
likely that the helical structure of bacteriorhodopsin will be a
feature common to the intramembranous sections of many integral
proteins. Energetic considerations dictate that the hydrogen bond
requirements of the polypeptide chain be fully satisfied, and in
the hydrophobic interior of the membrane this can only be done by
intramolecular bonding.

Rotational and Lateral Diffusion of Lipids

    It is clear from a variety of spectroscopic techniques that
biomembranes are dynamic not static structures. It is also known
that certain membrane functions depend critically on the fluidity
of the membrane lipids. Spin-labelled and fluorescent-labelled lipid
probes are found to perform rotational motions in the nanosecond
timescale in fluid lipid bilayers and membranes. For diffusive
rotation the characteristic correlation times are given by the
Debye equation ($\tau_R = \eta V/kT$) and correspond to effective viscosities in the range: $\eta \sim 0.1-1$ poise. A spin-labelled steroid analogue of cholesterol for instance rotates rapidly about its long

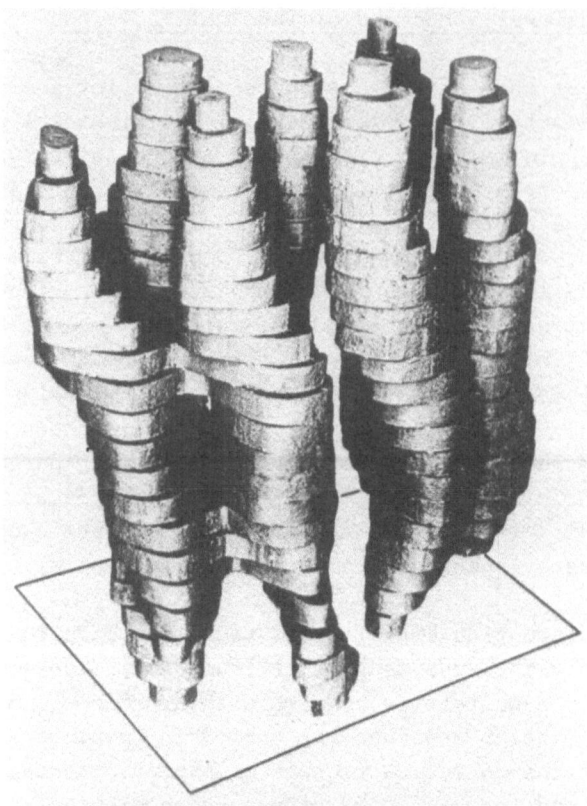

Fig. 1.5. Three-dimensional model of the bacteriorhodopsin molecule.
The model is viewed roughly parallel to the plane of the
membrane. The top and bottom of the model correspond to
the parts of the molecule in contact with the aqueous
phase, the rest being in contact with lipid (Henderson
and Unwin, 1975).

axis with a correlation time of 0.5-1 ns and performs angular os-
cillations of the long axis with amplitudes of 20°-45° depending
partly on the cholesterol content of the membrane. Such information
is deduced from the linesplittings and linewidths in the ESR spectra
of the spin-labelled lipids, which happen to be very sensitive to
angular motions in this particular time range (Marsh, 1975; 1981).

    The lipids not only undergo rapid rotational motion, but also
diffuse laterally within the plane of the membrane. This rapid

translational motion was first demonstrated by spin label methods
in which the spin-spin interaction between adjacent spin labels
was used to measure the bimolecular collision rate between lipids
in the plane of the membrane. Subsequently photobleaching experiments, of the type described for proteins in the following section,
have yielded similar results. The fluid membrane lipids have translational diffusion coefficients in the region of: $D_L \sim 10^{-8} cm^2 s^{-1}$,
corresponding to an effective hopping frequency of $\nu(=4 D_L / \lambda^2) \sim$
$10^7 s^{-1}$, or a mean distance of travel of $<x>(\sim \nu\lambda) \sim 1 \mu m/s$. Detailed references can be found in earlier publications (Marsh, 1975;
1981).

Rotational and Lateral Diffusion of Membrane Proteins

        The rates of rotational diffusion of a large variety of inte-
gral proteins have been measured from the decay in phosphorescence
anisotropy of long-lived triplet optical probes such as eosin or
erythrosin (Cherry, 1979). The method can be illustrated from a
similar experiment on rhodopsin (see Fig. 1.6), which was the first
membrane protein for which rapid rotational diffusion was demonstrated. In this case the intrinsic chromophore, retinal, is used
rather than a triplet probe, and photolysis rather than phosphorescence is detected. The retinal chromophores are oriented randomly
within the plane of the rod outer segment disc membrane (Fig. 1.6a).
A transient dichroism is induced in this population by photobleaching with polarized light. The induced dichroism is found to decay
very rapidly after the transient bleaching, due to rapid reorientation of the chromophores. This reorientation arises from the rotation of the rhodopsin molecule about an axis perpendicular to
the plane of the membrane. A rotational relaxation time of $\phi_{//} \simeq 20$
μs was calculated from the decay curve in Fig. 1.6b. Cross-linking
the proteins with glutaraldehyde stopped the rotation, causing a
long-lived photodichroism. In addition the static retinal dichroism
observed perpendicular to the plane of the disc membrane indicates
that rhodopsin performs only in-plane rotations.

        Rotational relaxation times measured for several membrane
proteins, mostly by the triplet phosphorescence method, are given
in Table 1.5. Clearly some integral proteins rotate rather rapidly
in membranes. The rotational relaxation time is defined by: $\phi_{//} =$
$1/D_{//}$ where $D_{//}$ is the rotational diffusion coefficient referred to

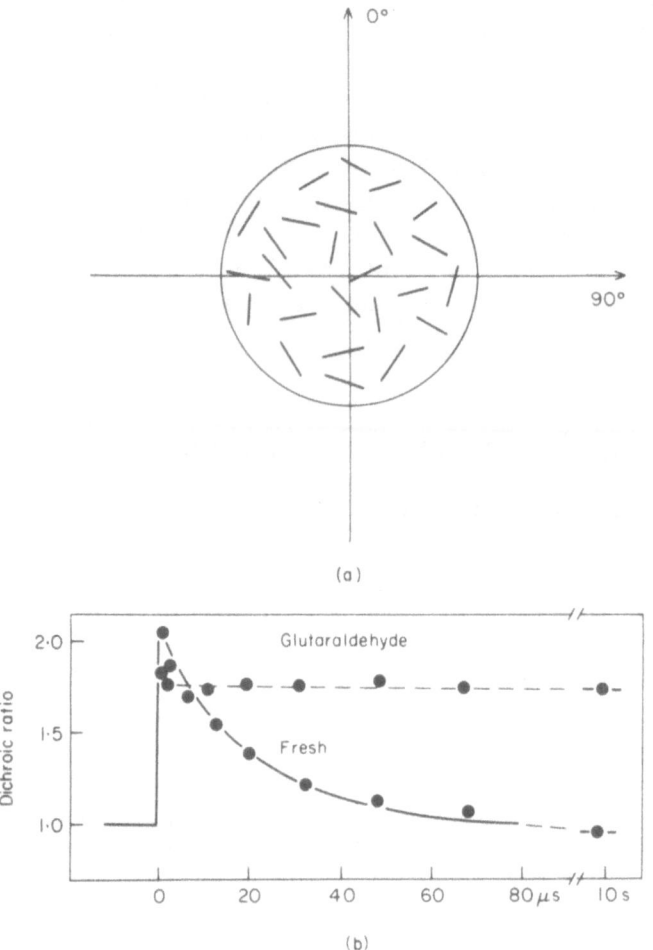

(a)

(b)

Fig. 1.6. Measurement of rotational relaxation of membrane proteins.
a) Random arrangement of chromophores in the plane of the
membrane. The short bars indicate the orientation of the
chromophores relative to the polarization direction of
the polarizing or bleaching light ($0°$ axis).   b) Decay
in induced photodichroism after a transient bleaching of
the retinal/rhodopsin chromophores in rod disc membranes
(Cone, 1972).

an axis perpendicular to the plane of the membrane. For a membrane
protein treated as a cylinder: $\phi_{//} = \eta\ 4\pi\ a^2 h/kT$, a being the radius
and h the intramembranous height. For the rapidly rotating species

Table 1.5. Rotational Relaxation Times of Membrane Proteins
         (from Cherry, 1979).

Membrane	Protein	$\phi_{/\!/}$ (µs)	T°C
Rod outer segment	Rhodopsin	20	20
Sarcoplasmic reticulum	Ca^{2+}-ATPase	100	22
Brown membrane (H. halobium)	Bacteriorhodopsin[a]	400	22
DMPC vesicles	Bacteriorhodopsin[a]	10-200	25
Erythrocyte	Band 3[b]	200	37
Rat liver microsomes	Cytochrome P-450	270	21
Lipid vesicles	Cytochrome oxidase[c]	500	22

[a] No rotation in purple membrane or DMPC in gel phase.

[b] Fast rotating component, slower component has $\phi_{/\!/}$ = 2-5 ms.

[c] Mobile component, immobile component also found at low lipid/protein ratio and
   in inner mitochondrial membrane (Kawato et al., 1981).

in Table 1.5 this gives effective membrane viscosities in the range
1-10 poise. These values are larger than those obtained from lipid
probes (presumably because of different rotational slip and fric-
tional coefficients) but nonetheless indicate a rather fluid membrane environment. Thus it appears that, in the absence of specific
inhibiting interactions, integral proteins are able to rotate freely as rigid bodies in the plane of fluid membranes. Such rotational
mobility could be important in adjusting the correct orientation
for reactions with substrates and protein components within the
plane of the membrane. On the other hand, the restriction of protein
in an otherwise fluid membrane is an indication for specific associations with other membrane components, and phosphorescence de-
polarization measurements can be used to study these supramolecular
interactions. A trivial example is that bacteriorhodopsin is found
not to rotate in the purple membrane since the protein is arranged
in a two-dimensional crystalline lattice. A more interesting case
is band 3 in the erythrocyte membrane which is found to exist in
different states of mobility (see Table 1.5). The slower rotating
component could arise from higher aggregates of band 3 dimers. The
faster component is known to correspond to band 3 dimers, since
covalent cross-linking of the dimers does not change the rotation
rate. A similar situation with regard to coexisting aggregates might
also be the case for cytochrome oxidase (see Table 1.5).

   The lateral mobility of membrane proteins can be investigated
by the fluorescence photobleaching method. This technique is illustrated schematically in Fig. 1.7. The protein is labelled with
a fluorophore such as fluorescein or rhodamine, either covalently
or via a fluorescent-conjugated antibody. A small patch of the membrane surface is photobleached by an intense light pulse destroying
all fluorophores in this region. The recovery of fluorescence due
to unbleached fluorophores diffusing into the bleached area is then
monitored. The lateral diffusion coefficient, D_L, is obtained from
the rate of fluorescence recovery, and the fraction of proteins
which is accessible to the bleached area from the percentage recovery. Typical results for a variety of different membrane proteins
are given in Table 1.6. The systems fall into two groups: the fast

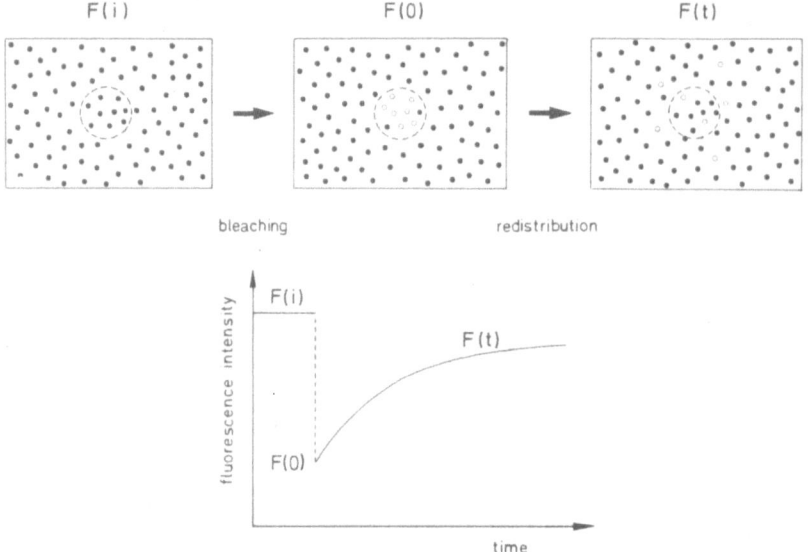

Fig. 1.7. Photobleaching experiment. The fluorescence intensity,
         F(i), arising from a small area of fluorophores is mea-
         sured. The laser beam intensity is then increased 10,000-
         fold to bleach all fluorophores in the indicated region.
         The time-course of the fluorescence recovery, F(t), is
         then recorded immediately following the bleaching pulse
         (Vaz et al., 1982.)

diffusing proteins with diffusion coefficients $D_L \sim 10^{-8}$-10^{-9} cm^2/s and complete recoveries, and the slow diffusing proteins with diffusion coefficients $D_L \sim 10^{-11}$-10^{-10} cm^2/s and only approximately 50% recoveries. For comparison lipid probes have $D_L \sim 10^{-8}$-10^{-7} cm^2/s as noted previously, and complete recoveries.

The theoretical diffusion coefficient for two-dimensional free diffusion is given by the Saffman and Dellbrück expression: $D_L = (kT/4\pi\eta h)\cdot ln[(\eta h/\eta_w a) - 0.5772]$, where h, a are the intramembranous height and protein radius and $\eta_w \simeq 10^{-2}$ poise refers to the aqueous phase. The expression is very insensitive to the particle radius which explains why there is not a strong dependence on molecular weight for the fast diffusing proteins, nor a very large decrease in rate as compared to the individual lipid molecules. Values of the membrane viscosity in the range $\eta = 1$-10 poise predict diffusion coefficients $D_L = 6\cdot10^{-8} - 6\cdot10^{-9}$ cm^2/s. Comparing with Table 1.6, this indicates that the group of faster moving proteins are performing essentially free lateral diffusion within the plane

Table 1.6. Lateral Diffusion Coefficients of Proteins Measured by Photobleaching (adapted from Cherry, 1979).

Cell	Protein	$D_L (cm^2/s)$	Recovery	T^oC
Rod outer segment	Rhodopsin	2-$6\cdot10^{-9}$	-	20°
DMPC vesicles[a]	Cyt. P-450	$7\cdot10^{-9}$	-	26°
DMPC vesicles[a]	Glycophorin	$1\cdot10^{-8}$	-	25°
Mast cells	F_c receptors	$2\cdot10^{-10}$	50-80%	-
Fibroblast	anti-P388 receptors	$3\cdot10^{-10}$	50%	-
"	S-ConA[c] receptors	1-$10\cdot10^{-11}$	36%	22°
Myotubes	S-ConA[c] receptors	$2\cdot10^{-11}$	-	22°
"	AChR[c]	$5\cdot10^{-11}$	75%	22°
Blebs[b]	AChR[c]	$2\cdot10^{-9}$	100%	-
Blebs[b]	ConA	$7\cdot10^{-9}$	≥97%	-

[a] Data from Vaz et al. (1982)

[b] Data from Wu et al. (1981) and Tank et al. (1981)

[c] S-ConA, succinylated concanavalin A; AChR, acetylcholine receptor

of the membrane, which is also in accordance with the 100% fluorescence recoveries. This fast group consists of purified proteins reconstituted in lipid bilayers, rhodopsin in the disc membrane and
proteins in membrane blebs. The rod outer segment discs contain no
cytoskeletal elements, and the blebs are sections of membrane which
have been artificially induced to bud off from the underlying cytoskeletal network. Thus all three systems correspond essentially to
free proteins in bilayers.

The group of much slower moving proteins with incomplete recoveries which are found in cell plasma membranes most probably
corresponds to membrane proteins whose diffusion is restricted by
interaction with the cytoskeleton and/or glycocalyx. The large decrease in diffusion coefficient cannot simply be accounted for by
an increase in membrane viscosity; this would have to be too large.
Similarly, because the free diffusion rate is very insensitive to
particle size the large decrease in rate cannot be due to oligome-
rization; the aggregates required would be very large. In this connection it is interesting to compare the results on protein rotation
and lateral diffusion. For free diffusion the lateral diffusion
coefficient is related to the rotational relaxation time by:
$D_L \simeq 4a^2/\phi_{//}$. For rhodopsin this would require a particle radius of
10-17 Å, which is reasonable. For band 3 on the other hand the
rotational relaxation time would predict a lateral diffusion coefficient of $D_L \sim 3.10^{-9}$ cm^2/s for the dimer whereas the measured
value is much smaller $D_L \sim 4.10^{-11}$ cm^2/s. Clearly band 3 does not
undergo unrestricted diffusion and most probably its diffusion is
limited in range by the spectrin-actin network on the inner surface
of the erythrocyte membrane, which nonetheless allows free rotational diffusion within the interstices.

Lipid Chain Dynamics

The dynamic properties of the membrane discussed in the previous two sections have their origin in rotation about the bonds
in the lipid hydrocarbon chains. It is this rotational isomerism
which is the fastest molecular motion and is the fundamental basis
for the membrane fluidity. The chain motions are best studied by
^{2}H NMR of lipids whose chains have been specifically deuterated.
The nuclear quadrupole splittings can be used to measure the order
parameter $S_n = 1/2 (3<\cos^2 \theta_n> - 1)$, or time-average angular am-

plitude of motion, of the chain segments (Seelig and Seelig, 1980).
The quadrupole splittings are dependent on the orientation of the
chains with respect to the static magnetic field of the NMR spectrometer. Angular rotations of the chain segments at frequencies greater
than 200 kHz will average these splittings, the extent of averaging
depending on the angular amplitude, θ_n. A typical ^{2}H NMR spectrum
is given in Fig. 1.8; this spectrum is from bilayers of dimyristoyl
phosphatidylethanolamine in which the sn-2 has been perdeuterated.
The quadrupole splittings from the various chain segments are indicated, the smaller splitting corresponding to smaller order parameter ($\Delta\nu_n = (3/8) \, (e^2qQ/h) \cdot S_n$)), i.e. greater amplitude of motion.
The angular amplitude of motion increases (quadrupole splitting decreases) on proceeding from the glycerol backbone to the terminal
methyl group of the chain. The C-2 group is the only position which
is out of order, the reason being that the bend in the sn-2 chain

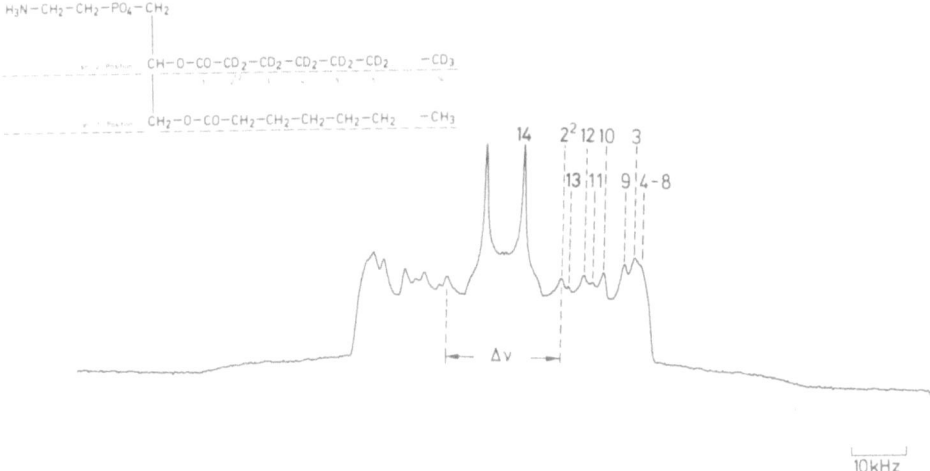

Fig. 1.8. ^{2}H NMR spectrum of dimyristoyl phosphatidylethanolamine
bilayer membranes with per-deuterated sn-2 chains, T =
50°C. The spectral peaks arising from the different methylene segments are indicated. The bilayers are randomly
oriented and the spectra from each $-CD_2$ group are a superposition of the individual spectra from each orientation.
The peaks arise from the axial degeneracy of all bilayers
with their planes containing the magnetic field direction
(Marsh et al., 1982).

found in the crystal structure (Fig. 1.3) is preserved in the hydrated bilayers, causing the C-2 group to have an entirely different
orientation from the rest of the chain.

The reason for the increase in angular amplitude of motion
down the chain lies in the rotational isomerism. If it is assumed
that the chain is effectively anchored at the glycerol backbone,
there are increasing numbers of bonds about which rotations can
take place as one proceeds down the chain. This is the intramolecular effect which allows a uniform population of rotational isomers
down the chain. There are also intermolecular effects which cause
a reduced population of gauche rotational isomers in the upper part
of the chain and a steadily increasing gauche population in the
lower half of the chain. It is seen from Fig. 1.8 that the quadrupole splittings for methylene segments C 3 - C 8 are all very
similar. This is the so-called order parameter plateau which has

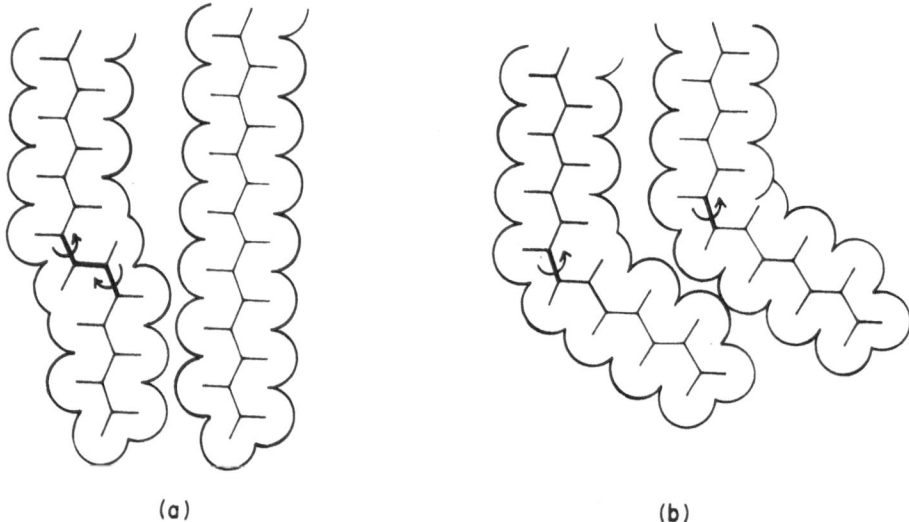

(a)   (b)

Fig. 1.9. Angular motion and packing of lipid chains arising from
          rotational isomerism from the linear all-trans chain.
          a) The all-trans chain and a gauche$^+$-trans-gauche$^-$ kink
          conformation. b) A single gauche$^+$ conformation caused
          by a +120° rotation about one C-C bond.

its origin in the chain packing restrictions. Since single <u>gauche</u>
conformations cause large angular deviations of the chain axis, only
those combinations of <u>gauche</u> conformations such as kinks (see Fig. 1.9)
which minimize lateral displacements and preserve the parallel
packing of the chains, are allowed in the upper parts of the chain.
Lower down the chain the lateral displacements caused by single
<u>gauche</u> conformations are much smaller and hence these are also al-
lowed. In addition bond rotations cause a shortening of the chains
(Fig. 1.9) which tends on average to create more free volume towards the centre of the bilayer, allowing increased rotational isomerism.

Deuterium NMR is thus capable of giving a rather detailed description of the dynamic configuration of the lipid chains. Quantitatively it is found that there are on average ~ 4 <u>gauche</u> conformations or ~ 2 kinks per chain. Spin lattice relaxation measurements which are sensitive only to fast motions, indicate that
the correlation times for the rotational isomerism are ~ 0.1 ns
for the plateau region in the first half of the chain, decreasing
to ~ 0.01 ns at the end of the chain. The cooperative interaction
of these chain rotations gives rise to the dynamic properties of
membranes discussed in the previous sections, and the equilibrium
populations of <u>gauche</u> isomers determine the structural and thermodynamic properties of lipid bilayers discussed in the following
chapter.

2. PHOSPHOLIPID BILAYER MEMBRANES

<u>Introduction</u>

The central role of the phospholipid bilayer in biological
membrane organization was outlined in the previous chapter. Bilayers
composed of single defined phospholipids provide an excellent model
system for in depth study of lipid membrane properties. In the
present chapter the structure and thermodynamics of phospholipid
bilayers, as determined by a variety of different physical techniques, will be reviewed. Particular attention will be paid to the ordered-fluid lipid phase transition and its use in studying bilayer
properties. The effects of hydration, surface electrostatics and
ion interactions will be considered. Discussion is also given of
the transformation to non-bilayer phases.

Lipid-Water Phase Behaviour

    Fully hydrated bilayers composed of a single phospholipid
species undergo a well-defined thermotropic phase transition (see
Fig. 2.1) in which the lipid chains change from an ordered, pseudo-
crystalline or gel state to a fluid or liquid crystalline state
which is similar to that found in biological membranes. The lipid
chain configuration in the high-temperature, fluid L_α phase is
that described in the final section of the previous chapter. In the
low temperature, ordered phase, the chains are essentially in the
parallel, all-<u>trans</u> state, and may be tilted relative to the bilayer
normal, the L_β' state as in phosphatidylcholines, or untilted the
L_β state as in phosphatidylethanolamines. Additionally, an intermediate phase, P_β', in which the bilayer is rippled with a period
of 100-200 Å is found in the gel phase of certain phospholipids,
notably phosphatidylcholine and phosphatidylglycerol. The occurrence of the rippled phase is associated with the pre-transition.
The structure of the various phases has been established by X-ray
diffraction (see below), and also by freeze fracture electron
microscopy, with which the rippled phase can be readily visualized.

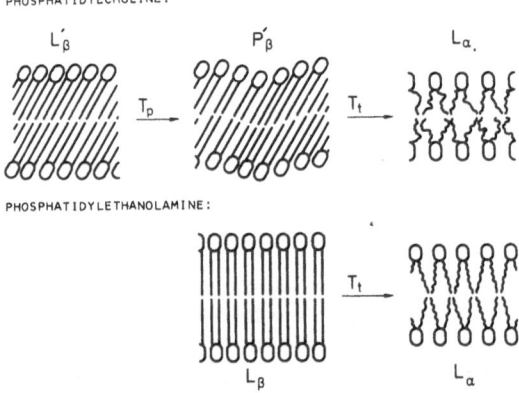

Fig. 2.1. Thermotropic phase transitions in phospholipid bilayers.
        T_t is the main transition temperature corresponding to
        the change from the gel phase to the fluid phase, L_α . T_p
        is the pretransition corresponding to the change from the
        tilted ordered phase L_β' to the ordered rippled phase
        P_β' .

The transition temperatures, T_t, and phase behaviour differ between the various phospholipid types but all depend strongly on the
degree of hydration. A typical phospholipid-water phase diagram is
given in Fig. 2.2 for dimyristoyl phosphatidylserine. The anhydrous
phospholipid forms bilayer crystals which undergo a transition to
a liquid crystalline state at high temperature. The transition
temperature decreases progressively as water is taken up by the
bilayer. The water molecules bind to the phospholipid headgroups

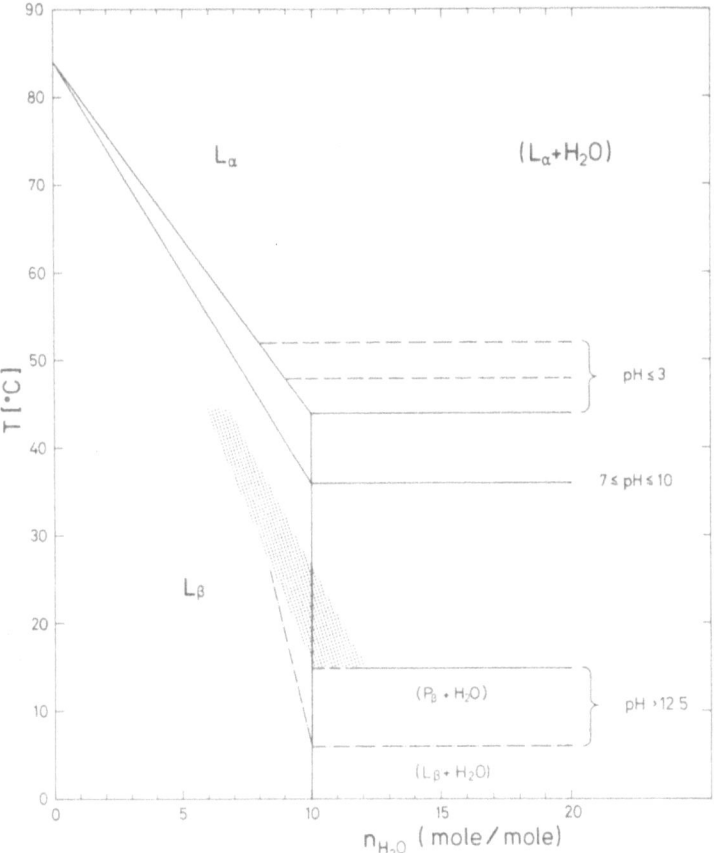

Fig. 2.2. Composite phase diagram for dimyristoyl phosphatidyl-
          serine bilayers in the three different states of proton-
          ation: at pH $\lesssim$ 3 (zwitterionic), at 7 $\lesssim$ pH $\lesssim$ 10 (singly
          charged) and at pH $>$ 12.5 (doubly charged). The phase
          boundaries were established by measuring the bilayer
          transition temperatures as a function of water content
          (Cevc et al., 1981).

and depress the transition temperature by approximately 4°C per
water molecule. The bound water molecules can be detected by the
quadrupole splitting of the ^{2}H NMR spectrum from D_2O and clearly
must mediate any ionic interactions with the phospholipid headgroups. Differences in transition temperature between different
lipids may also be due in part to different degrees of hydration.
The bilayers reach a limiting hydration at approximately 20% water
in the gel phase. Beyond this the transition temperature remains
constant and further water is not incorporated between the bilayers,
but is present as a free water phase in which the fully hydrated
bilayers are dispersed. Above the phase transition the bilayers can
accommodate more water and for phosphatidylcholine a free water
phase does not appear until $\sim$40% water content.

Calorimetric Properties

Differential scanning calorimetry is a basic method to study
the phospholipid bilayer phase transitions since these are accompanied by a positive enthalpy of fusion. The excess heat capacity,
ΔC_p, of the sample is measured (see Fig. 2.3) and thus the enthalpy
of the transition is given by the integrated area:

$$\Delta H_t = \int_{T_i}^{T_f} \Delta Cp \cdot dT.$$

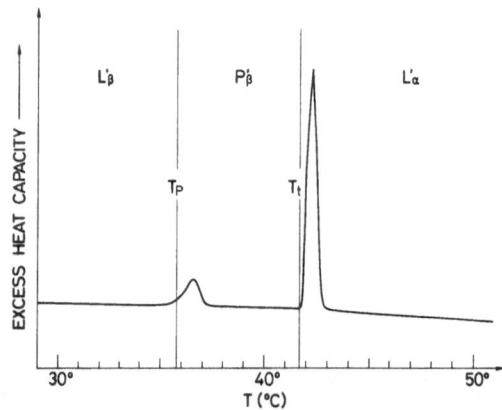

Fig. 2.3. Differential scanning calorimeter trace of dipalmitoyl
        phosphatidylcholine bilayers dispersed in water. The
        reference pan contained distilled water.

If the transition is first order: $\Delta G_t = 0$, and the entropy of transition is given by: $\Delta S_t = \Delta H_t / T_t$. The values of the transition entropy and of the transition enthalpy depend linearly on the lipid chain length and are very similar for different lipids of the same chainlength (see Table 2.1). Particularly interesting is the transition entropy, since this is dominated by the configurational disorder of the lipid chains in the fluid phase. The incremental transition entropy has a value $d\Delta S_t / dCH_2 \sim 1.54$ cal/mole K per chain, approximately the same for all phospholipids. This is approximately half the entropy of fusion for crystalline hydrocarbons (~ 2.61 cal/mole K/CH_2), indicating that the lipid chains in fluid bilayers are in a much more ordered state than in liquid hydrocarbons. This is in agreement with the results of deuterium NMR studies discussed in the previous chapter. The transition entropy can be directly related to the chain rotational isomerism by statistical mechanical theories (Marsh, 1974). The incremental transition enthalpy, $d\Delta H_t / dCH_2 = 0.58$ kcal/mole per chain, is also considerably smaller than the melting enthalpy of crystalline hydrocarbons (~ 1.0 kcal/mole/CH_2). This quantity contains both intramolecular (rotational isomerism) and intermolecular (van der Waals interaction) terms, but clearly again the chain packing must be more ordered in the fluid bilayers than in liquid hydrocarbons (Marsh, 1974).

Table 2.1. Thermodynamic Properties of the Phospholipid Bilayer Phase Transition

	T_t (^{o}C)	ΔH_t (kcal/mole)	ΔS_t (cal/mole K)	$\Delta \bar{V}_t$ (ml/g)
di C_{12}-PC	-1.0	4.3	15.8	–
di C_{14}-PC	24.0	6.3	21.1	0.027
di C_{16}-PC	41.4	8.7	27.6	0.037
di C_{18}-PC	58.2	10.8	32.7	0.045
di C_{12}-PE	30.5	3.6	11.9	0.016
di C_{14}-PE	49.6	6.4	19.8	0.0204
di C_{16}-PE	63.8	9.6	28.6	–

From Seelig (1981), Nagle and Wilkinson (1978), and Wilkinson and Nagle (1981).

The chainlength dependence of the transition temperature can be explained from the calorimetric measurements; it essentially comes from the end-effects in the chainlength dependence of ΔH_t and ΔS_t. If we write the linear dependence as

$$\Delta H_t = (n-n_o)\,\Delta h; \qquad \Delta S_t = (n-n_o')\,\Delta s,$$

where Δh and Δs are the incremental transition enthalpy and transition entropy per CH_2 group, and n is the chainlength, then

$$T_t = \Delta H_t / \Delta S_t = (\Delta h/\Delta s)\left[1-(n_o-n_o')/(n-n_o')\right] =$$

$$T_t^{\infty}\left[1-(n_o-n_o')/(n-n_o')\right].$$

Here $T_t^{\infty} = \Delta h/\Delta s$ is the transition temperature extrapolated to infinite chainlength. Normally $n_o > n_o'$, since the headgroup contributions to the end-effects are greater for ΔH_t than for ΔS_t, and this gives rise to the chainlength dependence. As the chains become longer the chainlength dependence of T_t progressively decreases $(T_t - T_t^{\infty} \sim 1/(n-n_o'))$, as is observed in Table 2.1.

Dilatometric Results

A discontinuous change in the lipid partial specific volume as expected for a first order transition, is observed at the bilayer phase transition. The net volume change however is relatively small: $\Delta V_t / V_t \sim +4\%$ (see Table 2.1). This indicates that the lipid chain disordering is highly cooperative and does not cause large disruptions in the chain packing. The absolute value of ΔV_t increases with increasing chainlength (Table 2.1), as should be the case if the dominant contribution is from chain isomerism. The volume changes are only 2/3 of those observed at the melting of paraffins however, again indicating that the lipid chains in fluid bilayers are more ordered than in liquid hydrocarbons. The free volume produced by an isolated kink is ~ 25 $\overset{o}{A}^3$ (Träuble and Haynes, 1971), as compared with $\Delta V_t \simeq 46$ $\overset{o}{A}^3$/molecule for dipalmitoyl phosphatidylcholine. This would imply a minimum value of ~ 2 kinks _per molecule_ created at the phase transition if the volume change is due solely to kink formation. If allowance is made for the cooperative packing of the defects, the estimated number of kinks would be higher. The value of ~ 2 kinks _per chain_ determined by ^{2}H NMR (see previous chapter) indicates that this cooperative packing of the chains with correlated motions is found in fluid bilayers.

X-Ray Diffraction Measurements

Multibilayer dispersions of phospholipids give X-ray reflections in the low-angle region (Fig. 2.4a) from the bilayer repeat spacing, and in the high-angle region (Fig. 2.4b) from the chain-chain packing. In the high-angle region at pH 1.5 there is a single sharp reflection at 4.12 $\overset{o}{A}$, characteristic of hexagonal packing of the lipid chains. The chain-chain spacing is a = $(2/\sqrt{3})\cdot s$ = 4.76 $\overset{o}{A}$, where s = 4.12 $\overset{o}{A}$ is the short spacing, and the area per chain: $f_o = (2/\sqrt{3})s^2 = 19.6 \overset{o}{A}^2$. At pH 8 there is a sharp reflection at 4.21 A with a broad shoulder at 4.13 $\overset{o}{A}$. This corresponds to a distorted hexagonal packing with the chain tilted in the direction of the 4.13 A spacing.

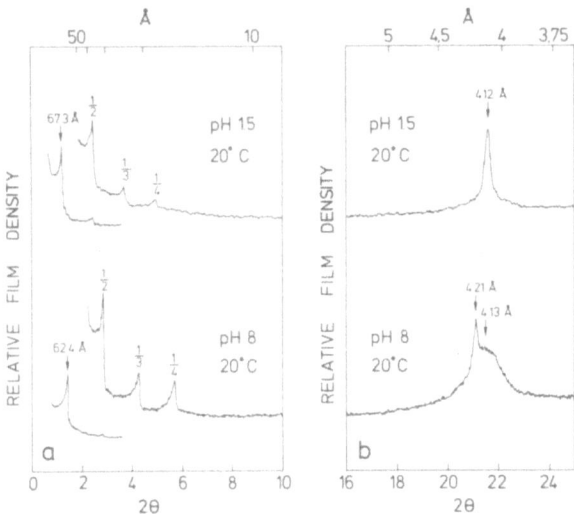

Fig. 2.4. Densitometer traces of powder X-ray diffraction intensities from fully-hydrated dipalmitoyl phosphatidylglycerol multibilayers at pH 8.0 (1.5 M KCl/50 mM Tris) and pH 1.5 (1.5 M KCl/HCl), at 20°C. (a) Low-angle, and (b) high-angle reflection regions, respectively. The diffraction intensity is plotted against the diffraction angle 2θ and the corresponding Bragg spacings (Watts et al., 1981).

The long spacings in Fig. 2.4a are in the simple ratio 1:2:3:4 indicating a lamellar structure consistent with lipid bilayers. The repeat distances, d = 67.3 Å at pH 1.5 and d = 62.4 Å at pH 8.0, correspond to the bilayer thickness, d_1, plus the thickness of the interlamellar water layer, d_w. This is seen clearly in Fig. 2.5 in which the lamellar repeat increases with increasing water content (c = weight fraction of lipid) until a limiting value of around 22–24% water. At this point no more water is taken up between the bilayers and further added water is present as an excess free water phase (cf. Fig. 2.2). Using lumped values $\bar{V}_1$ and $\bar{V}_w$ for the partial specific volumes of the bilayer and the interlamellar water, respectively, it is possible to calculate the effective bilayer thickness from the long spacing and the value of the limiting hydration

$$d_1 = d/\left[1 + (\bar{V}_w/\bar{V}_1)(1-c)/c\right],$$

where c is the lipid weight fraction at limiting hydration. In this way it is found that d_1 = 51.1 Å at pH 8.0 and 53.3 Å at pH 1.5. Since the bilayers are in the gel phase at both pHs, the difference corresponds to a tilting of the chains at pH 8.0. The value of d_1 at pH 1.5 is close to the bilayer thickness of 54.5 Å estimated[1]

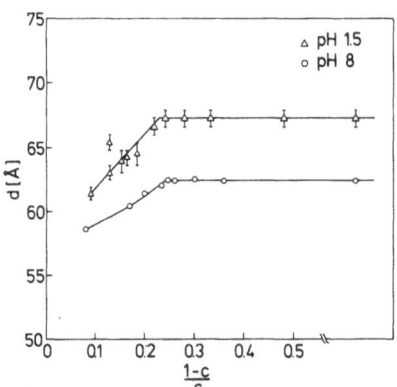

Fig. 2.5. X-ray long spacings at 20°C as a function of water/lipid weight ratio, (1–c)/c. Derived from the low-angle diffraction of dipalmitoyl phosphatidylglycerol multibilayers at pH 8.0, 1.5 M KCl/50 mM Tris (o——o) and pH 1.5, 1.5 M KCl/HCl (△——△) (Watts et al., 1981).

from molecular models with perpendicular all-<u>trans</u> chains.

The area per molecule can also be calculated from the bilayer thickness

$$F = 2 \, M \overline{V}_1 / (N \cdot d_1),$$

where M is the lipid molecular weight and $N = 6.023 \cdot 10^{23}$. The lipid molecular areas in the plane of the bilayer are 48 Å^2 at pH 8.0 and 37 Å^2 at pH 1.5. The chain tilt can be determined by comparison with the chain areas, f_o, determined from the short spacing

$$\cos \theta = 2 \, f_o / F .$$

At pH 8.0 $\theta = 32^o$ confirming that the bilayer is in the L_β' phase, and at pH 1.5 $\theta \simeq 0^o$ confirming that the bilayer is in the L_β phase.

A point of considerable interest is the structural changes which take place during the bilayer phase transition. It is found that the bilayer thickness decreases by $\Delta d_1 / d_1 \approx -10\%$ or $\Delta d_1 \simeq -4.5$ Å for dipalmitoyl phosphatidylcholine. Allowing for the tilt, the changes along the length of the chain are -15% and -5.7 Å, respectively. The origin of the bilayer thinning is the creation of rotational isomers in the lipid chains. For instance 1 kink reduces the chainlength by 1.27 Å, the projected length of 1 CH_2 group (cf. Fig. 1.8). This is in rough agreement with the estimate of 2 kinks per chain from ^{2}H NMR.

The bilayer thinning is accompanied by an expansion of the bilayer area. For dipalmitoyl phosphatidylcholine it is found that $\Delta E_t / F_t \simeq +14\%$, the expansion in membrane area more than compensating for the 10% decrease in bilayer thickness to give the +4% expansion in bilayer volume discussed in the previous section. The absolute increase in area/molecule is +8 Å^2. The actual increases in chain area, allowing for tilt, are $\Delta f_t / f_t \simeq 35$–40% or $\Delta f_t \simeq$ 18 Å^2, which indicates the considerable increase in rotational isomers. These area increases clearly will strongly affect surface binding properties, and are also important in discussing transition temperature shifts below.

Electrostatic and Ionic Interactions and Transition Temperature Shifts

The bilayer transition temperature depends on the phospholipid headgroup as indicated in Table 2.1. The negatively-charged phospholipids are interesting and biologically relevant since their headgroups can be modified by ionic interactions. In particular pH titration will protonate or deprotonate the headgroup and change the bilayer surface charge, hence modifying the electrostatic properties of the membrane.

The pH titration of the transition temperature of dimyristoyl and dipalmitoyl phosphatidylglycerol bilayers is given in Fig. 2.6.

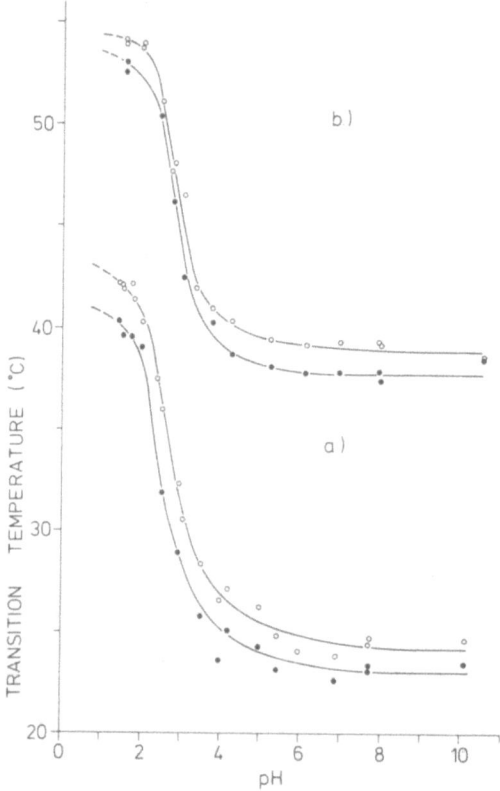

Fig. 2.6. Phase transition temperatures of (a) dimyristoyl and (b) dipalmitoyl phosphatidylglycerol bilayers (1.4 mM) in 0.1 M KCl as a function of pH. Open circles, heating; closed circles, cooling (Watts et al., 1978).

From this it is seen that the transition temperature increases by $18^{\circ}C$ for dimyristoyl phosphatidylglycerol and by $15^{\circ}C$ for dipalmi-toyl phosphatidylglycerol, on protonating the phosphate group which has an apparent pK_a of 2.9. The transition temperature shifts can be interpreted in terms of the increase in bilayer free energy, G^{Pr}, due to the protonation. Then the shift is

$$\Delta T_t = T_t^* - T_t = \Delta G_t^{Pr} / \Delta S_t^* ,$$

where the asterisk refers to the protonated state and ΔG_t^{Pr} is the change in the protonation contribution to the free energy at the phase transition (Träuble, 1976; Cevc et al., 1980). Consistent values of G_t^{Pr} = 320 cal/mole and 340 cal/mole are obtained for di-myristoyl and dipalmitoyl phosphatidylglycerol, respectively, pro-ving that the protonation is purely a headgroup interaction, not directly affecting the chains.

In principle there are several possible contributions to ΔG_t^{Pr}, and hence to the transition temperature shift. One contri-bution is from the bilayer surface charge and arises because the surface electrostatic energy is less in the more expanded fluid phase above the transition, hence causing a reduction in T_t. The shift depends on the increase in molecular area ΔF at the tran-sition and can be estimated from Gouy-Chapman electrostatic double layer theory (Träuble, 1976). At low ionic strength the shift be-tween charged and uncharged bilayers is given by

$$\Delta T_t^{max} = (2\ kTN / \Delta S_t^*) \cdot (\Delta F/F) .$$

Taking a value of $\Delta F/F \simeq 0.14$ (see the above section) gives an estimate for the maximum electrostatic shift of $\Delta T_t^{max} \simeq -10^{\circ}C$ and $-8^{\circ}C$ for dimyristoyl and dipalmitoyl phosphatidylglycerol, re-spectively. These values are considerably smaller than the experi-mentally determined shifts and are likely to be an upper limit since Gouy-Chapman theory tends to overestimate surface potentials. Clearly there are also non-electrostatic contributions to the total titration-induced shifts.

The salt dependence of the transition temperature shift gives a method of investigating and isolating the electrostatic contri-bution to the total shift. Increasing monovalent ion concentration shields the surface charges and hence should decrease the electro-

statically induced depression of T_t. Data for dimyristoyl and dipalmitoyl phosphatidylglycerol bilayers are given in Fig. 2.7.
As expected there is an increase in transition temperature with increasing ionic strength for both systems in the charged state at
pH 8.0. The transition temperature of bilayers in the uncharged
state at pH 1.1 does not change with ionic strength, confirming that
the salt affects essentially only the electrostatic properties. As
the salt concentration is increased the transition temperature of
the charged bilayers increases as $c^{1/2}$, the square root of the salt
concentration. This is in accordance with Gouy-Chapman theory which
predicts

$$\Delta T_t^{GC} = \Delta T_t^{max} - (4\,\epsilon\epsilon_o Nk^2 T^2 / e^2 \Delta S_t^{*})\,\Delta F \cdot \varkappa \quad ,$$

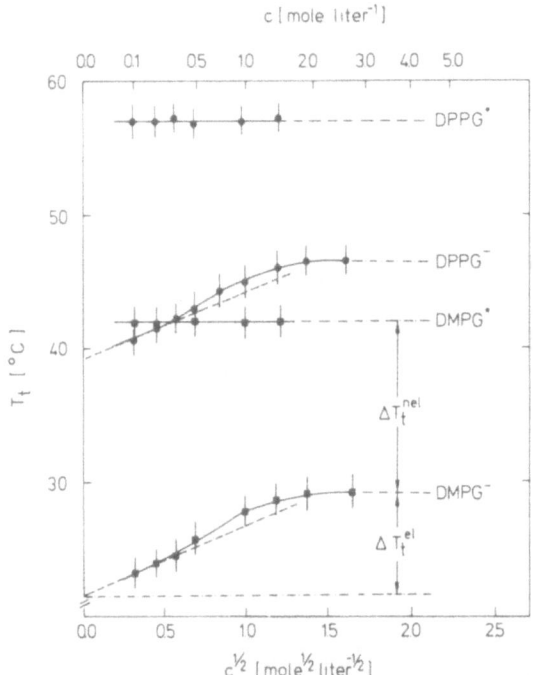

Fig. 2.7. Phase transition temperatures of dimyristoyl (DMPG) and
dipalmitoyl (DPPG) phosphatidylglycerol as a function
of the bulk NaCl concentration, c, and the headgroup
ionization. PG^o: 0.1 N HCl pH 1.08; PG$^-$: triethanol-
amine·HCl/NaOH pH 7.0 (Cevc et al., 1980).

with the reciprocal Debye screening length given by

$$\varkappa = (2000\ Ne^2 /\ \varepsilon\varepsilon_o kT)^{1/2}\ c^{1/2} \quad .$$

However, the initial rate of screening predicted for $\Delta F = 8\ \mathring{A}^2$: $d\Delta T_t^{GC}/dc^{1/2} \simeq 3.5^{o}C\ mole^{-1/2}\ litre^{1/2}$, is considerably smaller than that observed experimentally: $d\Delta T_t/dc^{1/2} \simeq 4.5 - 5^{o}C\ mole^{-1/2}\ litre^{1/2}$.

At higher salt concentration (c >0.5 M) the screening exceeds to an even greater extent that predicted by Gouy-Chapman theory, and the transition temperature flattens off to a plateau value at salt concentrations of c $\simeq$ 1.5 - 2 M. Independent measurements using charged partitioning spin labels indicate that the surface potential of the bilayers is completely screened at this salt concentration. Thus from Fig. 2.7 it appears that the electrostatic phase transition temperature shift is effectively screened in >2 M NaCl and the total electrostatic contribution to the transition temperature shift is $\Delta T_t^{el} \simeq 6.5^{o}$ and $5.5^{o}C$ for dimyristoyl and dipalmitoyl phosphatidylglycerol, respectively. These are again smaller than the values for ΔT_t^{max} predicted above, in line with the tendency of Gouy-Chapman theory to overestimate surface potentials. Thus the salt dependence in Fig. 2.7 provides clear evidence for a large non-electrostatic contribution, ΔT_t^{nel}, to the phase transition temperature shift caused by lipid titration. Likely sources for the non-electrostatic shift are changes in the head-group-water interactions and in the hydrogen bonding of the head-groups.

The results of this section illustrate the usefulness of phase transition titrations in studying ionic interactions with the bilayer surface and interactions between the phospholipid headgroups. For phosphatidylserine (Cevc et al., 1981) it has been found that the increase in transition temperature on titrating the carboxyl group can be accounted for solely by an electrostatic shift of $8^{o}C$ and shifts of $4^{o}C$ per water molecule involved in the loss of one or two waters of hydration. The much larger decrease of $\Delta T_t \simeq -22^{o}C$ on titrating the amine group contains only a small electrostatic contribution, $\Delta T_t^{el} \simeq -6^{o}C$, the remainder probably being due either directly or indirectly to hydrogen bonding.

Hexagonal-Bilayer Phase Transformations

Although a wide range of phospholipids spontaneously form bilayers when dispersed in water, non-lamellar structures such as the
inverted hexagonal (H_{II}) phase are also found under certain circum-
stances. Such non-bilayer phases could be especially relevant in
"activated" processes in membranes, such as membrane fusion, membrane synthesis and biogenesis, in which the normal lamellar membrane topology is interrupted. Studies of bilayer-nonlamellar phase
transformations thus give information about the non-bilayer phase
in much the same way as studies of the gel-fluid phase transition
gave information on the biologically-relevant fluid phase in the
previous sections. In addition the phase transformation itself is
interesting since this may be the basis for the "activated" membrane processes. This is particularly so in view of the fact that these
phase transformations can be of extremely low entropy. Consideration of the above section thus shows immediately that these transitions can be triggered by very weak energetic interactions.

The bilayer-inverted hexagonal phase transition is taken as
a model for non-bilayer transformations in this section. A calorimetric recording from an aqueous dispersion of the ether phospholipid dihexadecyl phosphatidylethanolamine is given in Fig. 2.8.
The main gel-fluid bilayer phase transition occurs at $68.5^{\circ}C$ with
an enthalpy of 7.9 kcal/mole, corresponding to a transition entropy
of $\Delta S_t \sim 23$ cal/mole K. At higher temperature, in the fluid phase
a lower enthalpy transition occurs which corresponds to the transition from multilamellar organization to one in which the lipids
are arranged in inverted cylinders which are packed in a hexagonal
array (see Fig. 2.8). This transition occurs at $T_h = 86.5^{\circ}C$ and
has an enthalpy of $\Delta H_h \simeq 1.4$ kcal/mole corresponding to a transition entropy of $\Delta S_h \simeq 4$ kcal/mole K, much lower than that for
the gel-fluid transition.

The structure of the inverted hexagonal phase is established
by X-ray diffraction. The long spacings are found to be in the
ratio $1:3^{1/2}:2$ which is diagnostic for the hexagonal packing of
the lipid-water cylinders. For ditetradecyl phosphatidylethanolamine
in water for example, the low-angle diffraction changes abruptly
at $T_h \sim 92^{\circ}C$ from a lamellar pattern with orders in the ratio 1:2:3
and long spacing $\sim 44 \AA$ to a hexagonal pattern with orders in the
ratio $1:3^{1/2}:2$ and long spacing of $\sim 55 \AA$ (Seddon et al., 1982).

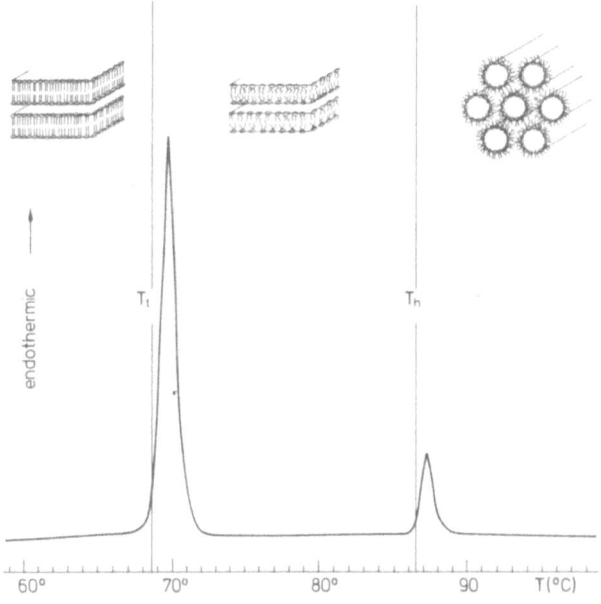

Fig. 2.8. Differential scanning calorimeter scan of dihexadecyl
phosphatidylethanolamine dispersed in water. T_t indicates
the ordered-fluid bilayer phase transition and T_h the
lamellar-hexagonal phase transition (Seddon, et al., 1982).

Phosphorus NMR provides a further diagnostic method for de-
tecting hexagonal phases in phospholipid dispersion. The ^{31}P NMR
spectra of a dihexadecyl phosphatidylethanolamine dispersions in
2.5 M NaCl in the lamellar and hexagonal phases, at 70°C and 80°C
respectively, are given in Fig. 2.9. The NMR spectra are powder
patterns consisting of a superposition of the spectra from bilayers
(or hexagonal cylinders) at all orientations with respect to the
spectrometer magnetic field. The lamellar powder pattern has axial
symmetry arising from rotations of the phospholipid molecule in the
bilayer. The peak at high field arises from the axial degeneracy of
all bilayers with their planes containing the magnetic field direction. The shoulder at low field corresponds to those (far fewer)
bilayers which are oriented with their plane perpendicular to the
magnetic field. The chemical shift anisotropy is given by the splitting between the peak and shoulder, and for the lamellar phase at
70°C is $\Delta \sigma = -50$ ppm.

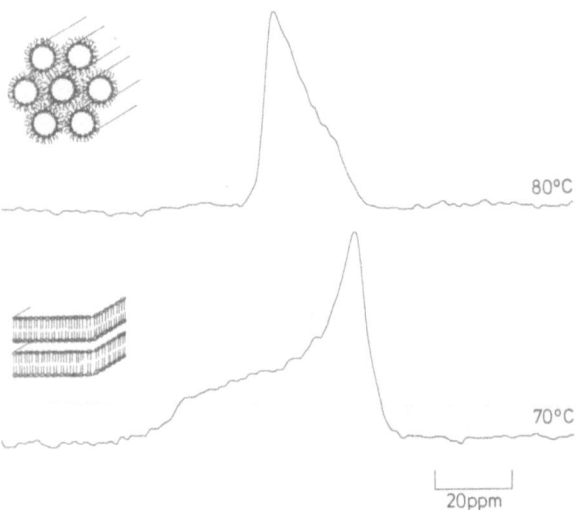

Fig. 2.9. 109 MHz ^{31}P NMR spectra of dihexadecyl phosphatidyletha-
         nolamine in 2.5 M NaCl in the lamellar phase at 70°C and
         in the inverted hexagonal phase at 80°C (Marsh and Seddon,
         1982).

In the hexagonal phase there is an additional rotational motion which gives rise to a further averaging of the chemical shift
anisotropy. This is rotation around the cylinder axis, arising from
rapid lateral diffusion of the lipid molecules. This also gives an
axial powder pattern, but in the opposite sense to that of the
lamellar pattern: the peak is now to low-field rather than high-field
of the shoulder. The additional motion around the cylinder axis
causes the shoulder to remain at the position of the peak in the
lamellar pattern and the peak to be half-way between the lamellar
peak and shoulder. This gives rise to the very characteristic hexagonal powder pattern in Fig. 2.9 in which $\Delta \sigma = +21$ ppm, at 80°C.

The bilayer-hexagonal transition can be used to determine which
structural and energetic features affect the stability of the hexagonal phase. A simple example is given in Fig. 2.10, which compares
the chainlength dependence of the bilayer-hexagonal and lamellar

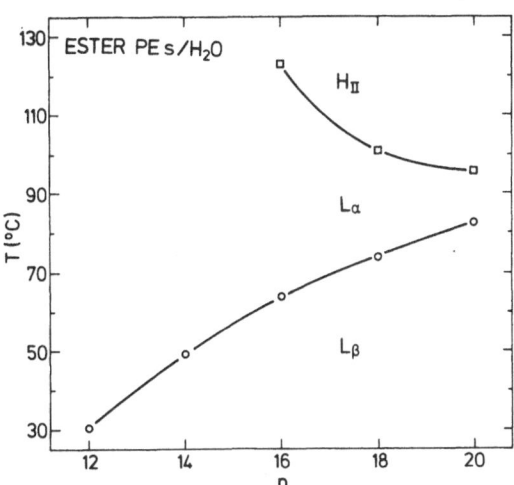

Fig. 2.10. Bilayer-hexagonal (□——□) and lamellar gel-fluid
          (o——o) phase transition temperatures of disaturated
          phosphatidylethanolamines dispersed in excess water, as
          a function of acyl chainlength (Seddon et al., 1982).

gel-fluid transition temperatures for disaturated phosphatidylethanolamines. The gel-fluid transition temperature increases with increasing chainlength as discussed previously, whereas the bilayer-
hexagonal transition temperature <u>decreases</u> steeply with increasing
chainlength. The reason the hexagonal phase is favoured by increasing chainlength can be seen from the "inverted-cone" configuration
of the lipid molecules in this phase (see Fig. 2.8). Any structural
or energetic effect which tends to increase the volume of the hydrophobic region relative to the polar headgroups will favour the hexagonal phase. Conversely increasing the headgroup size or charge dis-
favours the hexagonal phase and increases the bilayer-hexagonal
transition temperature. The water content, salt concentration and
pH titration all affect the hexagonal transition in the opposite
way to which they affect the lamellar-fluid transition, in accordance with the inverted nature of the hexagonal phase. In addition
the hexagonal transition is far more sensitive to all these perturbations than is the gel-fluid transition, as was predicted above
(Seddon et al., 1982).

3. LIPID-PROTEIN INTERACTIONS

Introduction

It is clear from chapter 1 that lipid-protein interactions are likely to play an important role in both the structure and function of biological membranes. The lipid-protein interface is involved not only in sealing the proteins into the membrane, thus maintaining the structural integrity of the membrane permeability barrier, but also in controlling the conformational stability of the protein and interfacing it with the fluid environment of the bulk bilayer lipids. This chapter is devoted to the study of lipid-protein interactions in membranes and reconstituted systems, using the electron spin resonance (ESR) spectra of spin-labelled lipids such as those indicated in Fig. 3.1. Such aspects as the stoichiometry, specificity and mobility of the protein-interacting lipids will be discussed. A more detailed and extensive treatment can be found in a contemporary review (Marsh and Watts, 1982).

Bilayer Lipid in Membranes

The chromaffin granule from the adrenal medulla has a membrane with high lipid/protein ratio, in line with its function as an exocytotic storage vesicle. The ESR spectra of various positional isomers of the stearic acid spin labels (cf. I(3,12) = 14-SASL, Fig. 3.1), at probe concentrations in these membranes, are given in Fig. 3.2. The total spectral anisotropy of these labels, given by the linesplitting: $2(A_\parallel - A_\perp)$, is seen to decrease as the spin label is stepped down the lipid chain from the polar headgroups towards the terminal methyl group. This corresponds to an increasing amplitude of segmental motion [the order parameter is given by $S = (A_\parallel - A_\perp)/(A_{zz} - A_{xx})$] towards the end of the chain, which is similar but not identical to that discussed for ^{2}H NMR in chapter 1. This spin label flexibility gradient is a characteristic feature of lipids organized in a bilayer form and indicates that a large part of the lipid in chromaffin granule membranes is arranged in bilayers. The spin label order parameters in the membranes are also very close to those in bilayers of the extracted lipids, indicating that any perturbation of the lipids by the protein must decrease very quickly with distance away from the protein.

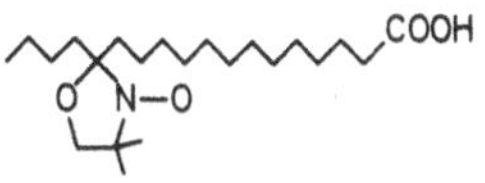

Fig. 3.1. Nitroxide lipid spin label derivatives: 14-SASL, stearic
          acid bearing the spin label group on the 14-C atom. 14-
          PCSL: phosphatidylcholine bearing the spin label on the
          14-C atom of the sn-2 chain. 14-CLSL: cardiolipin (diphos-
          phatidylglycerol) bearing the spin label on the 14-C atom
          of a single chain.

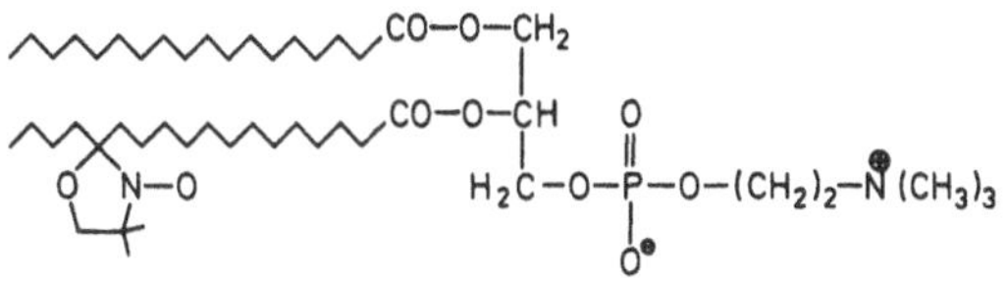

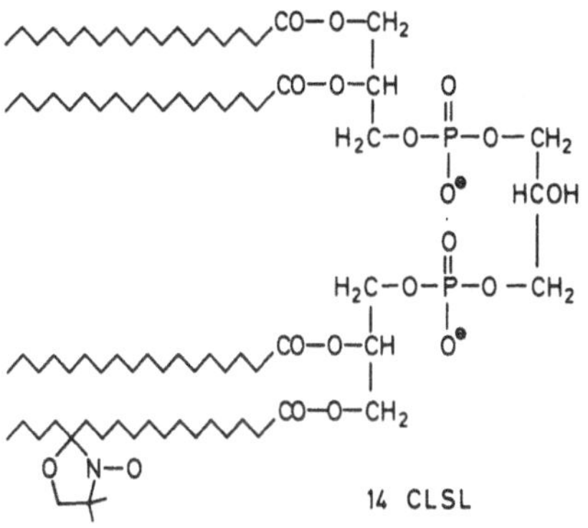

Motionally Restricted Lipid in Membranes

     Membranes with a relatively high protein content frequently
display a second component in the ESR spectra of lipid spin labels,
in addition to the fluid lipid bilayer component discussed in the
previous section. This component is best resolved with labels close
to the end of the chain, since a large degree of averaged spectral
anisotropy is available to detect any immobilization induced by the
protein. The ESR spectra of the 16-SASL stearic acid spin label in
acetylcholine receptor-rich membranes and in bilayers of the extracted lipids is given in Fig. 3.3. A motionally restricted spin
label component is seen in the outer wings of the spectrum from the
membranes which is not present in the spectrum from the lipids alone.

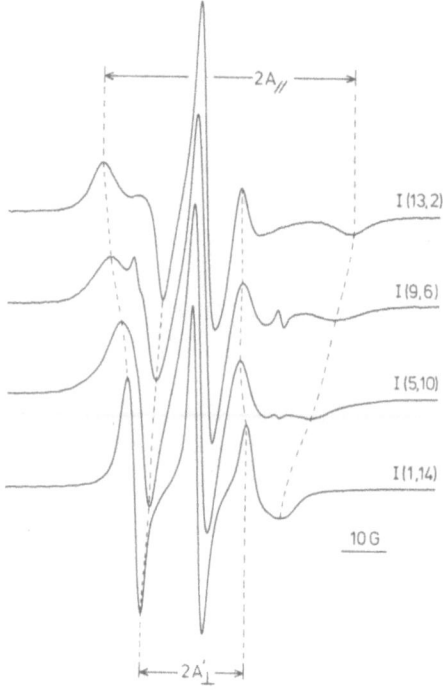

Fig. 3.2. ESR spectra of various positional isomers of the stearic
          acid spin label I(m,n) [= (n + 2) - SASL] (ca. 1 wt%)
          in chromaffin granule membranes at 30°C (Marsh and Watts,
          1982; Fretten et al., 1980; Marsh et al., 1976).

Quantitation of the two spectral components by difference spectroscopy (see Fig. 3.3), indicates that 45% of the total intensity
of the membrane spectrum is in the motionally restricted component
and 55% in the fluid component. The motionally restricted component
represents lipids which are interacting directly with the protein,
either by specific association or simply by direct proximity to
the protein.

Stoichiometry of Lipid/Protein Interactions

     The titration of cytochrome oxidase-dimyristoyl phosphati-
dylcholine complexes of different lipid/protein ratios, labelled
with the 14-PCSL spin label, is given in Fig. 3.4. The ratio of

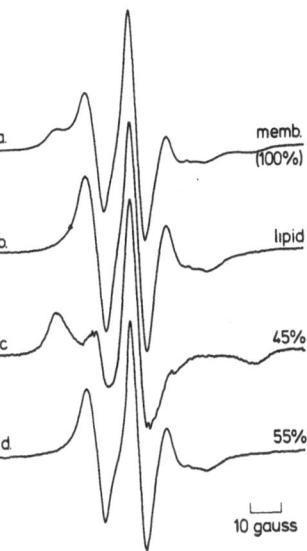

Fig. 3.3. ESR spectra of the stearic acid spin label, 16–SASL, in
         acetylcholine receptor-rich membranes from <u>Torpedo mar-
         morata</u>. a) Membranes. b) Aqueous dispersion of extracted
         membrane lipids. c) Immobilized component difference
         spectrum obtained by subtracting the lipid spectrum (55%
         of the total integrated intensity) from the membrane
         spectrum. d) Fluid component difference spectrum obta-
         ined by subtracting a purely immobilized spectrum (45%
         relative intensity) from the membrane spectrum (Marsh and
         Barrantes, 1978; Marsh et al., 1981).

the intensity of the fluid component to the intensity of the motionally restricted component, n_f^*/n_b^*, determined from the ESR spectra
as in Fig. 3.3, is plotted against the total lipid/protein ratio,
n_t. If the two spin label components are modelled in terms of an
exchange reaction in competition with the unlabelled lipids for n_1
sites on the protein

$$L^* + L_{n_1-m}L_m^*P \underset{}{\overset{K_r}{\rightleftharpoons}} L + L_{n_1-m-1}L_{m+1}^*P \; ,$$

then

$$(n_f^*/n_b^*) = n_t/(n_1 K_r) - 1/K_r \; ,$$

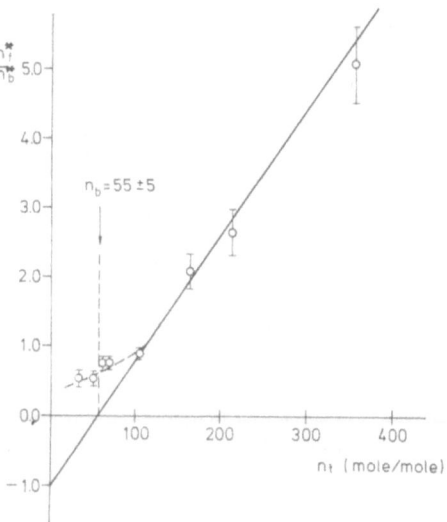

Fig. 3.4. Lipid-protein titration of cytochrome oxidase-dimyristoyl
         phosphatidylcholine complexes with total lipid/protein
         ratios, n_t; obtained from spin-label difference spectra.
         n_b^* is the motionally restricted part of the lipid/protein
         ratio which must be subtracted from the spectra of the
         complexes to yield the fluid part, n_f^* (cf. Fig. 3.3)
         (Knowles et al., 1979).

where K_r is the association constant of the spin label <u>relative</u> to
the host lipid (Marsh and Watts, 1982). Clearly from Fig. 3.4 $K_r = 1$
and there is no selectivity between the labelled and unlabelled
phosphatidylcholine, for sites on cytochrome oxidase. Since at high
lipid/protein ratios there are a constant number of lipids $n_1 =$
55 ± 5 associated with each protein, independent of the total amount
of lipid in the complex, the motionally restricted component is
operationally defined as the first shell of lipid interacting with
the protein. Based on presently available structural information
this is approximately sufficient to cover the intramembranous perimeter of the protein (Marsh and Watts, 1982).

A motionally restricted lipid spin label component has been
observed in several different membrane systems, including rod outer
segment disc membranes, and (Na^+, K^+)-ATPase-containing membranes.
Our results are summarized in Table 3.1. Only for the cytochrome

Table 3.1. Stoichiometries of the Motionally-Restricted Lipid
Spin Label Component in Various Lipid-Protein
Systems

Protein/Membrane	n_1^{exp} (mole/mole)	$n_1^{exp} / \sqrt{M.W.}$ $(\cdot 10^3)$	n_1^{calc} (mole/mole)
Cytochrome oxidase - DMPC	55 ± 5	0.123 ± 0.011	50
Bovine rod outer segment disc/rhodopsin	24 ± 3	0.125 ± 0.016	24
Frog rod outer segment disc/rhodopsin	22 ± 2	0.114 ± 0.010	(24)
Na^+/K^+-ATPase shark rectal gland	58 ± 4	0.112 ± 0.008	(~ 60)
Acetylcholine receptor-rich membrane/T. marmorata	45%	--	52-55

n_1^{exp} is the effective number of motionally restricted lipids per protein
deduced from the spin label experiments. n_1^{calc} is the estimated number of
lipids which can be accomodated around the intramembranous perimeter of the
protein. M.W. is the protein molecular weight (from Marsh et al., 1982).

oxidase-DMPC system has a lipid/protein titration been performed;
for the other membranes the values of n_1^{exp} are calculated from the
single lipid/protein ratio and for the lipid label of lowest affinity. The number of motionally restricted lipids scales approximately with the square root of the protein molecular weight (as
might be expected for proteins which protrude to roughly equal extents from the membrane) and also correlates with the limited amount
of dimensional information available on the proteins.

Lipid Specificity

The ESR spectra of similar lipid spin labels, but with dif-
fering headgroups, are given in Fig. 3.5 for three membrane systems.
These show a degree of specificity of the motionally restricted
component for various lipids, and also a different pattern of selectivity for the different proteins. The spectra indicate a greater
preference for stearic acid and for cardiolipin than for phosphatidylcholine in the first lipid shell of both cytochrome oxidase

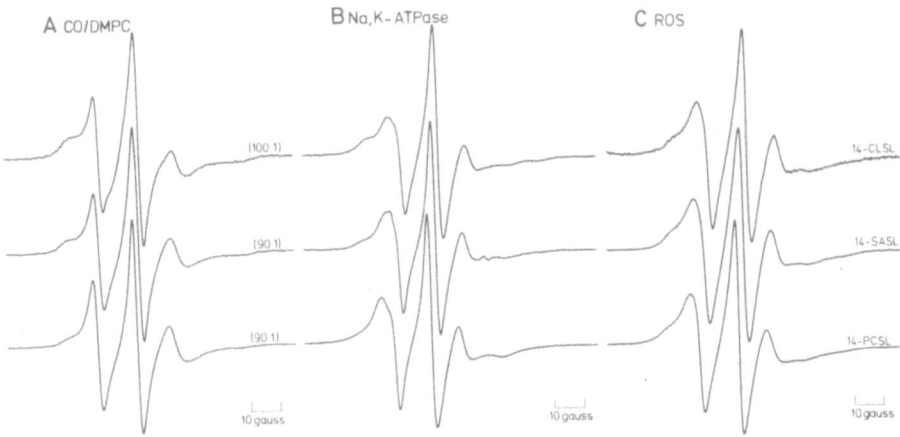

Fig. 3.5. ESR spectra of the cardiolipin spin label 14-CLSL (top
row), stearic acid spin label 14-SASL (middle row) and
phosphatidylcholine spin label 14-PCSL (bottom row), in:
A) cytochrome oxidase-dimyristoyl phosphatidylcholine
complexes of the indicated lipid/protein mole ratios, at
$T = 32^{\circ}C$; B) (Na^+, K^+)-ATPase membranes at $T = 8^{\circ}C$;
C) bovine rod outer segment disc membranes at $T = 5^{\circ}C$ (top
row), $T = 10^{\circ}C$ (middle row), $T = 7^{\circ}C$ (bottom row) (Marsh
et al., 1982).

(Knowles et al., 1981) and the (Na^+, K^+)-ATPase. The most marked
selectivity is for cardiolipin, but this is not an inevitable feature of the cardiolipin structure, since no selectivity is found
with rhodopsin, in agreement with results on this system using a
wide range of lipid labels (Watts et al., 1979). The lipid selectivities can be interpreted in terms of the cooperative binding
equation given in the previous section. In the case of there being
multiple sites with different effective association constants, one
measures the average relative association constant, K_r^{av}, for the
different sites (Brotherus et al., 1981). The values of K_r^{av} for
cardiolipin relative to phosphatidylcholine deduced from Fig. 3.4
are 4.3, 3.0 and 0.8 for the (Na^+, K^+)-ATPase, cytochrome oxidase
and rhodopsin, respectively. These are not very high affinities in

energetic terms, if they simply represent a uniform increase in affinity constant for all sites. However, if the selectivity is for a single specific site only, then the affinity constants for this site are: K_r^{sp} ($\simeq n_1 \cdot K_r^{av}$) ~ 250 and 170 for (Na$^+$, K$^+$)-ATPase and cytochrome oxidase, respectively. This indicates an appreciable e-nergetic interaction, and also that the sites would be nearly al-ways occupied by the specific lipid. Thus the lipid selectivities for the motionally restricted spin label component indicate a de-gree of specificity in the lipid-protein interactions, beyond that of simple occupancy of the first shell lipid sites next to the protein.

Mobility and Exchange of the Motionally Restricted Lipid Component

The ESR difference spectra of the motionally restricted lipid component (cf. Fig. 3.2) lie within the slow-motion regime of spin label spectroscopy, indicating that the effective rotational cor-relation times are $\gtrsim 10$ ns. Experiments on the temperature depen-dence in rod outer segment disc membranes show changes which are diagnostic for motion on this timescale (Watts et al., 1981; 1982). Thus the mobility of the motionally restricted lipid is significan-tly reduced relative to that of the fluid lipids ($\tau_R \sim 1$ ns), but lies considerably closer to this, than to the mobility of the pro-tein as a whole ($\tau_R \sim 20$ μs, cf. chapter 1). This motion of the first shell lipid must represent either segmental motion of the lipid relative to the protein or exchange of the protein with the fluid lipids. For a correlation time $\tau \sim 50$ ns the upper limit for the exchange rate is $\nu_{ex} \sim 2 \cdot 10^7$ s^{-1}, which is comparable with the lipid-lipid exchange rate in fluid lipid bilayers (cf. chapter 1). Exchange at this rate should cause an additional broadening of the fluid component. Linewidths in rod outer segment membranes are greater than in the extracted lipids, which may be due to exchange, or to direct perturbation by the protein of lipids beyond the first shell (see below). From the additional broadening of the rod outer segment spectra, an upper estimate for the exchange rate is ν_{ex} ($\sim g \beta \delta H/h$) $\simeq 2$-$3 \cdot 10^6$ s^{-1} (Marsh et al., 1982). Thus it seems that at least some of the lipids next to the protein have both segmental flexibility and exchange at appreciable rates with the fluid lipids. Most probably the first shell is at exchange equilibrium during a functional cycle of the membrane protein.

Conclusion: Range Dependence of Lipid-Protein Interaction

   In the previous section it was mentioned that the increased
linewidth of the fluid membrane component could represent perturbation of lipid beyond the first shell around the protein. The
range dependence of the interaction can be obtained from measurements of the perturbation of the fluid component as a function of
lipid/protein ratio. Results for the cytochrome oxidase-DMPC system
are given in Fig. 3.6, from which it can be seen that the perturbation drops off very rapidly away from the protein. This, together,

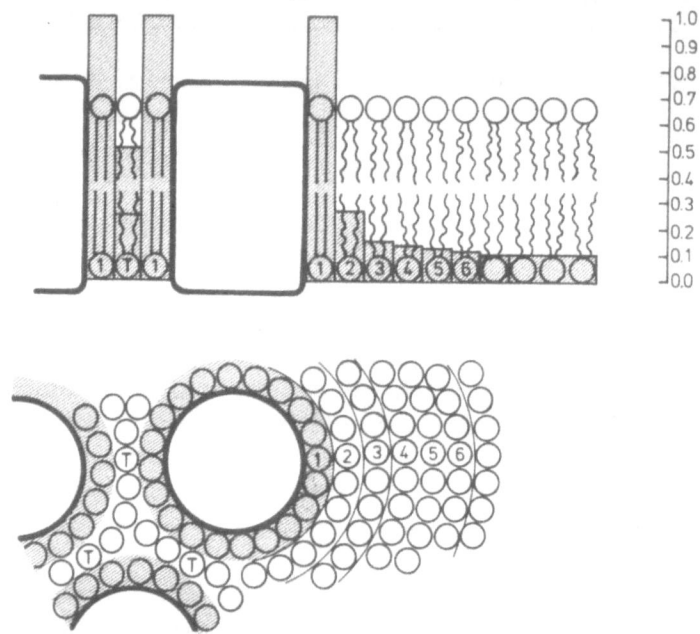

Fig. 3.6. Schematic indication of the degree of chain motional re-
         striction in the various shells of lipid surrounding
         cytochrome oxidase in dimyristoyl phosphatidylcholine com-
         plexes. The vertical scale corresponds to the normalized
         linesplittings in the ESR spectra of the 14-PCSL spin
         label (Knowles et al., 1979). T is a shared second shell
         which exists only in regions of high protein packing
         density (low lipid/protein ratios), where the pertur-
         bations from two adjacent proteins overlap (Marsh et al.,
         1978).

with the dynamic information of the previous section, illustrates how effectively the protein integrates into the fluid bilayer structure. The second lipid shell is much less perturbed than the first shell, and the third and fourth shells are very little perturbed relative to the pure lipid. Since many biological membranes, e.g. the rod outer segment disc membrane, have only two shells of lipid surrounding the protein this short correlation length of interaction is necessary if fluid bilayer, as required by the fluid mosaic model, is to exist. ^{2}H NMR studies (Seelig, 1982) have demonstrated that the first shell lipids are more disordered than, and in fast exchange with, the bulk fluid lipids. It may be that the combination of slower motion but greater disorder of the motionally restricted lipids provides the ideal interface between the protein and the fluid bilayer, allowing rapid exchange with the latter.

ACKNOWLEDGEMENTS

    I would like to thank Drs. R. Henderson and W. Vaz for providing the originals for Figs. 1.5 and 1.7, respectively.

    Much of the work reviewed in chapter 2 was carried out in collaboration with Gregor Cevc, Karl Harlos, John Seddon and Tony Watts. I would like to thank John Seddon for providing the DSC scan for Fig. 2.3.

    Much of the work reviewed in chapter 3 was carried out in collaboration with Pancho Barrantes, Mike Esmann, Pam Fretten, Steven Morris, Peter Knowles, Rob Pates, Igor Volotovski, Tony Watts and Rainer Uhl.

REFERENCES

Chapter 1

Cherry, R.J., 1979, Rotational and lateral diffusion of membrane
    proteins, Biochim. Biophys. Acta, 559:289.
Cone, R.A., 1972, Rotational diffusion of rhodopsin in the visual
    receptor membrane, Nature New Biol., 236:39.
Elder, M., Hitchcock, P., Mason, R., and Shipley, G.G., 1977, A

refinement analysis of the crystallography of the phospholipid, 1,2-dilauroyl-DL-phosphatidyl-ethanolamine, and some remarks on lipid-lipid and lipid-protein interactions, Proc. Roy. Soc. Lond. A, 354:157.

Engelmann, D.M., Henderson, R., McLachlan, A.D., and Wallace, B. A., 1980, Path of the polypeptide in bacteriorhodopsin, Proc. Natl. Acad. Sci. USA, 77:2023.

Hauser, H., Pascher, I., and Sundell, S., 1980, Conformation of phospholipids. Crystal structure of lysophosphatidylcholine analogue, J. Mol. Biol., 137:249.

Henderson, R., and Unwin, P.N.T., 1975, Three-dimensional model of purple membrane obtained by electron microscopy, Nature (London), 257:28.

Hitchcock, P.B., Mason, R., Thomas, K.M., and Shipley, G.G., 1974, Structural chemistry of 1,2-dilauroyl-DL-phosphatidylethanol-amine: Molecular conformation and intermolecular packing of phospholipids, Proc. Natl. Acad. Sci. USA, 71:3036.

Kawato, S., Sigel, E., Carafoli, E., and Cherry, R.J., 1981, Rotation of cytochrome oxidase in phospholipid vesicles, J. Biol. Chem., 256:7518.

Korn, E.D., 1966, Structure of biological membranes, Science, 153:1491.

Marsh, D., 1975, Spectroscopic studies of membrane structure, in: "Essays in Biochemistry", P.N. Campbell and W.N. Aldridge, eds., Vol. 11, Academic Press, New York, p. 139.

Marsh, D., 1981, ESR: Spin labels, in: "Membrane Spectroscopy", E. Grell, ed., Springer-Verlag, Berlin, p. 51.

Marsh, D., Watts, A., and Smith, I.C.P., 1982, to be published.

Pearson, R.H., and Pascher, I., 1979, The molecular structure of lecithin dihydrate, Nature, 281:499.

Seelig, J., and Seelig, A., 1980, Lipid conformation in model and biological membranes, Q. Rev. Biophys., 13:19.

Singer, S.J., and Nicholson, G.L., 1972, The fluid mosaic model of the structure of cell membranes, Science, 175:720.

Tank, D., Wu, E.S., and Webb, W.W., 1981, Enhanced mobility of acetylcholine receptor and membrane probes in muscle membrane blebs, Biophys. J., 33:74a.

Tomita, M., Furthmayr, H., and Marchesi, V.T., 1978, Primary structure of human erythrocyte glycophorin A. Isolation and characterization of peptides and complete amino acid sequence, Biochemistry, 17:4756.

Vaz, W., Derzko, Z.I., and Jacobson, K.A., 1982, Photobleaching

measurements of the lateral diffusion of lipids and proteins
in artificial phospholipid bilayer membranes, in: "Cell Surface
Reviews", G. Poste and G.L. Nicolson, eds., Vol. 8, Elsevier/
North-Holland, Amsterdam.

Wu, E.S., Tank, D., and Webb, W.E., 1981, Lateral diffusion of
concanavalin A receptors and lipid analog in normal and bul-
bous lymphocytes, Biophys. J., 33:74a.

Zwaal, R.A., Roelofsen, B., and Colley, C.M., 1973, Localization
of red cell membrane constituents, Biochim. Biophys. Acta,
300:159.

Reading List

Cherry, R.J., 1979, Rotational and lateral diffusion of membrane
proteins, Biochim. Biophys. Acta, 559:289.

Jain, M.K., and Wagner, R.C., 1980, "Introduction to Biological
Membranes", J. Wiley, New York.

Marsh, D., 1975, Spectroscopic studies of membrane structure, in:
"Essays in Biochemistry", P.N. Campbell and W.N. Aldridge, eds.,
Vol. 11, Academic Press, New York, p. 139.

Marsh, D., and Watts, A., 1982, Spin-labeling and lipid-protein interactions in membranes, in: "Lipid-Protein Interactions",
P.C. Jost and O. H. Griffith, eds., Vol. II, J. Wiley, New York.

Seelig, J., and Seelig, A., 1980, Lipid conformation in model and
biological membranes, Q. Rev. Biophys., 13:19.

Chapter 2

Cevc, G., Watts, A., and Marsh, D., 1980, Non-electrostatic contribution to the titration of the ordered-fluid phase transition in phosphatidylglycerol bilayers, FEBS Lett., 120:267.

Cevc, G., Watts, A., and Marsh, D., 1981, Titration of the phase
transition of phosphatidylserine bilayer membranes. Effects
of pH surface electrostatics, ion binding and headgroup hydration, Biochemistry, 20:4955.

Marsh, D., 1974, Statistical mechanics of the fluidity of phospholipid bilayers and membranes, J. Membrane Biol., 18:145.

Marsh, D., and Seddon, J.M., 1982, Gel-to-hexagonal ($L_\beta - H_{II}$) phase
transitions in phosphatidylethanolamines and fatty acid-phos-
phatidylcholine mixtures, demonstrated by ^{31}P NMR spectroscopy
and X-ray diffraction, Biochim. Biophys Acta, in press.

Nagle, J.F., and Wilkinson, D.A., 1978, Lecithin bilayers. Density
    measurements and molecular interactions, Biophys. J., 23:159.

Seddon, J.M., Cevc, G., and Marsh, D., 1982, Calorimetric studies
    of the gel-fluid and lamellar-inverted hexagonal (L_β - H_{II}) phase
    transitions in dialkyl and diacyl phosphatidylethanolamines,
    to be published.

Seelig, J., 1981, Thermodynamics of phospholipid bilayers, in:
    "Membranes and Intercellular Communication", R. Balian, M.
    Chabre, and P.F. Devaux, eds., North Holland, Amsterdam, p. 36.

Träuble, H., 1976, Membrane electrostatics, in: "Structure of Bio-
    logical Membranes", S. Abrahamsson and I. Pascher, eds., Plenum
    Press, New York, p. 509.

Träuble, H., and Haynes, D.H., 1971, The volume change in lipid bi-
    layer lamellae at the crystalline-liquid crystalline phase
    transition, Chem. Phys. Lipids, 7:324.

Watts, A., Harlos, K., and Marsh, D., 1981, Charge-induced tilt
    in ordered-phase phosphatidylglycerol bilayers. Evidence from
    X-ray diffraction, Biochim. Biophys. Acta, 645:91.

Watts, A., Harlos, K., Maschke, W., and Marsh, D., 1978, Control
    of the structure and fluidity of phosphatidylglycerol bilayers
    by pH titration, Biochim. Biophys. Acta, 510:63.

Wilkinson, D.A., and Nagle, J.F., 1981, Dilatometry and calori-
    metry of saturated phosphatidylethanolamine dispersions, Bio-
    chemistry, 20:187.

Reading List

Chapman, D., ed., 1968, 1973, 1976, "Biological Membranes. Fact and
    Function", Vols. 1-3, Academic Press, London.

Marsh, D., 1981, Electron spin resonsnce: Spin labels, in: "Mem-
    brane Spectroscopy", E. Grell, ed., Springer-Verlag, Berlin,
    p. 51.

Melchior, D.L., and Steim, J.M., 1976, Thermotropic transitions in
    biomembranes, Ann. Rev. Biophys. Bioeng., 5:205.

Träuble, H., 1976, Membrane electrostatics, in: "Structure of
    Biological Membranes", Plenum Press, New York, p. 509.

Chapter 3

Brotherus, J.R., Griffith, O.H., Brotherus, M.O., Jost, P.C.,
    Silvius, J.R., and Hokin, L.E., 1981, Lipid-protein multiple

binding equilibria in membranes, Biochemistry, 20:5261.

Fretten, P., Morris, S.J., Watts, A., and Marsh, D., 1980, Lipid-lipid and lipid-protein interactions in chromaffin granule membranes. A spin label ESR study, Biochim. Biophys. Acta, 598:247.

Knowles, P.F., Watts, A., and Marsh, D., 1979, Spin label studies of lipid immobilization in dimyristoyl phosphatidylcholine-substituted cytochrome oxidase, Biochemistry, 18:4480.

Knowles, P.F., Watts, A., and Marsh, D., 1981, Spin label studies of headgroup specificity in the interaction of phospholipids with yeast cytochrome oxidase, Biochemistry, 20:5888.

Marsh, D., and Barrantes, F.J., 1978, Immobilized lipid in acetyl-choline receptor-rich membranes from Torpedo Marmorata, Proc. Natl. Acad. Sci. USA, 75:4329.

Marsh, D., Radda, G.K., and Ritchie, G.A., 1976, A spin-label study of the chromaffin granule membrane, Eur. J. Biochem., 71:53.

Marsh, D., and Watts, A., 1982, Spin labelling and lipid-protein interactions in membranes, in: "Lipid-Protein Interactions", P.C. Jost and O.H. Griffith, eds., Vol. II, J. Wiley, New York.

Marsh, D., Watts, A., and Barrantes, F.J., 1981, Phospholipid chain immobilization and steroid rotational immobilization in acetyl-choline receptor-rich membranes from Torpedo Marmorata, Biochim. Biophys. Acta, 645:97.

Marsh, D., Watts, A., Maschke, W., and Knowles, P.F., 1978, Protein-immobilized lipid in dimyristoyl phosphatidylcholine-substituted cytochrome oxidase: Evidence for both boundary and trapped-bilayer lipid, Biochem. Biophys. Res. Commun., 81:397.

Marsh, D., Watts, A., Pates, R.D., Uhl, R., Knowles, P.F., and Esmann, M., 1982, ESR spin label studies of lipid-protein interactions in membranes, Biophys. J., 37:265.

Seelig, J., 1982, ^{2}H NMR studies of lipid-protein interactions, in: "Lipid-Protein Interactions", P.C. Jost and O.H. Griffith, eds., Vol. II, J. Wiley, New York.

Watts, A., Davoust, J., Marsh, D., and Devaux, P.F., 1981, Distinct states of lipid mobility in bovine rod outer segment disc membranes. Resolution of spin label results, Biochim. Biophys. Acta, 643:673.

Watts, A., Volotovski, I.D., and Marsh, D., 1979, Rhodopsin-lipid associations in bovine rod outer segment disc membranes. Identification of immobilized lipid by spin labels, Biochemistry, 18:5006.

Watts, A., Volotovski, I.D., Pates, R., and Marsh, D., 1982,
    Spin label studies of rhodopsin-lipid interactions, <u>Biophys.</u>
    <u>J.</u>, 37:94.

<u>Reading List</u>

Devaux, P.F., Davoust, J., and Rousselet, A., 1981, Electron spin
    resonance studies of lipid-protein interactions in membranes,
    <u>Biochemical Symposia</u>, 46:207.
Jost, P.C.,and Griffith, O.H., 1980, The lipid-protein interface
    in biological membranes, <u>Ann. N.Y. Acad. Sci.</u>, 38:391.
Marsh, D.,and Watts, A., 1982, Spin-labelling and lipid-protein
    interactions in membranes, <u>in</u>: "Lipid-Protein Interactions",
    P.C. Jost and O.H. Griffith,eds., Vol. II, J. Wiley, New York.
Marsh, D., Watts, A., Pates, R.D., Uhl, R., Knowles, P.F., and
    Esmann, M., 1982, ESR spin label studies of lipid-protein
    interactions in membranes, <u>Biophys. J.</u>, 37:265.

X-RAY AND NEUTRON SMALL-ANGLE

SCATTERING ON PLASMA LIPOPROTEINS

Peter Laggner

European Molecular Biology Laboratory
Hamburg Outstation, Hamburg, Fed. Rep. Germany
and
Institut für Röntgenfeinstrukturforschung der
Osterreichischen Akademie der Wissenschaften
Graz, Austria

INTRODUCTION

Plasma lipoproteins are macromolecular complexes of such
chemically diverse substances as proteins, various kinds of lipids
and, to minor proportions, carbohydrates. They serve primarily as
transport vehicles for lipids in the aqueous environment of blood
and lymph and represent an enormously broad spectrum of compostions
and particle sizes. Traditionally lipoproteins are classified upon
the operationally useful characteristic of buoyant density initiated by the ultracentrifugal separation studies by DeLalla and
Gofman (1954) (Table 1). A biochemically more specific nomenclature
was pioneered by Alaupovic (1972), whereby lipoprotein families
are defined by their specific apolipoproteins. This concept of
lipoprotein families has proven particularly useful in relating
certain biochemical functions of individual lipoprotein classes to
their apoproteins.

From a physico-chemical viewpoint, the problem of lipoprotein
structure can be broken down into four categories:
    Morphology and internal organization
    Structure-forming role of the components
    Selective interactions between components
    Structural dynamics.

Table 1. Classification and Properties of
          Human Plasma Lipoproteins

	Chylomicrons	VLDL[1]	LDL[1]	HDL[1]
Density g/ml	0.95	0.95–1.006	1.006–1.063	1.063–1.21
Mol. wt.	$>10^8$	$(0.5-5)\cdot10^7$	$(2-5)\cdot10^6$	$(1.8-5)\cdot10^5$
Protein weight-%	0.5–2.5	10–15	18–30	35–55
Major lipids[2]	TG	TG	CE	PL
Minor lipids[2]	PL,C	PL,C	PL,C,CE	CE,C,TG
Major Apoproteins	B,C	B,C	B	A

[1] VLDL, very low density lipoproteins; LDL, low density lipoproteins; HDL, high density lipoproteins.
[2] TG, triglycerides; CE, cholesteryl esters; C, cholesterol;
PL, phospholipids.

     The first three are classical categories of supramolecular
structure research in molecular biology, and as a result one might
expect a rationalization of lipoprotein structure by some kind of
static picture. However, from all we know about natural and artificial lipid – protein complexes, lipoproteins cannot be considered
as static particles: Mobility of components within the particles
is a well-established fact, and exchange of components among lipoproteins and with cell membranes is part of their biological function. In search of structure-function relationships, therefore,
considerations of structural dynamics are certainly of paramount
importance.

     This has to be borne in mind throughout the present chapter
which will be concerned primarily with the static picture of lipoproteins, that is, their time-average structural principles. Much
of this information has come from X-ray and neutron small-angle
scattering on lipoprotein solutions and therefore, these techniques
and their pertinent results will be discussed in some detail. For
the interested reader, a brief recollection of the principles of
the small-angle scattering method will be presented first, which

may be useful but is not essential to an appreciation of the results on lipoproten structure to be discussed in the subsequent
sections.

SMALL-ANGLE SCATTERING OF X-RAYS OR
NEUTRONS ON DILUTE SOLUTIONS

The classical diffraction theory treats X-rays and neutrons
similarly as plane waves scattered on atomic obstacles; their wavelengths (commonly in the range of a few Ångström units) are of the
same order of magnitude as the interatomic distances. When the
scattering centers (electrons for X-rays, atomic nuclei for neutrons) interact with the beam, they become the origins for isotro-
pically propagating secondary waves. The interference pattern
between these secondary waves determines the observable scattering
effect and is related in a well-defined way (through Fourier transformation) to the spatial arrangement of the scattering centers
within the molecules.

In diffraction on single crystals where the molecules are
fixed in regular spatial arrays, the diffraction pattern contains
information on distances and directions between scattering centers,
and the analysis of the scattering pattern eventually leads to a
set of three-dimensional coordinates for the atoms within a molecule.

What information can be gained from a similar experiment on a
dilute system of randomly oriented molecules? Here again, the diffraction pattern from a single molecule in a fixed orientation in
space is determined by the atomic coordinates; however, since the
distances between the molecules are large as compared to their
dimensions, interferences between different molecules become negligible and the total scattering from a dilute system is the sum of
the scattering effects from single particles ("particle scattering"). Due to the random orientation, this sum will be an average over all orientations. Obviously, this will result in total
loss of directional information, and the isotropic (about the direction of the incident beam) scattering pattern will contain information only about distances between the scattering centres
within a particle. Owing to the inverse relationship between object
size and diffraction angle (for a fixed wavelenth), the term "small-

angle scattering" is used to describe the scattering of X-rays and neutrons on macromolecules.

The exact mathematical formalism relating the experimental observable, i.e. the angular distribution of scattering intensity, to the "real-space" information, i.e. the distance distribution function p(r), is given by Fourier transformation according to

$$I(h) = 4\pi \cdot \int_0^\infty p(r) \cdot \frac{\sin (h \cdot r)}{h \cdot r} \cdot dr \; , \tag{1}$$

and conversely

$$p(r) = (1/2\pi^2) \cdot \int_0^\infty I(h) \cdot h^2 \cdot \frac{\sin (h \cdot r)}{h \cdot r} \cdot dh \; . \tag{2}$$

The argument h, by which the angular dependence is expressed, is related to the scattering angle, 2θ, and the wavelength, λ , by

$$h = \frac{4\pi \cdot \sin \theta}{\lambda} \; . \tag{3}$$

The distance distribution function p(r) is equal to the number of any distances r between two scattering centres within one particle, averaged over all orientations in space. If volume elements (with corresponding scattering densities) are considered rather than individual scattering centres, p(r) is the distance distribution of these volume elements, weighted by the product of the scattering densities; since the latter are conveniently expressed relatively to the solvent scattering density, p(r) can also acquire negative values if scattering densities lower than that of the solvent occur within the particle.

A special case of particular relevance to lipoproteins is represented by spherical particles, where orientational averaging is of no effect: here, the amplitudes F(h), equal in magnitude to the square root of the intensities, are the Fourier transforms of the radial scattering density distribution $\rho(r)$, according to the relation:

$$\rho(r) = (1/2\pi^2) \cdot \int_0^\infty F(h) \cdot h^2 \cdot \frac{\sin (h \cdot r)}{h \cdot r} \cdot dh \; . \tag{4}$$

The scattering function also contains information on several important integral structure parameters.

(i) Molecular weight: The intensity at angle zero, $I(0)$, is equal to the integral over $p(r)$ (see eqn. (1)) and hence related to the total number of scattering centres per particle. With the known chemical net composition the molecular weight can be calculated (Kratky, 1963).

(ii) Radius of gyration, R: The root-mean-square distance of all scattering centres from the common centre of gravity is equal to the normalized second moment of $p(r)$, according to the expression

$$R^2 = \frac{\int_0^\infty p(r) \cdot r^2 \cdot dr}{2 \cdot \int_0^\infty p(r) \cdot dr} \quad . \tag{5}$$

R can be computed either from the $p(r)$ function or, according to the Guinier approximation (Guinier and Fournet, 1955), from the slope of the scattering curve in its gaussian part close to the origin by

$$R^2 = -3 \cdot \frac{d(\ln I)}{dh^2} \quad . \tag{6}$$

(iii) Particle volume, V: This is obtained from the intensity at angle zero, $I(0)$, and an integral over the scattering curve by the relation

$$V = 2\pi^2 \cdot \frac{I(0)}{\int_0^\infty I(h) \cdot h^2 \, dh} \quad . \tag{7}$$

This simple form only holds for homogeneously filled particles, it can, however, also be extended in its use to particles with internal density fluctuations, provided the mean square fluctuation is known.

Contrast Variation

Further information on the internal structure of particles can be obtained by scattering measurements in solutions of different solvent scattering density ("Contrast variation", Stuhrmann, 1974;

1979). For its discussion it is practical to consider two extreme
situations:

(i) The net scattering density of the particle is equal to that of
the solvent (zero contrast): small-angle scattering only arises
from, and contains information about, internal density fluctuations.
This scattering curve is termed $I_S(h)$.
(ii) The difference between solvent and solute density approaches
infinity (infinite contrast): the particles essentially represent
"holes" within the dense solvent, internal density fluctuations
become negligible and the scattering function, $I_F(h)$ only reflects
the contours of the particle.

   The scattering curve, $I(h, \bar{\rho})$, at any given net contrast $\bar{\rho}$
can be expressed in terms of a combination between $I_F(h)$ and $I_S(h)$,
and a mixed term $I_{SF}(h)$:

$$I(h, \bar{\rho}) = I_F(h) \cdot \bar{\rho}^2 + I_{SF}(h) \cdot \bar{\rho} + I_S(h). \tag{8}$$

This equation allows to determine $I_S(h)$ and $I_F(h)$ from a number of
experiments carried out at different contrasts $\bar{\rho}$. An equally useful
relation holds for the contrast dependence of the radius of gyration:

$$R^2 = R_F^2 + \frac{\alpha}{\bar{\rho}} - \frac{\beta}{\bar{\rho}^2}. \tag{9}$$

The terms α and β are related to the second moment of the internal
density fluctuations, and the distance between the centres of gravity of different density regions. If $\beta = 0$, R and α can be determined directly from the remaining linear relationship.

   The most important benefit of contrast variation lies in the
isolation of the contour scattering function $I_F(h)$, which lends itself directly to shape analysis. It has to be critically ensured,
however, that the addition of contrast enhancing agents, as e.g.
salts or sucrose, does not by itself lead to any structural changes.
If D_2O is used in neutron experiments, the effects of H/D exchange
have to be taken into account.

Model Building

.Finally, some remarks on the strategies taken in shape analysis: They all constitute attempts to approximate the experimental results by theoretical curves of model bodies. This trial-and-error approach can be made more transparent by performing it at the level of the $p(r)$ function which, through its real-space nature, is better accessible to our physical imagination than its Fourier transform, the "reciprocal-space" function $I(h)$. It can also be performed in a systematic way by developing the model structure in the form of spherical harmonics of increasing order (Stuhrmann and Miller, 1978). In any case, on the basis of small-angle scattering results, any particle structure can strictly be discussed only in terms of integral parameters, as e.g. R, M, and V, and of scattering-equivalent models which have to be tested by other available chemical or physical information.

One way of testing a model is by defined perturbation of a structure. This can be achieved by physical, chemical or enzymatic modification, where the information about the possible loci of action can be set against the resulting small-angle scattering changes. Examples of such approaches will be discussed in the section dealing with LDL, where thermotropic effects, proteolytic cleavage, and selective deuteration have added important clues to the selection of models.

Effects of Polydispersity

Since the scattering curve from a solution is the sum of the individual particle scattering effects, it is clear that the results can only then be used to infer an unknown structure if the solution contains only one sort of identically sized particles (monodisperse system). It is instructive to consider also the other extreme case when all particles have identical and known shape but vary in size. There, the known shape function is convoluted with an unknown distribution function. In this case the results can be used to extract information about size distribution (Glatter, 1980). In intermediate cases, where neither structure nor degree of polydispersity are known, the method breaks down as a source for information on particle structure.

    Plasma lipoproteins are apparently a continuous, pauci-modal
spectrum of particles. By careful separation methods, subfractions
can be achieved which show symmetrical boundaries in the analytical
ultracentrifuge and appear as uniform by electron microscopy. To
be safe, however, one has to consider the limits of resolution of
these methods: even with all possible precautions taken, a size variation of about ±5% would appear as a homogeneous sample by these
techniques· Together with the above mentioned structural dynamics,
this notion has to be respected in the interpretation and discussion
of small-angle scattering data from lipoproteins.

Why Neutrons?

    The specific advantage of neutrons for biological studies has
its basis in the remarkable difference in scattering power, cor-
rectly expressed in coherent scattering lengths b, between hydrogen
and deuterium ($b_H = -0.374 \cdot 10^{-12}$ cm; $b_D = 0.667 \cdot 10^{-12}$ cm). This
opens two very attractive possibilities not offered to a comparable
degree by X-rays. First, selective replacement of ^{1}H by ^{2}H, being
ideally isomorphous, leads to a drastic change in scattering power
and hence allows to isolate the structure parameters of the deuterated domains. Secondly, the range of contrast variation that can
be covered by H_2O/D_2O mixtures is far beyond of what can be achieved
by salts or sucrose in the case of X-rays. This makes it convenient
to take full advantage of contrast variation methods. The main disadvantage of neutrons is the very small number of laboratories
around the world, where suitable high-flux reactors can be used for
work on biological structures.

HIGH DENSITY LIPOPROTEINS (HDL)

    Of all lipoprotein classes, HDL contain the greatest proportion
of polar constituents (protein, phospholipids and cholesterol)
which are able to stabilize lipid/water interfaces. Not surprisingly, therefore, this class comprises the smallest lipoprotein
particles, ranging in molecular weight from about $2 \cdot 10^5$ to $5 \cdot 10^5$.
Out of this density class, traditionally two major subfractions,
HDL_2 (density range 1.063-1.125 g/ml) and HDL_3 (1.125-1.21 g/ml),
are isolated by ultracentrifugation, but a recent systematic survey (Kostner, 1981) has shown that a considerably higher number of

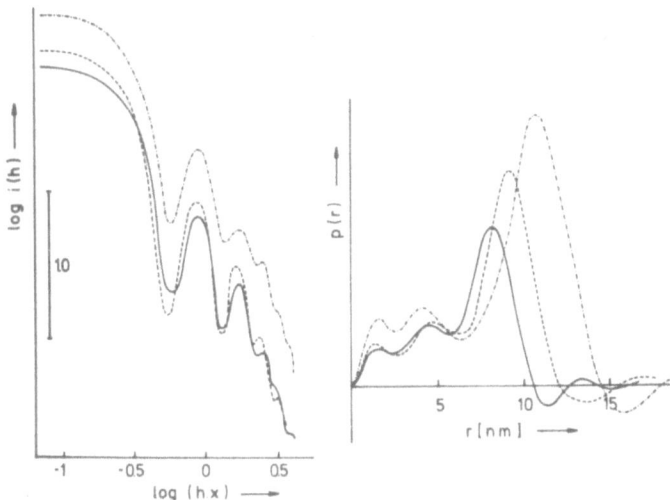

Fig. 1. Scattering curves (left) and distance distribution func-
        tions (right) of three subfractions of HDL. The angular
        scale is in units of h·x, where x equals 1.0, 0.85, and 0.76,
        respectively, in good agreement with the relations between
        the maximum particle diameters obtained from p(r) (from
        Laggner and Müller, 1978).

subfractions with different compositions can be resolved by careful
fractionation procedures.

   Detailed studies have so far been carried out on three selected
subfractions representing the extremes of this spectrum (Laggner et
al., 1973; Müller et al., 1974; Laggner et al., 1976). The X-ray
small-angle scattering curves and the corresponding electron-pair
distance distribution functions p(r) are shown in Fig. 1. In the logarithmic representation, the scattering curves can be brought to
almost perfect superposition of the side maxima by suitable linear
transformation of the angular argument on the abscissa. This already
indicates that the differences primarily relate to particle sizes
and reveals the existence of a common structural principle.

   Regarding particle shapes, both the well-defined maxima and minima in the scattering intensities and the steep decay of the p(r)
function towards the maximum dimension, show that the particles must

be highly isotropic. A further important piece of evidence comes
from a comparison of the radii of gyration of the contour volume,
R_F, with the maximum particle diameters derived from the p(r) functions. The values for R_F of HDL_3 and HDL_2, respectively are 42 Å and
47 Å; the smallest possible particle diameters consistent with these
values are those of spheres, and are calculated by $r = (5/3)^{1/2} \cdot R$:
$r = 54$ Å and 61 Å, for HDL_3 and HDL_2. On the other hand, the maximum
particle diameters from p(r) are 100-110 Å for HDL_3 and 120-130 Å
for HDL_2. This good agreement is evidence for highly isotropic,
quasi-spherical structure, and consequently the radial electron
density distributions can be evaluated by Fourier transformation of
the amplitudes (Fig. 2).

The results of this analysis show two characteristic features:
all three particle species contain a low-electron-density core of
radii between 37 Å (HDL_3) and 51 Å (LpC), and a high electron density shell of rather constant thickness of 15-20 Å. The electron
densities of the core (on average about 0.31 e/Å^3) correspond well
to the average of the cholesterol and hydrocarbon chain constituents.
The shell electron density (0.38 e/Å^3) is considerably lower than
that of unhydrated protein and phospholipid headgroups.

Considering the above parameters and the known molecular parameters of the constituents the following structure model is pro-

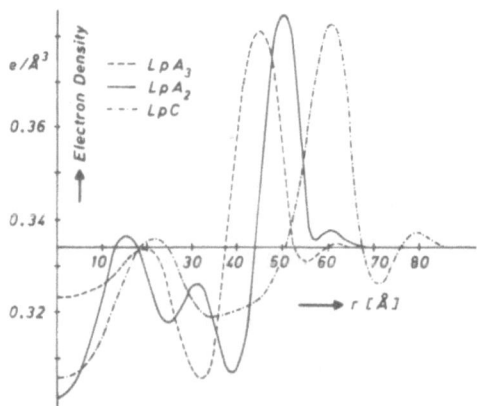

Fig. 2. Radial electron density distributions of three subfractions
        of HDL.

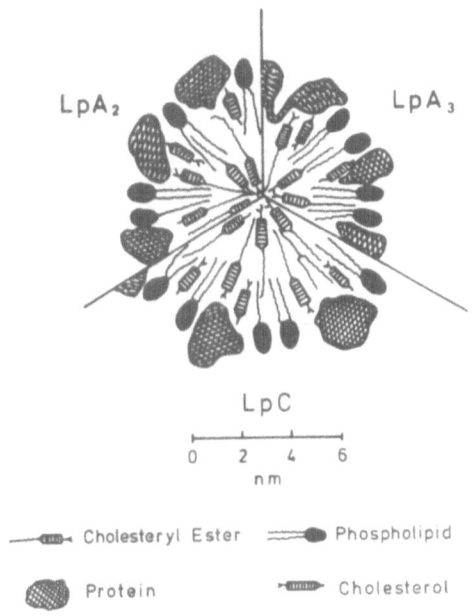

Fig. 3. Schematic cross-sectional view of the molecular arrangement
        in HDL particles (from Laggner and Müller, 1978).

posed (Fig. 3). The core is formed by a spherical micellar arrangement of the cholesteryl esters into which the phospholipid hydrocarbon chains and the unesterified cholesterol interdigitate from
the shell. The smallest core radius of 37 Å corresponds well to
the length of an extended cholesteryl oleate molecule, and in this
case full interdigitation is necessary. This model is supported by
the fact that no smaller HDL particle is found to any abundance in
the lipoprotein spectrum. As the particles grow larger (HDL$_2$ and
LpC), the overlap region for interdigitation decreases in length.
reaching the lowest value in LpC; there, the core radius (51 Å)
corresponds to the full length of a cholesteryl ester plus the length
of an all-trans $(CH_2)_{11}$ – chain, leaving about 10 Å corresponding
to about six methylene groups for overlap. Again, this is a limiting
case, since the lipoprotein spectrum shows a minimum at particle
sizes between 150 and 200 Å. Hence, according to this scheme, it is
the interdigitation between core and surface lipids which limits the
particle size in the HDL class.

What can these results tell about the surface structure? Since
the electron densities of protein and polar phospholipid headgroups
are rather similar (0.43-0.45 e/$\overset{\circ}{A}{}^3$) it is clear, that an individual
assignment of these components in the electron density profile is
impossible. A potential way to overcome this problem would be selective deuteration and neutron scattering (see below). Although such
experiments have not yet been reported, the present data still allow
some inference regarding the surface arrangement: the constant thickness of about 15-20 $\overset{\circ}{A}$ of the high electron density shell corresponds
well to the cross-section diameter of an α-helical protein. Since
all the major apolipoproteins of the HDL class contain considerable
portions of α-helical conformation when bound to phospholipids (see
reviews by Jackson et al., 1976; Morrisett et al., 1977; Laggner,
1981), it is postulated that they form a two-dimensional cage-like
arrangement at the particle surface into which the phospholipid headgroups are interspersed. A comparison of the available surface area
at the mean shell radii to the sum of the surface requirements calculated for the polar components shows, however, that only in LpA_3
(HDL_3 containing only apo-A proteins) the available surface can be
fully covered by the components; in LpA_2 and LpC, approximately 30%
of the surface area would be uncovered. This suggests, that the latter two species would contain considerable voids in the surface into
which solvent can penetrate. To prevent thermodynamically unfavourable hydrophobic contacts, it has to be postulated, that such crevi-
ces are formed by parts of the protein intruding at some points into
deeper regions of the particles. These calculations would predict
that a HDL particle could accomodate more than twice its original
phospholipid content without changing its absolute content in cholesteryl esters and its diameter. According to current concepts of
phospholipid transfer between lipoprotein subfractions this possibility is realized, at least in vitro (Tall and Green, 1981).

LOW DENSITY LIPOPROTEINS (LDL)

Studies on Native LDL

Chemically, LDL is distinguished from all other lipoproteins
mainly by two features: it contains the highest relative proportion
of cholesteryl esters (Table 1) and is characterized by the exclusive presence of one specific apoprotein, apo-B. Whereas most of
the other apoproteins are by now well characterized with respect

to primary and secondary structure, such information is still rather
fragmentary and slowly forthcoming in the case of apo-B. Although
still a disputed issue, a growing body of evidence points to the
existence of two copies of a unique polypeptide chain of 255,000
daltons within the LDL particle of molecular weight $(2-3) \cdot 10^6$
(Reynolds, 1980).

Similar to HDL, numerous physical and chemical studies including X-ray and neutron scattering methods have provided ample evidence for a highly isotropic particle shape, and for the surface location of most, if not all, of the polar lipids and the protein
(see the attached list of reviews). Early models for the internal
organization featuring a protein core and an extended phospholipid
bilayer "sandwiching" the cholesteryl esters (Mateu et al., 1972;
Finer et al., 1975) had to be revised on the grounds of more advanced analyses in favour of a model consisting of a cholesteryl ester
core surrounded by a monolayer shell of phospholipid, cholesterol
and protein (Fig. 4). In essence, the following pieces of evidence
have led to this proposal.

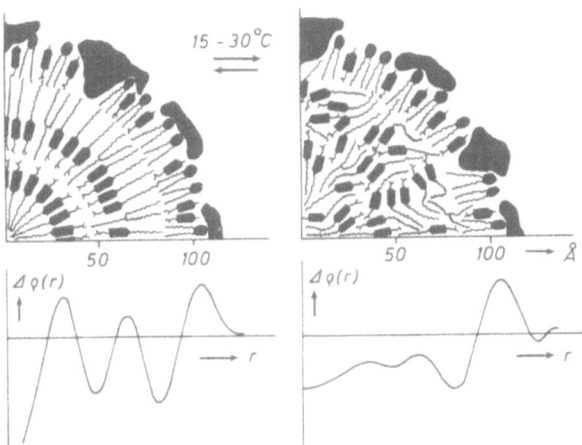

Fig. 4. Above: Schematic cross-sectional view of LDL below and above
        the thermotropic transition of the core.
        Below: Radial electron density distributions (from Laggner
        et al., 1977).

X-ray small angle scattering curves of LDL at temperatures below
$20\,^{\circ}C$ show a pronounced maximum around h = 0.17 $\overset{\circ}{A}{}^{-1}$ indicating a
structural regularity with a spacing of about 37 $\overset{\circ}{A}$ (Fig. 5).
Deckelbaum et al. (1975) first made the important observation that
this spacing disappears reversibly upon heating to about $40\,^{\circ}C$, by
analogy to the behaviour of isolated cholesteryl esters, concluded
that the cholesteryl esters in LDL undergo an order-disorder transition. Evaluation of the radial electron density profile from
quantitative X-ray studies (Laggner et al., 1977; Atkinson et al.,
1977) revealed three peaks around radii of 30 $\overset{\circ}{A}$, 65 $\overset{\circ}{A}$, and 105 $\overset{\circ}{A}$,
of which the first two disappear upon heating. At larger radii
(above 75 $\overset{\circ}{A}$, or distances of 150 $\overset{\circ}{A}$ in the p(r) function) no temperature effects could be observed. Calculation of the temperature-
sensitive spherical core volume of $1.8 \cdot 10^{6}$ $\overset{\circ}{A}{}^{3}$ shows excellent
agreement with the calculated volume requirements of the neutral
lipids (cholesteryl esters plus triglycerides) per particle.

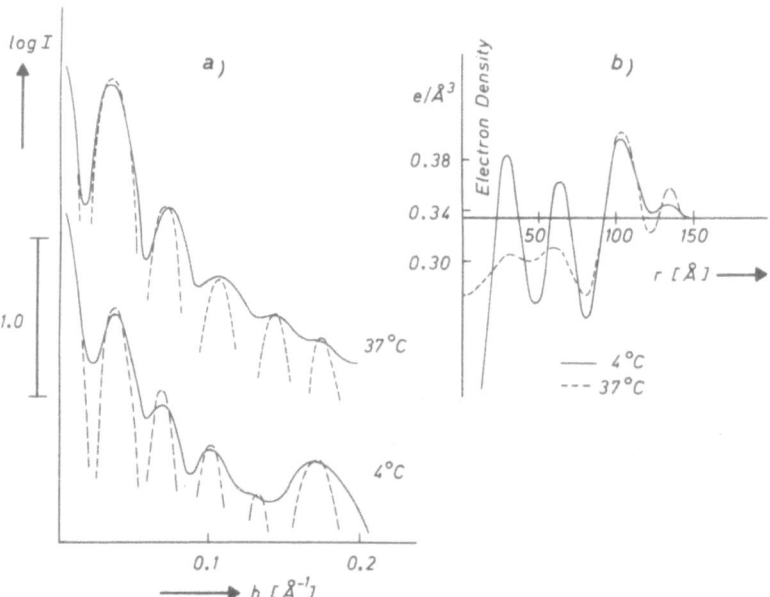

Fig. 5. X-ray small-angle scattering curves of LDL below and above
      the core-transition temperature (left). Radial electron
      density distributions corresponding to dotted scattering
      curves.

This leads to the conclusion that at low temperatures the cholesteryl esters are arranged in two concentric shells. The thicknesses of these shells correspond well to the molecular lengths
of cholesteryl esters and hence an arrangement similar to a smectic
liquid crystalline phase has been implied in the model of Fig. 5.
It is not yet quite clear, exactly which type of liquid-crystalline
order prevails in LDL, however, it appears likely that the spatial
restrictions within the particle cause certain differences to any
of the known phases of cholesteryl esters in model systems. It was
also found by Deckelbaum et al. (1977) that the temperature range
of the transition (in pooled LDL from normolipidemic donors this
lies between 15 and 30°C) shows a significant correlation to the
triglyceride/cholesteryl ester ratio, whereby an increased ratio
corresponds to a decreased transition temperature, i.e. it could
be discussed in terms of crystal impurities. However, also other
factors, like degree of fatty acid unsaturation (Kirchhausen et
al., 1979) and particle size (Jürgens et al., 1981) were found to
be of influence. In passing, it may be noted that as yet no physiological significance has been attributed to this transition phenomenon, although it seems remarkable, that controlled cholesterol
feeding experiments in swine (Pownall et al., 1980) have revealed
that dietary cholesterol leads to an increased transition temperature.

Regarding the surface structure of LDL, the same problems as
outlined above for HDL arising from the similarity in electron
density between protein and polar phospholipid headgroups make it
difficult to draw conclusive evidence from X-ray experiments on
the unperturbed structure. Two experimental approaches have been
taken so far to overcome this problem.

X-ray Scattering on Partially Trypsin-Digested LDL

In this approach (Laggner, Goldstein and Chapman, 1978),
partially trypsin-digested LDL preparations, in which about 20% of
the protein had been hydrolyzed, were compared to the native structure by X-ray small-angle scattering. The results have shown that
the maximum particle dimensions remained practically unchanged,
while the high-electron-density shell region around 105 Å decreased
in height. This was taken as an indication that the protein parts
removable by trypsin digestion are not standing proud of the parti-

cle surface to larger radii than about 110 to 120 Å, but that they
rather were parts of a uniformly spread layer at these radii. These
findings seem difficult to reconcile with models in which spikes
containing protein of about 80 Å length protrude from the particle
surface into the solvent (Tardieu et al., 1976), or in which the
protein is concentrated in four, tetrahedrically arranged globular
subunits (Gulik-Krzywicki et al., 1979; Luzzati et al., 1979).

Neutron Scattering on Selectively Deuterated LDL

    This experiment aimed at the structural arrangement of the
phospholipid headgroups in LDL and involved selective deuteration
and neutron scattering methods (Laggner et al., 1981). Due to the
novelty of this approach in this field and the high potential it
has revealed for further studies, it shall be outlined here in
some detail.

    In trials to find the most effective exchange procedure for
the endogenous phospholipids of LDL, it was found that artificial
complexes between apolipoproteins of HDL (mainly containing A-I
and A-II) and deuterated egg yolk lecithin, with buoyant densities
above 1.063 g/ml, were particularly useful as donnors of deuterated
lipids. The advantage of this system lies in the effective exchange
and in the easy and quantitative separation from LDL after exchange
by ultracentrifugation. By this method, samples containing more
than 80% of their phospolipids as $N(CD_3)_3$-phosphatidylcholine were
routinely obtained. Chemical analysis showed that practically all
of the lecithin molecules (including lysolecithin) but only approximately 50% of the sphingomyelin pool were readily exchangeable. By the criteria of X-ray small-angle scattering, analytical
ultracentrifugation, and negative-stain electron microscopy, the
deuterated samples were found to be closely similar to the native
material (Table 2). Contrast variation experiments using different
H_2O/D_2O buffers have been carried out both on native and deuterated
LDL at the D-11 camera at the high-flux reactor of the Institut
Laue-Langevin in Grenoble.

    The results have confirmed that both native and deuterated
LDL are highly isotropic, quasi-spherical structures and hence the
analysis aiming at the determination of the structure parameters
of the polar phospholipid headgroup shell could be carried out by

Table 2. Physical Parameters of Native and Deuterated LDL

Parameter	LDL	d-LDL
$M/10^{-6}$	2.3 ± 0.2	2.5 ± 0.2
$R/[\overset{\circ}{A}]$	77.3 ± 0.5	78.6 ± 0.5
$V/10^{-6} \overset{\circ}{A}{}^3$	3.8 ± 0.4	4.1 ± 0.4
$f = R_F/R_V$ *	1.04 ± 0.3	1.03 ± 0.3

*Anisotropy factor, by comparison of R_F with the radius of
gyration R_V of a sphere with the volume V.

amplitude subtraction according to the relation

$$F(h) = | \sqrt{I(h)_{d-LDL}} | - | \sqrt{I(h)_{LDL}} |, \qquad (10)$$

where $\Delta F(h)$ is the amplitude function of the deuterated headgroups
and $I(h)_{d-LDL}$ and $I(h)_{LDL}$, respectively, are the scattering curves
of deuterated and native LDL in the same solvent. Due to the higher
precision of the data with higher D_2O content in the solvent (the
incoherent scattering background of H_2O is much higher than of D_2O)
the data from 100% D_2O buffer were used for this purpose. Figure 6
shows the results of this subtraction in the Guinier plot, ·from
which a radius of gyration of 103 ± 5 Å was obtained for the deuterated headgroups. Towards larger angles, the $\Delta F(h)$ function could
be approximated roughly by the theoretical structure factor of an
infinitely thin, spherical shell of radius 90-100 Å. From the
knowledge of the radius of gyration of the polar headgroups alone,
and from that of the entire shell of 110 Å (containing protein plus
headgroups), the radius of gyration of the protein moiety could
also be calculated, according to the relation

$$R_{protein}^2 = \frac{(110)^2 - 0.09 \cdot (103)^2}{0.91}, \qquad (11)$$

where 0.09 and 0.91 are the respective scattering density contributions from deuterated headgroups and protein. This leads to the
value of 111 Å for the protein, indicating that the protein lies,
on average, about 8 Å above the level of the phospholipid headgroups.

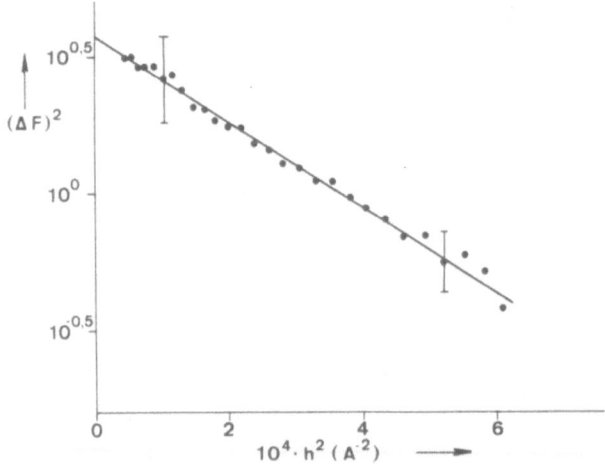

Fig. 6. Difference Guinier plot between deuterated and native LDL
        defining the radius of gyration of the deuterated phospho-
        lipid headgroups (from Laggner at al., 1981).

Model Considerations

    Regarding molecular models, the following considerations can
be made on the basis of the above parameters: The radius of the
neutral lipid core is approximately 75 Å; the phospholipid headgroups are located at about 103 Å. The difference between these
radii corresponds well to the thickness of a phospholipid monolayer.
It should be noted also, that the radius of gyration of the hydrocarbon core in native LDL, comprising apolar lipids plus phospholipid hydrocarbon chains and free cholesterol, found from measurements in 40% D_2O, corresponds to a sphere of a radius of 96 Å, which
is also in keeping with the above values.

    Further important clues arise from an estimation of the surface
area of the individual components. Assuming average molecular surface areas of 60 $Å^2$ and 38 $Å^2$ for phospholipids and cholesterol,
respectively, a total area of 6.9. 10^4 $Å^2$ is calculated for the
polar lipids; comparison with the surface area at 75 Å radius
($7.1 \cdot 10^4$ $Å^2$) shows that this lipid monolayer is able to completely
cover the neutral lipid core. On the other hand, at the radius of
the polar headgroups (103 Å) the total surface area of 1.33. 10^5

$\overset{\text{o}}{\text{A}}^2$ can only be covered to about 50%. If the rest is to be completely covered by protein, the available protein per LDL particle re-
presenting an estimated volume of $6.3 \cdot 10^4 \overset{\text{o}}{\text{A}}^2$, would have to be
spread in a layer of only 10 $\overset{\text{o}}{\text{A}}$ thickness. Although this seems an
implausible value, this calculation shows that in order to prevent
extensive hydrophobic contacts, the protein has to be dispersed in
a rather thin layer over the surface, rather than being concentrated in a few globular subunits, as has been suggested from freeze-
etching electron microscopy (Gulik-Krzywicki et al., 1979; Luzzati
et al., 1979).

Obviously it would be of particular interest to know whether
the surface structure reflects in any way the changes in the apolar
lipid core during the thermotropic transition. By X-ray methods no
significant changes have been observed, and as yet no neutron scattering study has been addressed specifically to this problem. An
ESR spectroscopic study using spin labels probing different sites
of the surface monolayer has shown pronounced spectral changes in
the temperature range of the core transition (Laggner and Kostner,
1978). This would indicate, that the information of the core transition is mediated to the surface and thus, in principle, would be
recognizable to specific receptors or enzymes. In a recent investigation using the same technique on two subfractions of porcine LDL
with different transition temperatures (Nöthig-Laslo et al., 1982),
however, it has been reported that there are differences between
the core and surface transition temperature. Thus, although these
findings agree with regard to the existence of structural changes
in the polar surface region close to the transition temperature of
the core, the underlying molecular details remain to be elucidated.

LIPOPROTEINS LACKING CHOLESTERYL ESTERS

According to the above developed scheme of HDL and LDL structure, the molecular structure of cholesteryl esters is an important
determinant to the size and internal arrangement of lipoproteins.
This is further demonstrated by the structures of lipoprotein
particles which lack cholesteryl esters. A natural system, where
this is the case, is given by the abnormal lipoprotein Lp-X occuring abundantly in the serum of patients suffering from obstruc-
tive jaundice or from a familial deficiency in the enzyme lecithin-
cholesterol-acyltransferase, which controls the conversion of cho-

lesterol into cholesteryl esters. By X-ray small-angle scattering
it was found that the structure of Lp-X resembles closely that of
artificial phospholipid/cholesterol bilayers (Laggner, Glatter, Müller, Kratky, Kostner and Holasek, 1977). A similar situation is encountered in reconstitution experiments when apoproteins (A-I and C-III
have been studied in considerable detail) are recombined in vitro
with phospholipids. There too, the lipid arrangement is characterized by phospholipid bilayer structure (Laggner et al., 1979;
Atkinson et al., 1980).

CONCLUSION

    The methods of X-ray and neutron small-angle scattering have
disclosed important information on the molecular arrangement within
lipoproteins which is not directly available from any other source.
A crucial prerequisite for the successful application of these
methods has been the preparation of well-defined fractions out of
the broad spectrum of particle sizes and compositions naturally
occurring in serum. Therefore, up to now, an analysis by these
methods has been possible only for the subfractions of HDL and LDL
which have proven to be sufficiently close to monodispersity.

    The structural picture emerging from these studies is remar-
kably simple and can be related in a logical way to the molecular
dimensions of the lipid constituents: In all cases, the spherical
particle core is built by the cholesterol esters with their long
molecular axes radially arranged. In HDL, where the polar constituents dominate, the particle sizes are limited by the degree of
interdigitation between the cholesteryl esters from the core and
the phospholipid hydrocarbon chains from the surface. In LDL, the
cholesteryl esters (and the minor amounts of triglycerides) form
a separate phase in the particle core consisting, in the low-tem-
perature state, of two concentrical layers in liquid crystalline
arrangement, which undergoes a thermotropic transition to a more
disordered state slightly below physiological temperature. This
core is covered by a monolayer of phospholipids, cholesterol and
protein.

    Regarding the arrangement of the protein within the polar
surface layer, the molecular details are still largely unclear.
However, the existing evidence suggests that both in LDL and HDL,

the protein is spread out in a disordered network at the surface
into which the polar headgroups of the phospholipids are interspersed. In HDL, the existing data are consistent with the hypothetical model of lipid-protein interaction developed by Segrest
(1977) featuring the protein largely as amphipathic helix. Direct
and compelling evidence, however, is not yet available.

ACKNOWLEDGEMENT

The original work cited from the authors laboratory has been
supported under grants no. 3524 and 4102 by the Österreichischer
Fonds zur Förderung der Wissenschaftlichen Forschung.

REFERENCES

Alaupović, P., 1972, Classification system of plasma lipoproteins,
    in "Protides of the Biological Fluids", H. Peeters, ed.,
    Pergamon Press, Oxford, p. 9.
Atkinson, D., Deckelbaum, R., Small, D.M., and Shipley, G.G.,
    1977, Structure of human plasma low-density lipoproteins:
    molecular organization of the central core, Proc. Nat. Acad.
    Sci. U.S.A., 74:1042.
Atkinson, D., Small, D.M., and Shipley, G.M., 1980, X-ray and
    neutron scattering studies of plasma lipoproteins, Ann. N.Y.
    Acad. Sci., 348:284.
Deckelbaum, R.J., Shipley, G.G., Small, D.M., Lees, R.S., and
    George, P.K., 1975, Thermal transitions in human plasma LDL,
    Science, 190:392.
Deckelbaum, R.J., Shipley, G.G., and Small, D.M., 1977, Structure
    and interations of lipids in human plasma density lipoproteins,
    J. Biol. Chem., 252:744.
DeLalla, O., and Gofman, J.W., 1954, Ultracentrifugal analysis of
    serum lipoproteins, Methods Biochem. Anal., 1:459.
Finer, E.G., Henry, R., Leslie, R.B., and Robertson, R.N., 1975,
    NMR studies of pig low and high-density serum lipoproteins.
    Molecular motions and morphology, Biochim. Biophys. Acta,
    380:320.
Glatter, O., 1980, Determination of particle size distribution
    functions from small-angle scattering data by means of the
    indirect transformation method, J. Appl. Cryst., 13:7.

Guinier, A., and Fournet, G., 1955, "Small-Angle Scattering of X-Rays", Wiley, New York, and Chapman-Hall, London.

Gulik-Krzywicki, T., Yates, M., and Aggerbeck, L.A., 1979, Structure of serum low-density lipoprotein. II. A freeze-etching electron microscope study, J. Mol. Biol., 131:475.

Jürgens, G., Knipping, G.M.J., Zipper, P., Kayushina, R., Degovics, G., and Laggner, P., 1981, Structure of two subfractions of normal porcine (sus domesticus) serum low density lipoproteins. X-ray small-angle scattering studies, Biochemistry, 20:3231.

Kirchhausen, T., Untracht, S.M., Fless, G.M., and Scanu, A.M., 1979. Atherogenic diet and neutral lipid organization in plasma low density lipoproteins, Atherosclerosis, 33:59

Kostner, G.M., 1981, Isolation subfractionation and characterization of human serum high-density lipoproteins, in: "High Density Lipoproteins", C.E. Day, ed., Marcel Dekker, New York, p. 1.

Kratky, O., 1963, X-ray small-angle scattering with substances of biological interest in diluted solutions, Progr. Biophys., 13:105.

Laggner, P., Degovics, G., Müller, K.W., Glatter, O., Kratky, O., Kostner, G., and Holasek, A., 1977, Molecular packing and fluidity of lipids in human serum low density lipoproteins, Hoppe Seyler´s Z. Physiol. Chem., 358:771.

Laggner, P., Glatter, O., Müller, K., Kratky, O., Kostner, G., and Holasek, A., 1977, The lipid bilayer structure of the abnormal human plasma lipoprotein-X. An X-ray small-angle scattering study, Eur. J. Biochem., 77:165.

Laggner, P., Goldstein, S., and Chapman, M.J., 1978, X-ray small-angle scattering study on the structure of partially trypsin-digested human plasma low-density lipoprotein, Biochem. Biophys. Res. Commun., 82:1332.

Laggner, P., Gotto, A.M., and Morisett, J.D., 1979, Structure of the dimyristoylphosphatidylcholine vesicle and the complex formed by its interaction with apolipoprotein C-III. X-ray small-angle scattering studies, Biochemistry, 18:164.

Laggner, P., and Kostner, G., 1978, Thermotropic changes in the surface structure of lipoprotein B from human plasma. A spin label study, Eur. J. Biochem., 84:227.

Laggner, P., Kostner, G., Rakusch, U., and Worcester, D., 1981, Neutron small-angle scattering on selectively deuterated human plasma low-density lipoproteins. The location of polar phospholipid headgroups, J. Biol. Chem., 256:11832.

Laggner, P., and Müller, K., 1978, The structure of serum lipopro-
    teins as analyzed by X-ray small-angle scattering, Quart. Rev.
    Biophys., 11:371.

Laggner, P., Müller, K., Kratky, O., Kostner, G., and Holasek, A.,
    1973, Studies on the structure of lipoprotein A of human high
    density lipoproteins HDL: the spherically averaged electron
    density distribution, FEBS Lett., 33:77.

Laggner, P., Müller, K., Kratky, O., Kostner, G., and Holasek, A.,
    1976, X-ray small-angle scattering of human plasma lipopro-
    teins, J. Coll. Interf. Sci., 55:102.

Luzzati, V., Tardieu, A., and Aggerbeck, L.P., 1979, Structure of
    serum low-density lipoprotein. I. A solution X-ray scattering
    study of a hyperlipidemic monkey low-density lipoprotein,
    J. Mol. Biol., 131:435.

Mateu, L., Tardieu, A., Luzzati, V., Aggerbeck, L., and Scanu, A.
    M., 1972, On the structure of human serum low density lipopro-
    tein, J. Mol. Biol., 70:105.

Müller, K., Laggner, P., Kratky, O., Kostner, G., Holasek, A., and
    Glatter, O., 1974, X-ray small-angle scattering of human
    plasma high density lipoprotein LpA from HDL: application of
    a new evaluation method, FEBS Lett., 40:213.

Nöthig-Laslo, V., and Knipping, G., 1982, Surface structure of the
    two porcine low density lipoprotein subclasses. A spin label-
    ling study, Submitted to Biochemistry.

Pownall, H.J., Jackson, R.L., Roth, R.I., Gotto, A.M., Patsch, J.
    R., and Kummerow, F.A., 1980, Influence of an atherogenic diet
    on the structure of swine low-density lipoproteins, J. Lipid
    Res., 21:1108.

Reynolds, J.A., 1980, Binding studies with apolipoproteins, Ann.
    N.Y. Acad. Sci., 348:174.

Segrest, J.P., 1977, Amphipathic helixes and plasma lipoprotein.
    Thermodynamic and geometric considerations, Chem. Phys. Lipids,
    18:7.

Stuhrmann, H.B., 1974, Neutron small-angle scattering of biological
    macromolecules in solution, J. Appl. Cryst., 7:173.

Stuhrmann, H.B., 1979, Ein neues Verfahren zur Bestimmung der Ober-
    flachenform und der inneren Struktur von geloesten globula-
    eren Proteinen aus Roentgenklein-winkelstreuungsmessungen,
    Z. phys. Chem. (Frankfurt/Main), N.F. 72:177.

Stuhrmann, H.B., and Miller, A., 1978, Small-angle scattering of
    biological structures, J. Appl. Cryst., 11:325.

Tall, A.R., and Green, P.H.R., 1981, Incorporation of phosphatidyl-

choline into spherical and discoidal lipoproteins during in-
cubation of egg yolk phosphatidylcholine vesicles with isolated high density lipoproteins or with plasma, J. Biol. Chem.,
256:2053.

Tardieu, A., Mateu, L., Sardet, C., Luzzati, V., Aggerbeck, L., and
Scanu, A.M., 1976, Structure of human serum lipoproteins in
solution. II. Small angle X-ray scattering study of HDL and
LDL, J. Mol. Biol., 101:129.

Selected Monographs and Reviews on Small-Angle Scattering and
Lipoprotein Structure

Guinier, A., and Fournet, G., 1955, "Small-Angle Scattering of X-
Rays", Wiley, New York, and Chapman-Hall, London.

Jackson, R.L., Morrisett, J.D., and Gotto, A.M.,Jr., 1976, Lipoprotein structure and metabolism, Physiol. Rev., 56:259.

Jacrot, B., 1976, The study of biological structures by neutron
scattering from solution, Rep. Progr. Phys., 39:911.

Kratky, O., 1963, X-ray small-angle scattering with substances of
biological interest in diluted solutions, Progr. Biophys.,
13:105.

Laggner, P., 1976, Physicochemical characterization of low density
lipoproteins, in "Low Density Lipoproteins" C. Day and R.S.
Levy, eds., Plenum Press, New York, p. 49.

Laggner, P., 1981, Physicochemical characterization of high density
lipoproteins, in "High Density Lipoproteins" C. Day, ed.,
Marcel Dekker, New York, p. 43.

Laggner, P., and Müller, K., 1978, The structure of serum lipoproteins as analyzed by X-ray small-angle scattering, Quart.
Rev. Biophys., 11:371.

Morrisett, J.D., Jackson, R.L., and Gotto, A.M.,Jr., 1977, Lipid-
protein interactions in the plasma lipoproteins, Biochim.
Biophys. Acta, 427:93.

Osborne, J.C., and Brewer, H.B., 1977, The plasma lipoproteins,
Advan. Protein Chem., 31:253.

Pilz, I., Glatter, O., and Kratky, O., 1979, Small angle X-ray
scattering, Methods Enzymol., 61:148.

Scanu, A.M. Edelstein, C., and Keim, P., 1975, Serum lipoproteins,
in "The Plasma Proteins", Vol. I., F. Putnam, ed., Academic
Press, New York, p. 317.

Stuhrmann, H.B., and Miller, A., 1978, Small-angle scattering of
    biological structures, J. Appl. Cryst., 11:325.

STRUCTURE AND DYNAMICS OF HUMAN PLASMA LIPOPROTEINS

Louis C. Smith, John B. Massey, James T. Sparrow
Antonio M. Gotto, Jr., and Henry J. Pownall

Department of Medicine
Baylor College of Medicine and
The Methodist Hospital
Houston, Texas 77030, USA

INTRODUCTION

Fatty acids have three distinct and important metabolic roles in animals. The fatty acyl moieties of phospholipids are the components of cellular membranes that provide the hydrophobic barriers between individual intra- and extracellular aqueous compartments. This structural role of fatty acids is the basis of cellular integrity. In addition, the degradation of fatty acyl chains is a highly efficient mechanism for the generation of metabolic energy. Moreover, fatty acids are precursors for prostacyclin, thromboxane, and leucotrienes, all of which are highly potent cellular effectors. Some of the cellular demand for phospholipid fatty acids for membrane synthesis can be met by intracellular synthesis. Most of the fatty acids needed for energy metabolism and all of the essential fatty acids for autocoid synthesis must be supplied exogeneously. Plasma lipoproteins are the vehicles for transport of these water insoluble lipids in the circulation. Lipoproteins also function as circulating reservoirs for the lipid soluble vitamins and environmentally derived contaminants, such as pesticides and polynuclear aromatic hydrocarbons (Smith et al., 1982b).

Lipoprotein metabolism involves three distinct processes (Smith et al., 1978). Physical exchange and transfer processes transform the nascent forms of the lipoproteins, secreted from the intestine

205

and the liver, to mature lipoproteins. Spontaneous transfer of
cholesterol, of fatty acid and of some of the protein components
of lipoproteins occurs simultaneously. In addition, there is protein mediated transfer of phospholipid, cholesteryl ester and triglyceride.

The second phase of lipoprotein metabolism is enzymatic and
involves initially the hydrolysis of lipoprotein triglyceride by
lipoprotein lipase. This catalytic event at the endothelial cell
surface releases the fatty acid and other lipid soluble components
for uptake by individual tissues (Robinson, 1970; Scow et al.,
1976; Nilsson-Ehle et al., 1980; Cryer, 1981; Smith et al., 1982b).
Lipoprotein lipase is involved in lipoprotein metabolism in four different ways. Lipoprotein lipase catalyzed hydrolysis of triglyceride is required to release fatty acids for entry into the cellular
metabolic pathways. Formation of triglyceride-depleted remnants
from chylomicrons and very low density lipoproteins[*], VLDL, is necessary before these lipoproteins are recognized by specific cell
surface receptors for subsequent endocytosis (Cooper and Yu, 1978;
Sherrill and Dietschy, 1978; Catapano et al., 1979a). Interconversions of high density lipoproteins, HDL, subclasses involve
lipoprotein lipase (Patsch et al., 1978). Finally, transfer of
hydrophobic minor components such as vitamins and environmental
contaminants from lipoproteins to specific tissues depends on lipoprotein lipase (Grubbs and Moon, 1973).

Lecithin:cholesterol acyltransferase, LCAT, is the second enzyme of major importance in the enzymic phase of lipoprotein metabolism. Cholesteryl ester formation by this enzyme significantly
changes the dynamics of the plasma cholesterol pool. Unlike the
spontaneous rapid equilibration of cholesterol between lipoproteins,

---

[*]Abbreviations are: VLDL, very low density lipoproteins; IDL, intermediate density lipoproteins; LDL, low density lipoproteins;
HDL, high density lipoproteins; HDL_2 and HDL_3, subclasses of HDL
that have densities of 1.063-1.125 and 1.125-1.120 respectively;
DNS, dimethylaminonaphthyl; LCAT, lecithin:cholesterol acyltransferase; DMPC, dimyristoylphosphatidylcholine; PPOPC, 1-palmitoyl-2-
palmitoleoyl phosphatidylcholine; GdmCl, guanidium chloride; LAP,
lipid associating peptide and PNA, 9(3´-pyrenyl)nonanoic acid.

cholesteryl ester transfer between lipoproteins is protein mediated and relatively slow.

The third and final phase of plasma lipoprotein metabolism is the cellular uptake. The principal routes involve receptor mediated endocytosis and fluid phase endocytosis in a variety of tissues and cell types (Brown et al., 1981). The interrelationship of these three aspects of lipoprotein metabolism are summarized in Fig. 1.

The principal interest in lipoproteins is associated with the importance of certain types of hyperlipoproteinemia as one of the major risk factors in atherosclerosis and premature cardiovascular disease (Fredrickson et al., 1978; Havel et al., 1980). An understanding of the role of lipoproteins in this complex disease pro-

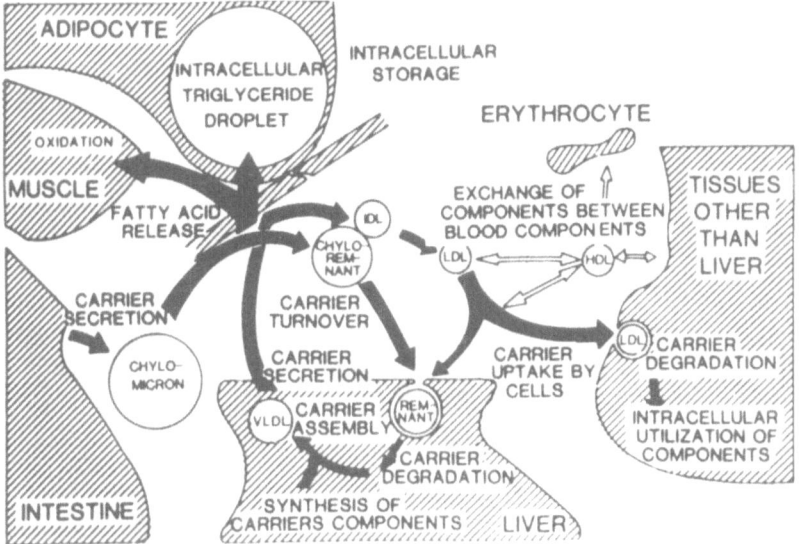

Fig. 1. Lipoprotein metabolism. The flow of the fatty acyl moieties of the triglyceride-rich lipoproteins originates in the intestine and in the liver with the synthesis of carrier components and the assembly of these components into lipoproteins. After secretion, exchange of lipoprotein components produce mature lipoproteins which are then metabolized by lipoprotein lipase at the capillary endothelium. Lipoprotein turnover eventually produces smaller remnant lipoproteins that are removed from the circulation for intracellular processing of the molecular components.

cess requires considerably more detailed information about the dynamics of these lipoproteins than is presently available. From a
biochemical perspective, human plasma lipoproteins are interesting
because they are a reasonably well defined biochemical system that
can be used to study the lipid:protein interactions that are also
important in membrane structure and function. The information derived from these relatively simple lipoprotein systems provides important insights for the more complex cellular membranes (Scanu
and Landsberger, 1980).

PHYSICAL PROPERTIES AND CHEMICAL COMPOSITION OF HUMAN PLASMA
LIPOPROTEINS

     The cammon features of plasma lipoprotein structure are shown
in Fig. 2. The interior of the lipoproteins contains the neutral
lipids, cholesteryl ester and triglyceride. The exterior surface
is a monomolecular film of specific proteins, termed apolipoproteins, and the polar lipids, phosphatidylcholine and cholesterol.
One possible arrangement (Edelstein et al., 1979) of the phosphatidylcholine, cholesterol and apolipoprotein A-1 (apoA-1) in HDL_3,
the most abundant of the plasma lipoproteins, is illustrated schematically in Fig. 3. In this model, there are no lipid domains in
the surface of HDL . The phospholipid molecules are widely dispersed so that intermolecular associations can involve only apopro-
tein:lipid and apoprotein:apoprotein interactions. By contrast,
with increasing size and a greater proportion of hydrophobic core
volume, the structure of the larger lipoproteins more closely re-

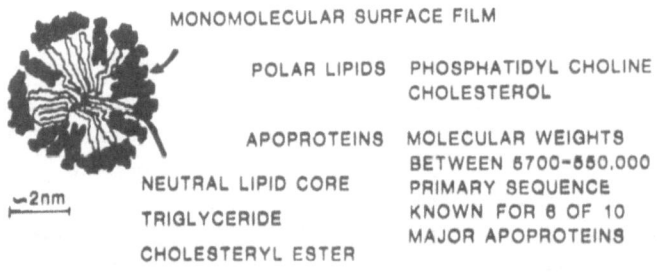

Fig. 2. Common features of plasma lipoprotein structure. The solid
        figures are surface film components, while the open figures
        are core components.

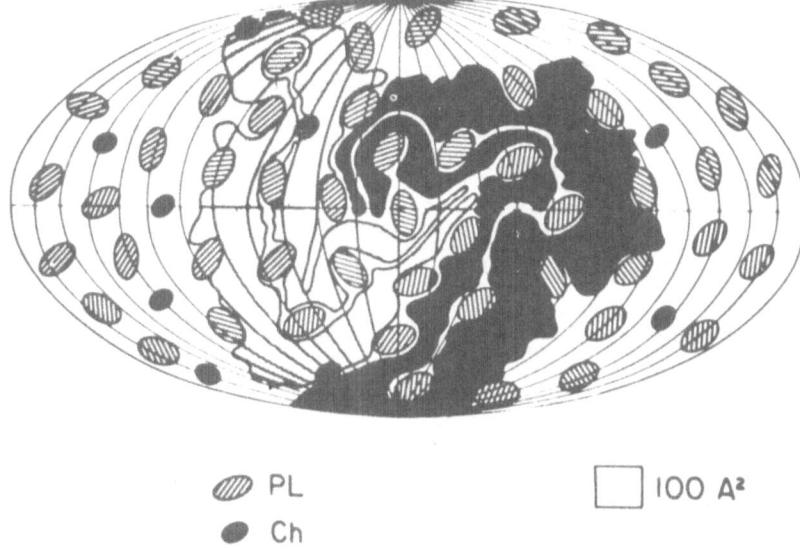

PL

Ch

100 A²

Fig. 3. Topology of surface components of HDL$_3$. The cross-hatched
areas represent the phospholipid polar head group, while
the solid eliptical areas represent cholesterol. This
scheme depicts two adjacent molecules of apoA-1 spread as
a monomolecular film on the lipoprotein surface with phospholipid molecules interspersed between the structural
domains of the apolipoprotein. (Edelstein et al., 1979,
used with permission).

sembles that of microemulsions. The surface film of the larger
lipoproteins probably contains authentic lipid domains. The topography of the surface components of the surface film of the larger
lipoproteins is unknown but is fundamentally important. For example
the recognition of some, but not all VLDL, (Gianturco et al., 1978),
by cell surface receptors can be attributed, as one possibility,
to lipid induced differences in the conformation of apolipoproteins
in the surface film.

    The lipoproteins are frequently classified according to the
density at which they are isolated (Havel et al., 1980). Even
though there is rapid exchange of components between lipoproteins,
the physical properties (Table 1) and chemical compositions of the
major lipoprotein families are essentially constant when isolated

Table 1. Physical Properties of Human Plasma Lipoprotein
        Families

	Electrophoretic Definition	Particle Size nm	Mol. Weight daltons	Density g ml^{-1}
Chylomicrons	remains at origin[*]	75-1200	>400,000,000	<0.93
VLDL	pre-β-lipoproteins	30-80	10-20,000,000	0.93-1.006
IDL	slow pre-β--lipoproteins[**]	25-35	5-10,000,000	1.006-1.019
LDL	β-lipoproteins	18-25	2,300,000	1.019-1.063
HDL$_2$	α-lipoproteins	9-12	360,000	1.063-1.125
HDL$_3$	α-lipoproteins	5-9	175,000	1.125-1.210

[*]On paper;   [**]On geon pevikon

by a variety of methods. The thermodynamic basis of this stability
is not known.

The concentrations of plasma lipoproteins in normal fasting
humans (Havel et al., 1980) are shown in Table 2. On a molar basis,
HDL$_3$ is the most abundant lipoprotein class. In both males and females, this HDL subclass accounts for about 80 and 70 mol percent,
respectively, of lipoproteins in the circulation. The most striking
difference between the lipoprotein concentrations in males and
females is the amount of HDL$_2$. The mol percent of HLD$_2$ in females is about three times that of males. Increased levels of HDL
have been associated with decreased risk of heart disease (Nikkila,
1978; Havel, 1979, Nicoll et al., 1980; Tall and Small, 1980). According to one view, increased amounts of HDL$_2$ may be protective
per se (Witztum and Schonfeld, 1979; Verdery, 1981). Alternatively,
accumulation of HDL$_2$ may be just one manifestation of the enhanced
action of lipoprotein lipase on triglyceride-rich lipoprotein rem-

Table 2. Concentrations of Major Plasma Lipoproteins in Normal
Fasting Humans

	Males		Females	
	μM	mol %	μM	mol %
VLDL	0.1	0.7	0.04	0.2
IDL	0.04	0.3	0.03	0.1
LDL	1.6	10	1.3	6
HDL_2	1.5	9	4.8	23
HDL_3	12.7	80	15.1	71

nants (Patsch et al., 1982). In the latter hypothesis, the lower
risk would be related to the decreased levels of cholesteryl ester
rich lipoprotein remnants (Zilversmit, 1980). A comprehensive
volume on HDL has appeared (Day, 1981).

The surface area is an important determinant of lipoprotein
function. When compared on the basis of the number of lipoprotein
particles ml^{-1}, about 90% of the total lipoprotein particles in the
plasma are HDL (Table 3). By contrast, in males, HDL have only about
half of the total lipoprotein surface area. In females, HDL contributes about 75% of the total lipoprotein surface area because
of the relative abundance of HDL_2. In terms of the core volume of
lipoproteins, LDL contains about half the lipoprotein core volume
in both males and females.

It should be noted that values for lipoprotein concentrations,
lipoprotein surface areas and lipoprotein core volumes in fasting
plasma are misleading with respect to the normal physiological conditions. Triglyceride levels are increased 2 to 10 fold postprandially and remain elevated 4-6 hrs, depending on the physical con-
ditioning of the individual (Patsch et al., 1982). Both the fat
content and the time interval between meals determines which of
the lipoprotein classes dominate the plasma lipoprotein distribution. Because of their transient nature and heterogeneity with
respect to size and chemical composition, the triglyceride-rich
chylomicrons and their remnants remain poorly characterized, although they may be, fundamentally, the most important lipoproteins
in the pathogenesis of atherosclerosis (Zilversmit, 1980; Wolinsky,
1980).

Table 3. Comparison of Lipoprotein Distribution in Normal
         Fasting Humans

	M a l e		
	Number of Particles $\times 10^{-12}$ ml^{-1}	Surface Area	Core Volume
		% of total	
VLDL	7	10	30
IDL	3	3	8
LDL	100	35	45
HDL$_2$	90	9	4
HDL$_3$	800	44	13

	F e m a l e		
	Number of Particles $\times 10^{-12}$ ml^{-1}	Surface Area	Core Volume
		% of total	
VLDL	2	3	14
IDL	1	1	4
LDL	60	24	46
HDL$_2$	230	26	15
HDL$_3$	710	20	20

The chemical composition of plasma lipoproteins is different
and characteristic for each density class (Table 4). With smaller
size and increasing density, the relative proportion of the cholesteryl ester content increases at the expense of triglyceride content
as the lipoprotein core volume becomes smaller (Shen et al., 1977).
Depending on the ratio of triglyceride to cholesteryl ester, there
can be substantial organization to the core region, as shown by thermal transitions and small-angle X-ray scattering (Atkinson et al.,1977)
and by induced circular dichroism (Chen and Kane, 1974; Sklar et
al.,1981). The surface components, except for the relative distribution of individual apoproteins, do not show substantial variation
in the molar proportions with lipoprotein size, although LDL, IDL
and VLDL contain about 40 mol percent cholesterol compared to about
20 mol percent for HDL. The reasons for these differences in surface
film composition are not known. This finding is intriguing since
lipoprotein cholesterol exists as a single, rapidly exchanging pool

Table 4. Chemical Composition of Normal Human Plasma Lipoproteins

	Surface Components			Core Lipids	
	Chole- sterol	Phospho- lipids	Apo- proteins	Trigly- cerides	Chole- steryl esters
	mol %			mol %	
Chylomicrons	35	63	2	95	5
VLDL	43	55	2	76	24
IDL	38	60	2	78	22
LDL	42	58	0.2	19	81
HDL$_2$	22	75	2	18	82
HDL$_3$	23	72	5	16	84

(Goodman and Noble, 1968; Miller et al., 1976), most of which is
associated with HDL.

On a molar basis, the content of total apoproteins is about
the same for lipoproteins in all density classes except for LDL, in
which the molar content of the apoprotein is about 10-fold lower.
The molecular weight of the protein component of LDL, apoB, has
long been disputed. Chemical evidence favors a molecular weight of
about 25,000 (Bradley et al., 1978). Physical methods provide molecular weights approximately 10 times greater (Smith et al., 1972;
Steele and Reynolds, 1979a; 1979b). The higher value for the molecular weight is used for these calculations since none of the biological experiments that have been described require the smaller
molecular weight for apoB.

PHYSIOLOGICAL ROLES OF PLASMA APOLIPOPROTEINS IN LIPOPROTEIN
METABOLISM

Most of plasma apolipoproteins in fasting plasma are associated
with HDL with the exception of apoB, which is associated with LDL
(Table 5). The concentrations of apoproteins range from 46 μM for
apoA-1 to about 2 μM for the combined isoproteins of apo-E (Havel
et al., 1980). It is believed that all apoproteins except apoB exchange between the various lipoprotein classes. For example, transfer of apoC from HDL to lymph chylomicrons has been demonstrated

Table 5. Distribution of Plasma Apolipoproteins in Normal
Fasting Humans

	Plasma Concentration		Lipoprotein Class			
			HDL	LDL	IDL	VLDL
	µM	mol %[*]		mol %[**]		
apoA-I	46	38	100			
apoA-II	23	20	100			
apoB	5	4		82	8	2
apoC-I	18	16	97		1	2
apoC-II	3	3	60		10	30
apoC-III	13	12	60	10	10	20
apoD	5	4	100			
apoE-II,III,IV	2	2	50	10	20	20

[*]Based on total plasma concentration as reported by Havel et
al., 1980.

[**]For each apoprotein.

when the newly secreted particles enter the circulation (Havel et
al., 1973; Imaizumi et al., 1978). The transfer of apoproteins to
triglyceride-rich lipoproteins is reversed by lipoprotein lipase
action in the transformation of HDL_3 to HDL_2 (Patsch et al., 1978;
Redgrave and Small, 1979) that accompanies IDL and chylomicron remnant formation.

Further metabolism of lipoprotein triglyceride of VLDL is accompanied by the loss of the smaller molecular weight apolipoproteins. Compared to VLDL, the proportions of apoE and apoB in IDL
are increased. The most abundant apoprotein is apoC-III (Table 6).
The proportion of apoC-I in IDL is increased about 3-fold compared
to most VLDL. The metabolism of IDL is too poorly understood to appreciate the significance, if any, associated with its apoprotein
composition.

In LDL, about 75 mol percent of the protein is apoB, based on
a molecular weight of 264,000. Small amounts of apoE are present,
while the remainder of the protein associated with LDL mass is
apoC-III.

Table 6. Apoprotein Composition of Human Plasma Lipoproteins

| | Mol % of Total Protein | | | |
	VLDL	IDL	LDL	HDL
apoA-I				45
apoA-II				22
apoB	2	13	75	
apoC-I	8	26		17
apoC-II	20	9		2
apoC-III	60	41	22	7
apoD				5
apoE-II,III,IV	9	13	3	1

The major protein components of human HDL are apoA-I and apoC-I. ApoA-II is present in approximately the same molar concentration as apoC-I. Differences in the relative proportions of the apoprotein components present in the HDL subclasses have not been quantified.

Certain plasma apolipoproteins activate lipolytic enzymes and are ligands for receptor mediated cellular uptake of lipoproteins. The role of plasma apolipoproteins are summarized in Table 7 and in recent reviews (Schaefer et al., 1978; Nester and Fidge, 1981). ApoA-I (Fielding et al., 1972) and apoC-I (Soutar et al., 1975) both active LCAT although the identity of the physiological substrate for LCAT remains unknown. The primary substrates for LCAT may be short-lived chylomicron surface remnants, such as those that appear postprandially in individuals with the familial absence of LCAT (Gjone et al., 1978). The decreased HDL containing abnormally low amounts of cholesteryl ester found in patients with an abnormal cysteine containing apoA-I (Weisgraber et al., 1980) may be the result of inhibition of LCAT (Fransceschini et al., 1981). A similar inhibition of LCAT has been suggested for the isoelectric variant of apoA-I that contains isoleucine (Nestruck et al., 1980).

ApoC-II functions as an activator for lipoprotein lipase (La-Rosa et al., 1970; Havel et al., 1970). The functional importance, if any, of the isoelectric heterogeneity of apoC-II (Havel et al., 1979) is not known. The precise mechanism of activation of lipopro-

Table 7. Metabolic Roles of Plasma Apolipoproteins

Apoprotein	Tissue source	Mol. Weight (daltons)	Metabolic importance
A. Enzyme Activators			
apoA-I	liver intestine	28,000	LCAT activator
apoC-I	liver	5,800	LCAT activator
apoC-II	liver	9,100	lipoprotein lipase activator
B. Ligands for Receptor-Mediated Cellular Uptake			
apoB-100	liver	550,000	non-hepatic receptors
apoE-III	liver	35,000	liver receptors
C. Function Controversial or Unknown			
apoA-II	intestine	17,000	? activator of hepatic lipase
apoA-IV	intestine	46,000	? remnant clearance
apoB-48	intestine	264,000	? remnant clearance
apoC-III	liver	8,750	? modulates apoE clearance
apoD	unknown	22,000	? cholesteryl ester transfer
apoLp(a)	unknown	unknown	unknown
β_2-glycoprotein-I (apoH)	liver	54,000	? lipoprotein lipase activator

tein lipase by apoC-II has not been established. Since lipoprotein lipase hydrolyzes triglyceride in the absence of apoC-II, a true coenzyme function for the apoprotein can be excluded. Earlier reports that apoC-I is a specific activator for lipoprotein lipase (Ganeson, et al., 1975) have not been confirmed.

ApoB and apoE are ligands for receptor-mediated cellular uptake. LDL are the principal reservoir of apoB in plasma. The principal tissue locations for removal of LDL from circulation of normal individuals are thought to be the non-hepatic tissues (Brown et al., 1981). Tissues that secrete steroids and those cells that are rapidly turning over require exogenous cholesteryl ester to maintain either the secretory activity of the cell or adequate rates of cell proliferation. A receptor that recognizes specifically apoE has been demonstrated in the liver (Sherrill et al., 1980) and appears to be important in the rapid removal of chylomicron remnants from the plasma. Whether or not apoE is involved in cellular uptake of HDL has not been established unambiguously. The different effects of HDL subclasses on suppression of HMG-CoA reductase activity in cultured fibroblasts are consistent with a role for apoE as a ligand for receptor mediated cellular uptake of HDL (Daerr et al., 1980). The plasma distribution of apoE and of individual isomorphic forms (Uterman et al., 1975) is uncertain and, like that of other apolipoproteins, depends on the levels of chylomicrons and chylomicron remnants. Under the usual lipoprotein isolation conditions, apoE is found primarily in HDL and VLDL, with significant amounts present in IDL. Substantial amounts of apoE are also found in the $\rho > 1.21$ fraction (Fainaru et al., 1977). A possible genetic and molecular basis for apoE polymorphism has been presented (Zannis and Breslow, 1982).

In vitro, abnormal lipoproteins are taken up by macrophages (Goldstein et al., 1979). This receptor mediated process, termed the scavanger pathway, is an efficient route for the removal of chemically modified lipoproteins in vitro. The nature of the lipoprotein ligand necessary for recognition of abnormal lipoprotein in vivo is not known. Identification of the physiologically important modified lipoproteins will provide the necessary context in which to evaluate the functional importance of the scavanger pathway in vivo (Steinberg, 1980: Brown et al., 1981).

Various roles for other apolipoproteins have been suggested, but remain uncertain because of conflicting reports in the literature. It has been reported that apoA-I is a ligand for receptor-mediated cellular uptake of HDL (Ose et al., 1981; Wandel et al., 1981) although other studies (Sigurdsson et al., 1979) indicate only minor amounts of HDL mass are degraded directly by the liver. ApoC-III is present in substantial amounts, at a concentration of 13 μM in fasting plasma, and exists in several forms that differ in sialic acid content (Albers and Scanu, 1971). The ratio of apoC-III to apoE may modulate hepatic regulation of apoE in triglyceride-rich lipoproteins (Windler et al., 1980). Recent evidence (Maeda et al., 1981a) suggests that the forms of apoC-III containing sialic acid and apoC-III-0 are under separate genetic control. No significant differences in polymorphic forms of apolipoproteinC-III are observed between patients with sialidase deficiency and normal individuals (Maeda et al., 1981b). ApoA-II has been reported to activate hepatic lipase (Jahn et al., 1981b). A minor lipoprotein component, β_2-glycoprotein-I (Polz et al., 1979) or apoH, has also been reported to activate lipoprotein lipase (Nakaya et al., 1980).

In these studies involving hepatic and lipoprotein lipase, respectively, the stimulating effects of the apoproteins were less than
2-fold in assay systems that were not defined; therefore these
findings may not have physiological importance.

The functions of several plasma apolipoproteins are not known.
ApoA-IV, which originates from the intestine, is associated initially with chylomicrons (Weisgraber et al., 1978). In rats, apoA-IV
is found in HDL. By contrast,in humans, it exists in a lipid free
form in fasting plasma. ApoB-48, a species of apoB that originates
from the intestine, is a component of chylomicrons (Krishnaiah et
al., 1980) and is not a precursor to apoB in LDL (Van´t Hooft et
al., 1982). While their roles are not known, it may be reasonably
speculated that both apoA-IV and apoB-48 are involved in removal
of chylomicron remnants by the liver.

Lp(a), identified initially as an antigenic component in a
subclass of LDL (Berg, 1963) is genetically controlled and has a
significant association with coronary heart disease (Berg et al.,
1974). The molecular basis for its antigenic behavior is not known.

Zilversmit et al. (1975) initially identified an activity in
plasma that catalyzed the transfer of cholesteryl ester between
lipoproteins (Pattnaik et al., 1978; Pattnaik and Zilversmit, 1979).
On the basis of a different response pattern to covalent inhibitors
(Barter et al., 1979; Barter and Jones, 1980) and partial purification (Rajaram et al., 1980), the existence of a separate triglyceride transfer protein has been established. A phospholipid transfer protein in plasma has also been reported (Brewster et al., 1978).
ApoD has been reported to function as a cholesteryl ester transfer
protein (Chajek and Fielding, 1978; Chajek et al., 1980: Fielding
and Fielding, 1980). This assignment is disputed by Morton and
Zilversmit (1981). The latter workers report that the cholesteryl
ester transfer activity and apoD do not copurify under several conditions. Moreover, monospecific antisera to apoD have no effect on
cholesteryl ester transfer in their experimental system. It has
been proposed that the cholesteryl ester transfer and phospholipid
transfer are facilitated by the same protein (Ihm et al., 1980).
These questions remain open since none of these proteins has been
adequately characterized.

The importance of these transfer activities may be understood
from consideration of data presented in Fig. 4. The fatty acid com-

positions of the molecular species of phospholipid in HDL, LDL, and VLDL are virtually identical (Morrisett et al., 1977a). This constant fatty acid composition is maintained even on diets that differ significantly in the polyunsaturated to saturated ratio of dietary fatty acids. Identical fatty acid composition of cholesteryl ester in VLDL, IDL and HDL and of triglyceride in HDL and LDL has been documented in the same study. One apparent function of these transfer proteins may be the maintenance of identical lipid composition and, therefore, equivalent physical properties of the plasma lipoproteins.

FUNCTIONAL REGIONS OF PLASMA APOLIPOPROTEINS

        Analysis of the amino acid sequence of plasma apolipoproteins led to the proposal (Segrest et al., 1974) that the apolipoprotein

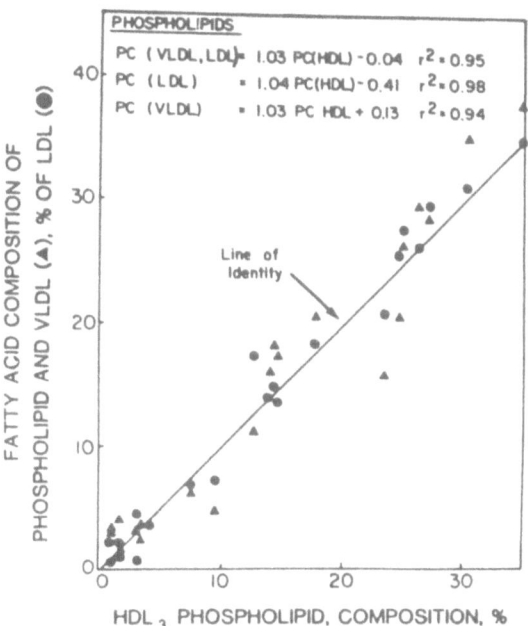

Fig. 4. Identical fatty acid composition of cholesteryl ester and phospholipids in individual lipoprotein classes. The solid circles and triangles represent individual samples as described in the paper by Morrisett et al. (1977a).

in the lipoprotein surface contains regions that are amphipathic
and α-helical (Figs.5 and 6). In this putative structure, the po-
sitively and negatively charged amino acids are oriented so that
they comprise a polar side of the α -helix. This hydrophilic face
extends into the aqueous solution and contributes to the structured
aqueous interface through hydrogen bonding with water molecules.

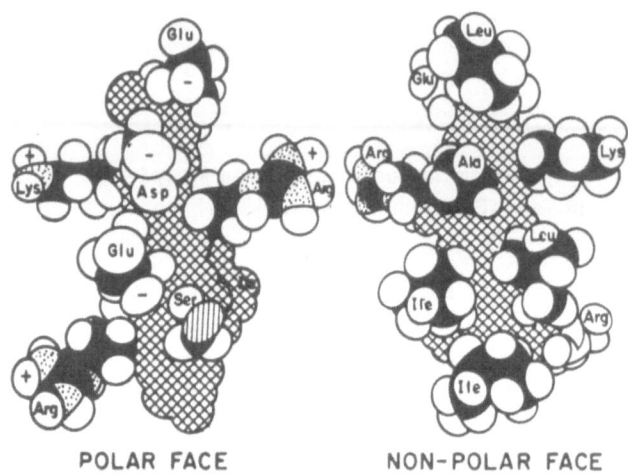

POLAR FACE   NON-POLAR FACE

Fig. 5. Bipolar nature of amphipathic α-helix in apolipoprotein
        C-I (from Jackson et al., 1974).

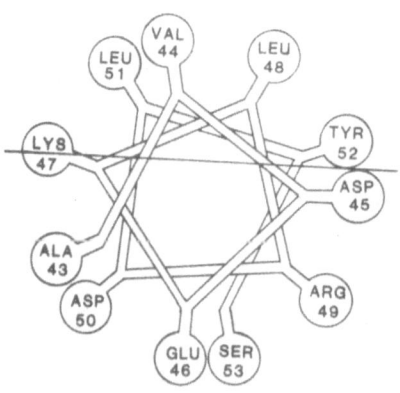

Fig. 6. α-Helical wheel of apoC-II (43-52).

The reverse side of this bipolar peptide is covered by hydrophobic amino acids that interface directly with a non-polar core lipids and with the fatty acyl side chains of phospholipids.

Apolipoproteins undergo changes in secondary protein structure when combined with phosphatidylcholine (Morrisett et al., 1977b). The circular dichroic spectrum and the blue-shifted tryptophan fluorescence spectrum are consistent with an amphipathic structure. Since lipoprotein lipase also can undergo hydrophobic association with phospholipids (Voyta et al., 1980), as indicated by blue-shifted tryptophan fluorescence, it seems probable that the inter-action of the enzyme with the apoprotein activator at the lipid-water interface involves extensive lateral protein:protein inter-actions (Smith and Scow, 1979).

Direct interaction of lipoprotein lipase with the activator has been shown with synthetic fragments of apoC-II that contain a single dansyl (DNS) group on the amino terminal residue of the peptide (Smith et al., 1982a). Resonance energy transfer from the tryptophan residues of lipoprotein lipase to the apoC-II-DNS pep-tides that do not contain tryptophan, demonstrates a specific en-zyme:activator association. This interaction is observed both in the absence of a lipid substrate and in the presence of a nonhydroly-zable substrate, 1-oleyl-2-palmityl-3-phosphorylcholine glyceryl ether containing 2 mol % trioleylglyceryl ether and 5 mol % choles-terol. The same chain length dependence both of energy transfer between the enzyme and DNS peptides at the lipid:water interface and of the peptide activation of lipoprotein lipase catalysis strongly supports this mechanism involving specific protein:protein association.

Studies of the activation of lipoprotein lipase by synthetic fragments of apoC-II have allowed a clear identification of distinct functional regions of the apolipoprotein (Kinnunen et al., 1977). The sequence regions in apoC-II that contain the high affinity phospholipid binding determinant are separate from those required for activation of lipoprotein lipase (Sparrow and Gotto, 1978). A discrete α-helical lipid binding region in apoC-II has been iden-tified in studies (Catapano et al., 1979b) of lipoprotein lipase action on VLDL from an individual with genetically determined ab-sence of apoC-II (Breckenridge et al., 1978). Normal rates of ester cleavage by lipoprotein lipase could be achieved in vitro with either

apoC-II (55-78)[*], apoC-II (50-78) or apoC-II (43-78). By contrast,
only apoC-II (43-78) associated strongly with the apoC-II deficient
VLDL, as shown by separation of unbound ^{125}I-peptides and VLDL on
Sephadex G-100. These results, along with conformational analysis
(Mantulin et al., 1980), indicate that residues 43-50 of apoC-II
are part of a helical lipid binding region. Since the synthetic
peptides differ significantly from each other and from apoC-II in
their abilities to bind to the apoC-II deficient VLDL, high affinity hydrophobic binding of the apoprotein to the lipoprotein
surface is not required for stimulation of lipoprotein lipase dependent hydrolysis of triglyceride.

Enzyme activation occurs through a direct protein:protein interaction between enzyme and the sequence region that extends from
residue 50 through 75 (Smith et al., 1980). An involvement of residues 50-59 in apoC-II interaction with lipoprotein lipase is indicated by experiments with apoC-II (60-68), apoC-II (55-68), apoC-
II (50-68), apoC-II (43-69) and apoC-II (35-68), which gives 0%,
10%, 25%, 100%, and 100% activation of the enzyme, respectively
(Smith et al., 1980). These results, compared to earlier studies
with apoC-II (55-78), which gives 100% activation, show that enhancement of enzymic catalysis can be achieved through association
of a putative β-sheet of the peptide with a corresponding β-sheet
of the enzyme. Alternatively, if this β-sheet region of the peptide
encompasing residues 62-75 is shortened, as is the case with the
68 carboxyl terminal series, inclusion of a lipid binding region
gives complete activation, as observed with apoC-II (43-68) and
apoC-II (35-68). Thus, activation depends on structural features
in apoC-II that concentrate the apoprotein at the interface with
the enzyme and the substrate. Both protein-protein and lipid-protein
interactions apparently contribute independently to this process.
The structure-function relationships of the various sequence regions in apoC-II are summarized in Fig. 7.

---

[*]For nomenclature, a synthetic peptide identified as apoC-II (55-78)
contains only those amino acids between residues 55 and 78 of the
primary sequence of apolipoproteinC-II.

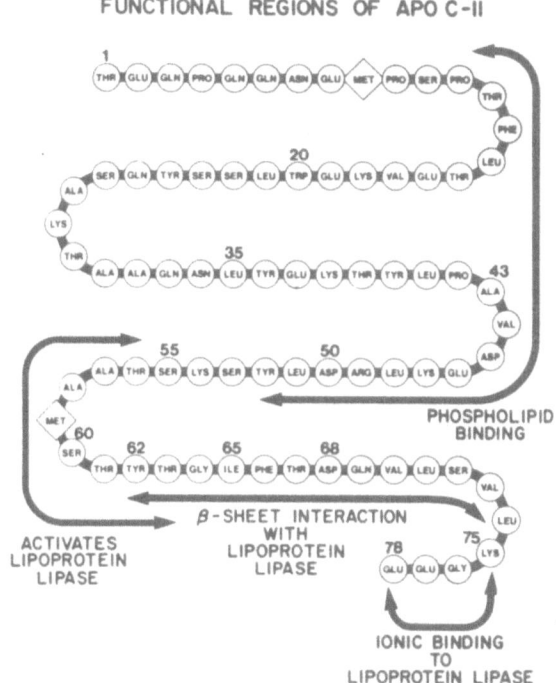

Fig. 7. Structure-function relationships in apoC-II.

DYNAMICS OF PHOSPHOLIPID:APOPROTEIN ASSOCIATION

The dynamic nature of lipoprotein structure and the importance of the physical processes of spontaneous transfer of lipid and apoprotein components in lipoprotein catabolism is shown in Fig. 8. This conception of a dynamic structure originates from elementary thermodynamic considerations. All components of isolated lipoproteins exist in equilibrium between the lipoprotein phase and the surrounding aqueous phase. Historically, lipoproteins have been viewed as static stoichiometric complexes of lipids and apoproteins, with emphasis on the fixed structural features. The lipoprotein structure has been considered to be the result of noncovalent interactions of oriented apoproteins in lipids with fixed arrangements, stoichiometries and distances in these complexes. This conceptualization of lipoproteins and membrane structure does not, however, recognize the dynamic aspects of lipoprotein and membrane structure.

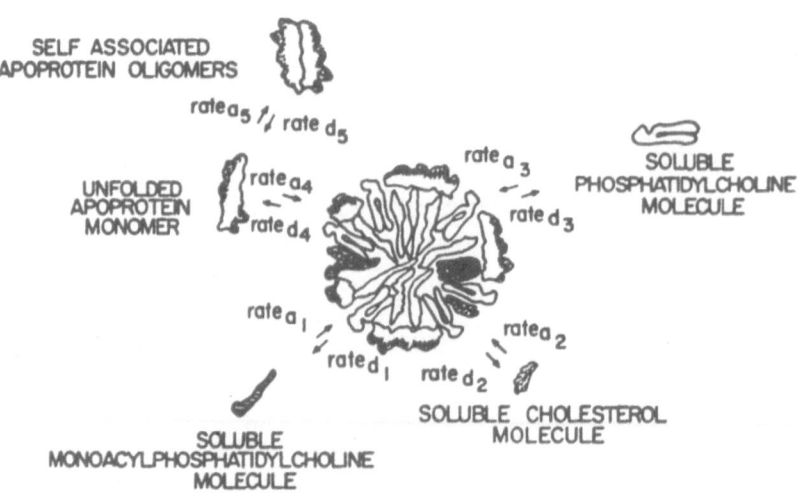

Fig. 8. Dynamics of lipoprotein structure. The individual rate constants for absorption and desorption of the lipoprotein components from the lipoprotein surface are designated d_1, $d_2 \ldots$, and $a_1, a_2 \ldots$, respectively.

From a metabolic viewpoint, the more appropriate conceptualization of the lipoprotein is as a fluid lipid droplet surrounded by second immiscible liquid, the aqueous solvent. According to this view, the lipoprotein surface has the dynamic properties of a two-dimensional fluid that separates the mobile lipid core from an equally dynamic aqueous phase. In this microemulsion model, the important processes are noncovalent interactions. The properties of the system are derived from the motion of individual components as they undergo rotational diffusion and translational movements (Israelachvili et al., 1980; Quinn, 1981). An additional important determinant is the partitioning of individual components between the macromolecular lipoprotein aggregate and the respective monomeric form in the aqueous solution. The rate at which these components transfer between lipoproteins is inversely related to their solubility in the aqueous solution (Charlton and Smith, 1982; Massey et al., 1982a). Cholesterol (McLean and Phillips, 1981), fatty acid (Doody et al., 1980), lysolecithin (Illingsworth and Portman, 1972) and some apoproteins spontaneously transfer in an extremely short time, as compared to the time required for other metabolic events. Some lipoprotein components, such as phospholipids, cholesteryl ester and triglyceride, are too poorly soluble in water to have a significant spontaneous transfer rate. As noted previously,

specific transfer proteins promote the equilibration of these compounds among lipoproteins on a metabolically important   time scale.

   Because of their relatively large hydrophobic surface area,
apolipoproteins, in the absence of lipids, readily self-associate
in aqueous solution (Stone and Reynolds, 1975; Vitello and Scanu,
1976). The rate of desorption of apolipoproteins from lipoprotein
surfaces has not been studied systematically. Extensive studies of
the reverse process, which is the assembly of lipid:apoprotein complexes, have been conducted in considerable detail. The dynamic of
lipid-protein interactions have been studied primarily with in vitro
model systems. Analysis of the association of apolipoproteins with
various phospholipid aggregates have provided important clues about
the nature of the kinetically important steps in the transfer of
apolipoproteins between lipoproteins (Pownall et al., 1977; 1978a;
Massey et al., 1981a; Mantulin et al., 1981).

   The disappearance of turbidity that accompanies formation of
lipid:protein complexes of apoA-I and dimyristoylphosphatidylcholine, DMPC, a synthetic lipid with a transition temperature at
23.9^{o}C, is temperature dependent (Pownall et al., 1978a) (Fig. 9).
The rate of reaction is approximately 10,000-fold higher at the
transition temperature than it is at either lower or higher tempe-

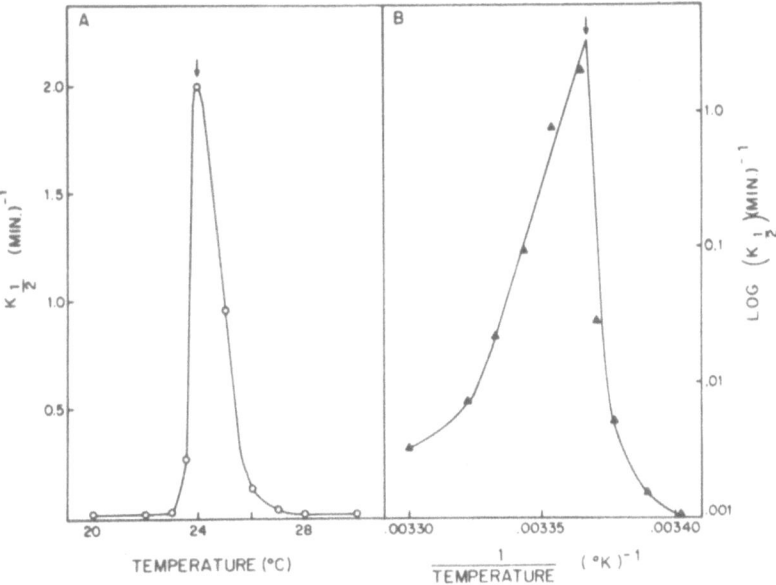

Fig. 9. Temperature dependence of apoA-I binding to DMPC (from
       Pownall et al., 1978a).

ratures. A theory developed to account for bilayer permeability
predicts that the permeability is a function of temperature and is
different for small molecules and for large molecules (Kaneshisa
and Tsong, 1978) (Fig. 10). Thus, permeation of small molecules
through a bilayer is predicted to be relatively rapid and to have
relatively little temperature dependence. For large molecules,
below the transition temperature, the rate of permeation is predicted to be highly temperature dependent. In addition, there should
be a relatively sharp maximum in the temperature dependence and
have significant decreases in rate of interaction both below and
above the transition temperature. Experimental confirmation (Pownall et al., 1981b) of this theory and its applicability to lipid:
protein association is shown in Fig. 11. The apoprotein ranged from
2,280 daltons for a model apoprotein, LAP-20, to 28,000 daltons
for apoA-I. The changes in temperature profile with increasing molecular weight shows excellent agreement with the results predicted
by theory.

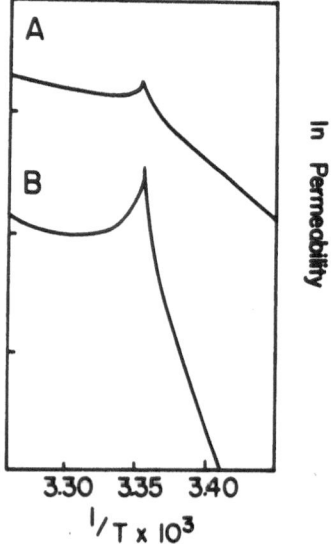

Fig. 10. Theoretical curves for phospholipid bilayer permeability.
         Curve A is that predicted for large molecules. Curve B is
         that predicted for small molecules (from Pownall et al.,
         1981a).

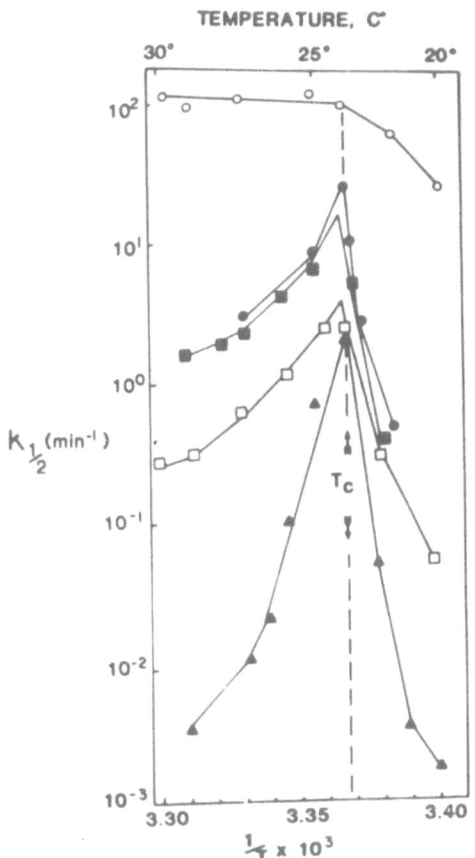

Fig. 11. Molecular weight dependence of apoprotein association with
DMPC. The proteins are (o) LAP-20, 2,280 daltons; (●) RCM-
apoA-II, 8,700 daltons: (■) apoC-III, 9,300 daltons;
(□) apoA-II, 17,400 daltons and (▲) apoA-I, 28,400
daltons (from Pownall et al., 1981a).

    The structural feature in the lipid surface that is required
for apoprotein associaton and for permeability through a lipid bilayer is the formation of defects, which are irregularities in the
packing of individual lipid molecules in the phospholipid surface
(Pownall et al., 1979). A schematic representation of the apoprotein:
phospholipid association is shown in Fig. 12. The irregularities
created by cholesterol prevent structuring of the interface, allow
water penetration between individual phospholipid molecules and
apparently disrupt the cooperative interaction between the phospho-

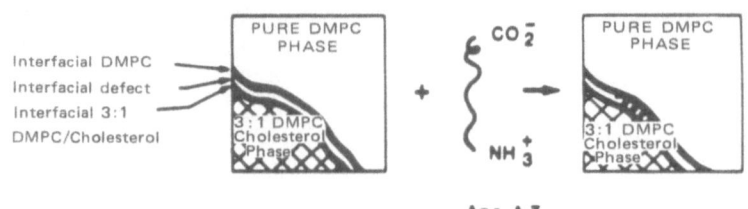

ISOLATED SECTION OF LIPOSOME SURFACE

Apo A-I INSERTED INTO LIPID MATRIX
AT THE BOUNDARY OF COEXISTING GEL
(DMPC:CHOLESTEROL = 3:1) AND LIQUID
CRYSTALLINE (PURE DMPC) PHASES

Fig. 12. Model for apolipoprotein:phospholipid association. In the
left panel, the surface of a phospholipid:cholesterol vesicle is depicted as a biphasic system in which homogeneous
DMPC and 1:3 cholesterol-DMPC phases coexist. Each phase
is bonded by interfacial lipid, and the interfacial lipid
of each phase is separated from that of the other phase
by a hole or channel defect. ApoA-I may insert into this
defect to produce the initial lipid:apoprotein intermediate on the right. The apoprotein is envisioned as a
helical structure in the lipid matrix (from Pownall et al.,
1979a).

lipid phase. The apoprotein then inserts the apolar residues into
lipid matrix at the boundary that exists between the gel and liquid
crystalline phases.

    Differences in the affinity of apoproteins for phospholipid
surfaces has been demonstrated by the displacement of apoA-I from
HDL by apoC (Pownall et al., 1978b). This observation has been confirmed by the addition of human apoA-II to canine HDL, which contain
no apoA-II, displaces apoA-I (Lagocki and Scanu, 1980). More quantitative studies of apoprotein association with lipid indicate that
the free energy of binding contains a large enthalpic component.
The excess enthalpy associated with helical formation (Fig. 13) is
a linear function of the increase in the helicity in the peptide
in the apoprotein that occurs on association with lipid (Massey et
al., 1981b; 1979). The distribution of reduced carboxymethylated

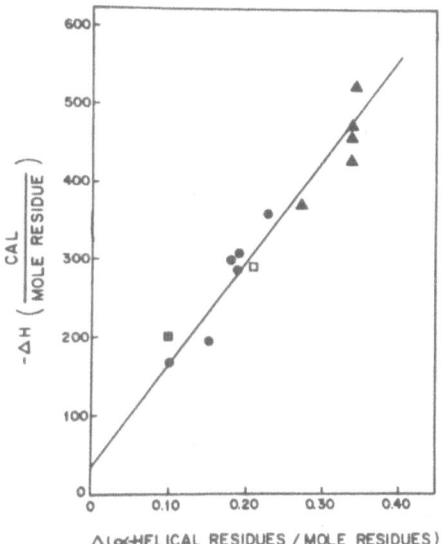

Fig. 13. Energetics of apolipoprotein:phospholipid binding and
helical formation. The apolipoproteins are (●) apoA-II;
(▲) apoC-III; (□,■) apoA-I). The curve is a linear re-
gression fit to the data from Massey et al. (1979). The
slope corresponds to -1.3 kcal/helical residue formed, an
x-intercept of 39 cal and a correlation coefficient of
0.94.

apoA-II (RCM-A-II) between the aqueous phase and DMPC:RCM-A-II complexes (Fig. 14) has shown that the free energy of association is
approximately 8 kcal mole^{-1} (Pownall et al., 1981a). The shift in
tryptophan fluorescence of apolipoproteins as the result of transfer to the lipid matrix from the aqueous solution (Pownall et al.,
1977) also gives a measurement of the free energy of association
(Table 8). With increasing peptide length, in a series of lipid
associating peptides (LAP) that differ in their content of leucine
residues, LAP-16, LAP-20, and LAP 24, there is an incremental increase in the free energy of association.

val-ser-ser-leu-lys-glu-tyr- trp -ser-ser-leu-lys-
glu- ser -phe-ser
LAP-16

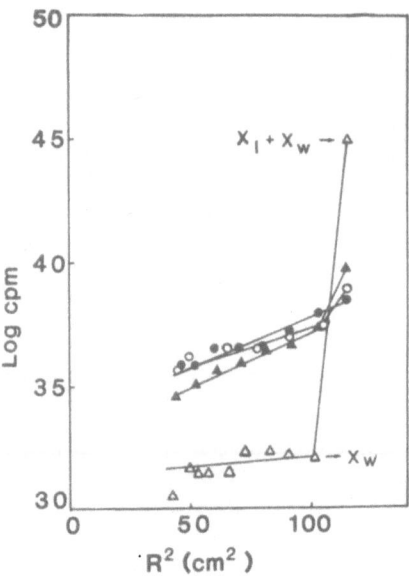

Fig. 14. Partitioning of reduced carboxymethylated apoA-II between
solution and DMPC by ultracentrifugation. The amount of
apolipoprotein is constant. The lipid/protein ratio is:(Δ)
no lipid; ($\blacktriangle$) 105; (o) 350 and ($\bullet$) 1050. x is the mol
fraction of the apoprotein in the aqueous phase and (x_1 +
x_w) is the mol fraction of the apolipoprotein associated
with the phospholipid multilayers. The direction of the
centrifugal force field is from left to right in the figure
(from Pownall et al., 1981a).

val-ser-ser-leu-leu-ser-ser-leu-lys-glu-tyr-trp-
        ser-ser-leu-lys-glu-ser-phe-ser
                    LAP-20

val-ser-ser-leu-leu-ser-ser-leu-leu-ser-ser-leu-
lys-glu-tyr-trp-ser-ser-leu-lys-glu-ser-phe-ser
                    LAP-24

The underscored residues denote the substitutions that distinguish
the peptides.The driving force for this lipid:protein association
resides largely in the transfer of the hydrophobic amino acid side
chains from the aqueous solution into the hydrophobic lipid matrix.

Table 8. Free Energy of Association, ΔG_a, of Lipid-Associating
        Peptides with DMPC

	ΔG_a, kcal mole^{-1}
ApoC-III	-10
+0.3 M GdmCl	-9.3
RCM-apoA-II	-8.0, -7.2[*]
+0.3 M GdmCl	-6.2
LAP-16	-6.5[**]
LAP-20	-8.9
LAP-24	-9.5

[*]First value is for association with single bilayer vesicles
and the second was obtained with multilayers.

[**]Measured in 1.0 M NaCl.

These synthetic lipid associating peptides are also activators
for LCAT (Pownall et al., 1980). Association of the peptide with
the lipid substrate is prerequisite for LCAT activation. While lipid
binding is necessary, lipid binding per se is not sufficient for
LCAT activation. The dependence of cholesteryl ester formation on
the association of the apoprotein cofactor with the substrate complex is shown in Fig. 15. When LAP-16 does not associate with the
phospholipid-cholesterol matrix, as indicated by the light scattering and the wavelength of the fluorescence maximum, there is no
cholesterol esterification by LCAT. With increasing concentration
of salt, the lipid:peptide complex forms. The tryptophan fluorescence maximum shifts to shorter wavelengths and there is a decrease
in the light scattering of the solution. Assembly of the macromolecular substrate then allows LCAT to form cholesteryl ester. The
loss of LCAT activity upon serial dilution of the DMPC-apoA-I complex also shows that phospholipid binding is reversible (Pownall
et al., 1982). Under conditions in which the apolipoprotein exists
in the aqueous solution, there is also a correpsonding loss of
cholesteryl ester formation.

A specific amino acid sequence of either apoA-I or apoC-I is
not necessary for activation of LCAT by an apolipoprotein (Pownall

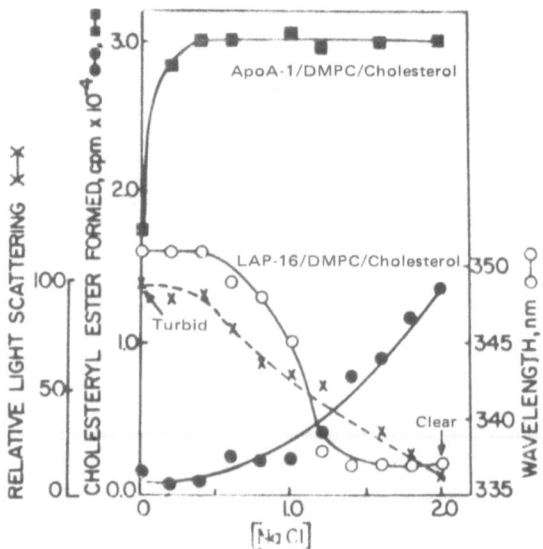

Fig. 15. Dependence of cholesteryl ester formation and phospholipid
binding by LAP-16 (from Pownall et al., 1981b).

et al., 1982). There are, however, exact requirements for enhancement of cholesteryl ester formation. For example, of the sequence
Lys_{16} and Glu_{17} is reversed, the resulting peptide LAP-20 (Glu_{16},
Lys_{17}) still associates with DMPC efficiently. However, its ability
to enhance cholesteryl ester formation by LCAT is reduced about 75%.
Subsequent studies will provide additional insight into the molecular basis of LCAT activation by the native and synthetic activator
apolipoproteins.

LIPID TRANSFER BETWEEN LIPOPROTEINS

     Two possible mechanisms of lipid transfer between lipoproteins
that do not involve proteins are the following. In the first mechanism, lipid molecules escape slowly from the lipoprotein surface
in the rate-limiting step and are taken up rapidly from the aqueous
medium. In the second mechanism, transfer requires the formation
of a collision complex of the donor and acceptor lipoproteins that

exists long enough for redistribution of the hydrophobic molecules between lipoprotein to occur without direct contact of the individual lipids with the aqueous environment.

The mechanism of lipid transfer between lipoproteins was demonstrated initially with fluorescent probes (Charlton et al., 1976). The fluorescence properties of pyrene and its derivatives depend on the microscopic concentration of the fluorophore in the lipoprotein, thus allowing both continuous monitoring of the kinetics of transfer and quantification of changes in mass, as well as obviating physical separation of reactants and products. The excimer/monomer ratio is linear with concentration (Pownall and Smith, 1973) and decreases with probe transfer from a donor to an acceptor lipoprotein (Fig. 16).

Pyrene and related fluorescent hydrocarbons (Smith and Doody, 1981) have served prototypes of lipids in the studies of the kinetics and mechanism of this transfer process. Other work with natural and fluorescent lipids (Illingsworth and Portman, 1972; Bjorrson et al., 1975; Martin and McDonald, 1976; Sengupta et al., 1976;

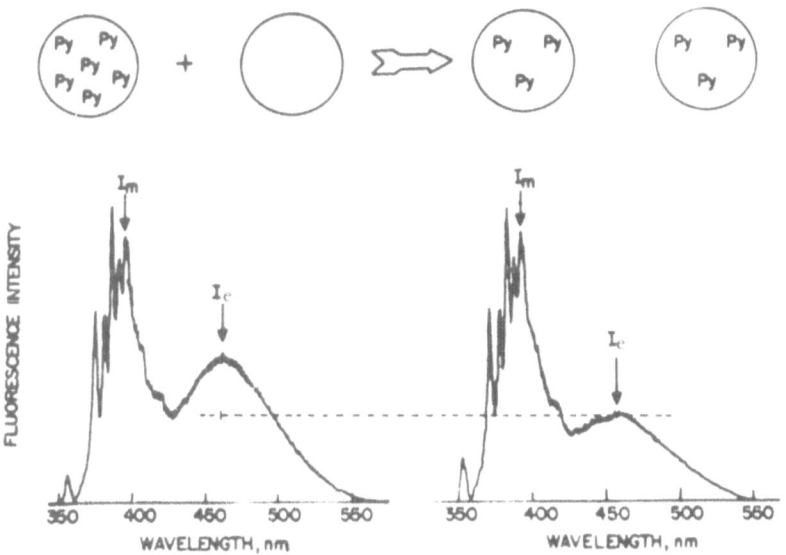

Fig. 16. Spectroscopy of pyrene transfer. The monomer and excimer fluorescence are designated I_m and I_e, respectively.

Kao et al., 1977; Duckwitz-Peterlein et al., 1977; Bell, 1978; Duckwitz-Peterlein and Moraal, 1978; Charlton et al., 1978; Galla et
al., 1979; Backer and Dawidowicz, 1979; Doody et al., 1980: Gottlieb, 1980; Roseman and Thompson, 1980; McLean and Phillips, 1981;
Massey et al., 1982a; 1982b) have demonstrated that the mechanisms
described with the fluorescent hydrocarbons are those that occur
with the naturally occurring lipids.

The actual transfer mechanism is a rate limiting dissociation
of the lipophilic material into the aqueous phase. In all instances, when the donor containing the fluorescent probe is mixed with
an unlabeled acceptor, the halftime of excimer fluorescence decrease is invariant over at least a 100-fold range of unlabeled acceptor concentrations, a 10-fold range of probe concentrations, and
exhibits first order kinetics. When low concentrations of labeled
donor are diluted further, the fluorescent lipids partition from
donor into water. Transfer rates are identical for movement of the
probe from the labeled donor either (a) to an identical acceptor
or (b) to the aqueous phase, i.e., no macromolecular acceptor. These
observations, coupled with invariant exchange rate with concentration and first order kinetics behavior, is compelling evidence
that the rate-limiting step in the transfer is the movement of the
hydrophobic molecule from the carrier into the aqueous solvent.

Studies with the 9(3'-pyrenyl)nanonoic acid, PNA, (Doody et
al., 1980) suggest that the formation of the transition state in
the transfer reaction involves the interfacial water at the surface of the phospholipid matrix. Studies with a series of synthetic
phosphatidylcholines that contain pyrene-labeled fatty acids have
shown that the rate constants for transfer between lipoproteins are
determined by the number of methylene units in the phosphatidylcholine. (Massey et al., 1982a) (Fig. 17). The fluorescent phosphatidylcholines contain PNA in the SN-2 position and myristic,
palmitic, stearic, oleic, or linoleic acid in the SN-1 position.
The model lipoproteins are recombinants of apoA-II and the synthetic
phospholipids, DMPC or 1-palmitoyl-2-palmitoleoylphosphatidylcho-
line, PPOPC. The rates of transfer (see Table 9) are influenced by
the properties of the lipid-protein surface of the recombinant
donor. Thus, the addition of two methylene units to the acyl chains
decreases the rate of desorption from DMPC by a factor of 8-10,
whereas, the addition of a double bond increases the rate by a
factor of 10. When a PPOPC matrix is used, the change is about a

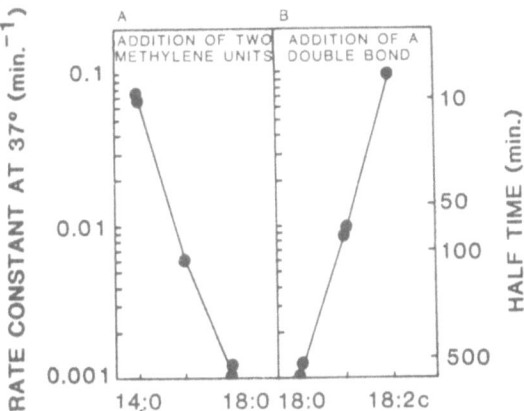

Fig. 17. Dependence of phospholipid transfer rate on fatty acyl
composition. In Panel A, the fatty acyl composition in the
SN-1 position of the pyrene labeled PC contains a different number of methylene units which are 14:0, myristic;
16:0, palmitic; and 18:0, stearic acids, respectively. In
Panel B, the fatty acid differ in the number of double
bonds which are 18:0, stearic; 18:1, oleic and 18:2c, linoleic acids, respectively (from Massey et al., 1982a).

factor of 5 to 7 per two methylene units and about 3 to 4 per
double bond. Because the spontaneous transfer of phospholipids can
produce a net flux, depending on the structure of the lipid and
the lipid-protein surface from which the lipid dissociates, this
mechanism of equilibration can be quantitatively important for unsaturated or short chain phospholipids. The transfer of a series of
phospholipids that differ in the head group is shown in Fig. 18.
The transfer rates are pH dependent for those phospholipids that
have titratable headgroups (Massey et al., 1982b).

In summary, plasma lipoproteins are dynamic structures that
transport lipids in the circulation. They also contain important
minor components such as vitamins and environmentally derived contaminants. The studies of lipoprotein dynamics suggest that many
aspects of the metabolism of the triglyceride-rich lipoproteins are
kinetically limited. For the smaller lipoproteins, such as HDL and
LDL that have relatively long life times in the circulation, the
structure of these lipoproteins reflect a thermodynamic equili-

Table 9. Rate Constants for Lipid Transfer Between Identical
Donor and Acceptors

Transfering Molecule	Donor	Rate Constant		Halftime	
		sec^{-1}	min^{-1}	sec	min
Pyrene	HDL	250		0.003	
	LDL	67		0.01	
	VLDL	13		0.05	
Benzo(a)pyrene	HDL	17		0.04	
	LDL	3.9		0.18	
	VLDL	1.8		0.39	
PNA	DMPC, pH 7.4	5.4		0.042	
	pH 2.8	0.08		4.2	
Cholesterol	C:EPC		0.005		138
PMC	HDL		0.58		1.2
	LDL		0.09		7.5
	VLDL		0.03		22
MPNPC	DMPC/apoA-II (45/1)		0.071		9.8
PPNPC	DMPC/apoA-II (45/1)		0.0061		114
MPNPC	PPOPC/apoA-1 (100/1)		0.0895		7.8
PPNPC	PPOPC/apoA-I (100/1)		0.0126		54.9
SPNPC	PPOPC/apoA-I (100/1)		0.00257		269
OPNPC	PPOPC/apoA-I (100/1)		0.00960		72
LPNPC	PPOPC/apoA-1 (100/1)		0.0373		18.6
MPNPA	DMPC/apoA-II (45/1)		0.0545		12.7
MPNPE	DMPC/apoA-II (45/1)		0.0192		36.1
MPNPG	DMPC/apoA-II (45/1)		0.0468		14.8
MPNPS	DMPC/apoA-II (45/1)		0.0425		16.3
MPNG	DMPC/apoA-II (45/1)		0.00378		183

* The abbreviations are PNA, 9(3′-pyrenyl)nonanoic acid; EPC, egg yolk phosphatidylcholine; PMC, 3′-pyrenemethyl-3 β-hydroxy-22,23-bisnor-5-cholenate; MPNPC, PPNCP,
SPNPC, OPNPC, LPNPC, phosphatidylcholines containing myristoyl, palmitoyl, stearoyl,
oleoyl and linoleoyl moiety at the SN-1 position and 9(3′-pyrenyl)nonanoyl group at
the SN-2 position, respectively; MPNPA, MPNPE, MPNPG, MPNPS, SN-1-myristoyl-2[9(3′-
-pyrenyl)] nonanoyl phosphatidic acid, phosphatidyl ethanolamine, phosphatidyl
glycerol and phosphatidyl serine, respectively; and SN-1-myristoyl-2-[9(3′-pyrenyl)]
nonanoylglycerol. The molar ratio of lipid to apolipoprotein is either 45 to 1 or
100 to 1, as indicated.

brium. Consequently, the structure is based primarily on thermodynamic principles. The parameters that account for the relatively
constant composition of these lipoproteins and for their dynamic
nature have not yet been elucidated.

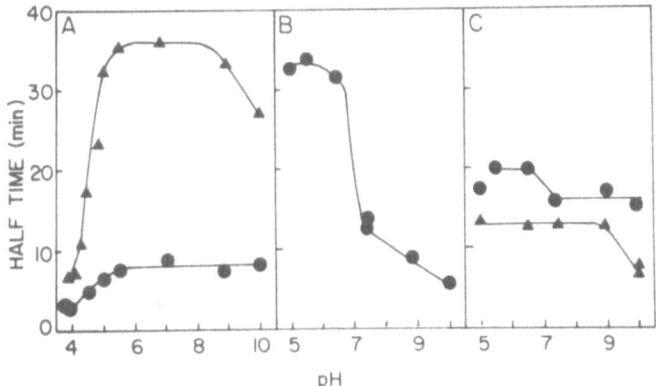

Fig. 18. pH Dependence of spontaneous phospholipid transfer between
DMPC:apoA-II complexes. The data are for (Panel A) MPNPC
(●), MPNPE (▲), (Panel B) MPNPA (●), (Panel C) MPNPG (●)
and MPNPS (▲) (from Massey et al., 1982b).

ACKNOWLEDGEMENTS

     Support for the research was provided by the Robert A. Welch
Foundation Q-343, Human Health Service grants HL-07282, HL-17269,
HL-15648, HL-27341, HL-19459 and the Office of Research and Development, Environmental Protection Agency, under grant number R-80-
9773. Additional funding was provided by Baylor College of Medicine. The Environmental Protection Agency does not necessarily
endorse any commercial products used in the study. The conclusions
represent the views of the authors and do not necessarily represent
the opinions, policies, or recommendations of the Environmental
Protection Agency. Carol Coreathers provided excellent assistance
in the preparation of the manuscript.

REFERENCES

Albers, J.J., and Scanu, A.M., 1971, Biochim. Biophys. Acta,
     236:29.
Atkinson, D., Deckelbaum, R.J., Small, D.M., and Shipley, G.G.,
     1977, Proc. Natl. Acad. Sci. USA, 74:1042.

Backer, J.M.,and Dawidowicz, E.A., 1979, Biochim. Biophys. Acta, 551:260.

Barter, P.J., Gooden, J.M., and Rajaram, O.V., 1979, Atheroscl., 33:165.

Barter, P.J.,and Jones, M.E., 1980, J. Lipid Res., 21:238.

Bell, F.P., 1978, Prog. Lipid Res., 17:207.

Berg, K., 1963, Acta Path. Microbiol. Scand., 59:369.

Berg, K., Dahlen, G., and Frick, M.H., 1974, Clin. Genet., 6:230.

Bjornson, L.K., Ghiewkowski, C., and Kayden, J.J., 1975, J. Lipid Res., 16:39.

Bradley, W.B., Rhode, M.F., Gotto, A.M., Jr., and Jackson, R.L., 1978, Biochem. Biophys. Res. Comm., 81:928.

Breckenridge, W.C., Little, J.A., Steiner, G., and Poapst, M., 1978, New Engl. J. Med., 298:1265.

Brewster, M.E., Ihm, J., Brainard, J.R., and Harmony, J.A.K., 1978, Biochim. Biophys. Acta, 529:147.

Brown, M.S., Kovanen, P.T., and Goldstein, J.L., 1981, Science, 212:628, and references therein.

Catapano, A.L., Gianturco, S.H., Kinnunen, P.K.J., Eisenberg, S., Gotto, A.M., Jr., and Smith, L.C., 1979a, J. Biol. Chem., 254:1007.

Catapano, A.L., Kinnunen, P.K.J., Breckenridge, W.C., Gotto, A.M., Jr., Jackson, R.L., Little, J.A., and Smith, L.C., 1979b, Biochem. Biophys. Res. Commun., 89:951.

Chajek, T.,and Fielding, C.J., 1978, Proc. Natl. Acad. Sci. USA, 75:3445.

Chajek, T., Aron, L., and Fielding, C.J., 1980, Biochemistry, 19:3673.

Charlton, S.C., Hong, H.Y., and Smith, L.C., 1978, Biochemistry, 17:3304.

Charlton, S.C., Olsoń, J.S., Hong, K.Y., Pownall, J.J., Louie, D.D., and Smith, L.C., 1976, J. Biol. Chem., 251:7951.

Charlton, S.C. and Smith, L.C., 1982, submitted for publication.

Chen, G.C., and Kane, J.P., 1974, Biochemistry, 13:3330.

Cooper, A.D., and Yu, P.Y.S., 1978, J. Liquid Res., 19:635.

Cryer, A., 1981, Intl. J. Biochem., 13:525.

Daerr, W.H., Gianturco, S.H., Patsch, J.R., Smith, L.C., and Gotto, A.M., Jr., 1980, Biochim. Biophys. Acta, 619:287.

Day, C.E., 1981, "High Density Lipoproteins", C. Day, ed., Marcel Dekker, New York.

Doody, M.C., Pownall, H.J., Kao, Y.K., and Smith, L.C., 1980, Biochemistry, 19:108.

Duckwitz-Peterlein, G., Eilenberger, G., and Overath, P., 1977, Biochim. Biophys. Acta, 469:311.

Duckwitz-Peterlein, G., and Moraal, H., 1978, Biophys. Struct. Mechanism, 4:315.

Edelstein, C., Kezdy, F.J., Scanu, A.M., and Shen, B.W., 1979, J. Lipid Res., 20:143.

Fainaru, M., Havel, R.J., and Imaizumi, K., 1977, Biochem. Med., 17:347.

Fielding, P.E., and Fielding, C.J., 1980, Proc. Natl. Acad. Sci. USA, 75:3445.

Fielding, C.J., Shore, V.G., and Fielding, P.E., 1972, Biochem. Res. Comm., 46:1493.

Franceschini, G., Sirtori, M., Gianfranceschi, G., and Sirtori, C. R., 1981, Metabolism, 30:502.

Fredrickson, D.S., Goldstein, J.L., and Brown, M.S., 1978, in: "The Metabolic Basis of Inherited Disease", J.B. Stanbury, J.B. Wyngarden, and D.S. Fredrickson, eds., 4th Edn., McGraw-Hill, New York, p. 604.

Galla, J.J., Thilen, V., and Hartman, W., 1979, Chem. Phys. Lipids, 23:239.

Ganesan, D., Bradford, R.H., Ganesan, C., Alaupovic, P., and Bass, H.G., 1975, J. Appl. Physiol , 39:1022.

Gianturco, S.H., Gotto, A.M., Jr., Jackson, R.L., Patsch, J.R., Sybers, H.D., Taunton, O.D., Yeshurun, D.L., and Smith, L.C., 1978, J. Clin. Invest., 61:320.

Gjone, E., Norum, K.R., and Glomset, J.A., 1978, in: "The Metabolic Basis of Inherited Diseases,J.B. Stanbury, J.B. Wyngarden, and D.S. Fredrickson, eds., 4th Edn., McGraw-Hill, New York, p. 589.

Goldstein, J.L., Ho, Y.K., Basu, S.K., and Brown, M.S., 1979, Proc. Natl. Acad. Sci. USA, 76:333.

Goodman, D.S., and Noble, R.P., 1968, J. Clin. Invest., 47:231.

Gottlieb, M.H., 1980, Biochim. Biophys. Acta, 600:530.

Grubbs, C.J., and Moon, R.C., 1973, Cancer Res., 33:1785.

Havel, R.J., 1979, Circulation, 60:1.

Havel, R.J., Goldstein, J.L., and Brown, M.S., 1980, in: "The Metabolic Control and Disease", P.K. Bondy and L.E. Rosenberg, eds., 8th Edn., Saunders, Philadelphia, p. 393.

Havel, R.J., Kane, J.P., and Kashyap, M.L., 1973, J. Clin Invest., 52:32.

Havel, R.J., Kotite, L., and Kane, J.P., 1979, Biochem. Med., 21:121.

Havel, R.J., Shore, V.G., Shore, B., and Bier, D.M., 1970, Circ.
    Res., 27:595.
Ihm, J., Harmony, J.A.K., Ellsworth, J., and Jackson, R.L., 1980,
    Biochem. Biophys. Res. Commm., 93:1114.
Illingsworth, D.R., and Portman, O.W., 1972, Biochim. Biophys.
    Acta, 280:281.
Imaízumi, K., Fainaru, M., and Havel, R.J., 1978, J. Lipid Res.,
    19:712.
Israelachvili, J.N., Marčelja, S., and Horn, R.C., 1980, Quart.
    Rev. Biophys., 13:121.
Jackson, R.L., Morrisett, J.D., Sparrow, J.T., Segrest, J.P.,
    Pownall, H.J., Smith, L.C., Hoff, H.F., and Gotto, A.M., Jr.,
    1974, J. Biol. Chem., 249:5314.
Jahn, C., Osborne, J.C., Jr., Schaefer, E.J., and Brewer, H.B.,
    Jr., 1981, Atheroscl., 1:381a.
Kaneshisa, M.I., and Tsong, T.Y., 1978, J. Am. Chem. Soc., 100:424.
Kao, Y.J., Charlton, S.C., and Smith, L.C., 1977, Fed. Proc.,
    56:936.
Kinnunen, P.K.J., Jackson, R.L., Smith, L.C., Gotto, A.M., Jr.,
    and Sparrow, J.T., 1977, Proc. Natl. Acad. Sci. USA, 74:4848.
Krishnaiah, K.V., Walker, L.F., Borensztajn, J., Schonfeld, G.,
    and Getz, G.S., 1980, Proc. Natl. Acad. Sci. USA, 77:3806.
Lagocki, P.A., and Scanu, A.M., 1980, J. Biol. Chem., 255:3701.
LaRosa, J.C., Levy, R.I., Herbert, P., Lux, S.E., and Fredrickson,
    D.S., 1970, Biochem. Biophys. Res. Comm., 41:57.
Maeda, H., Uzawa, H., and Kamei, R., 1981a, Biochim. Biophys. Acta,
    665:578.
Maeda, H., Uzawa, H., Nakamura, N., Nakashima, Y., and Goto, I.,
    1981b, Life Sci., 29:2065.
Mantulin, W.W., Massey, J.B., Gotto, A.M., Jr., and Pownall, H.J.
    1981, J. Biol. Chem., 256:10815.
Mantulin, W.W., Rhode, M.F., Gotto, A.M., Jr., and Pownall, H.J.,
    1980, J. Biol. Chem., 255:8185.
Martin, F.J., and MacDonald, R.C., 1976, Biochemistry, 15:321.
Massey, J.B., Gotto, A.M., Jr., and Pownall, H.J., 1979, J. Biol.
    Chem., 49:9359.
Massey, J.B., Gotto, A.M., Jr., and Pownall, H.J., 1981b, Bioche-
    mistry, 20:1575.
Massey, J.B., Gotto, A.M., Jr., and Pownall, H.J., 1982a, Bioche-
    mistry, 21: in press.
Massey, J.B., Gotto, A.M., Jr., and Pownall, H.J., 1982b, J. Biol.
    Chem., 257: in press.

Massey, J.B., Rhode, M.F., Van Winkle, W.B., Gotto, A.M., Jr., and Pownall, H.J., 1981a, Biochemistry, 20:1569.

McLean, L.R., and Phillips, M.C., 1981, Biochemistry, 20:2893.

Miller, N.E., Nestel, P.J., and Clifton-Bligh, P., 1976, Atherosclerosis, 23:535.

Morrisett, J.D., Jackson, R.L., and Gotto, A.M., Jr., 1977b, Biochim. Biophys Acta, 472:93.

Morrisett, J.D., Pownall, H.J., Jackson, R.L., Segura, R., Gotto, A.M., Jr., and Taunton, O.D., 1977a, in: "Polysaturated Fatty Acids", W.-H. Kunau and R.T. Holman, eds., Am. Oil Chem. Society Press, Champaign, Ill., p. 139.

Morton, R.E., and Zilversmit, D.B., 1981, Biochim. Biophys Acta, 663:350.

Nakaya, V., Schaefer, E.J., and Brewer, H.B., Jr., 1980, Biochem. Biophys. Res. Comm., 95:1168.

Nestel, P.J., and Fidge, N.H., 1981, in: "Lipoproteins, Atherosclerosis and Coronary Heart Disease", N.E. Miller and B. Lewis, eds., Elsevier, Amsterdam, p. 3.

Nestruck, A.C., Suzueand, G., and Marcel, Y.L., 1980, Biochim. Biophys. Acta, 617:110.

Nicoll, A., Miller, N.E., and Lewis, B., 1980, Adv. Lipid Res., 17:53.

Nikkila, E.A., 1978, Eu. J. Clin. Invest., 8:111.

Nilsson-Ehle, P., Garfinkel, A.S., and Schotz, M.C., 1980, Ann. Rev. Biochem., 49:667.

Ose, L., Roken, I., Norum, K.A., Drevon, C.A., and Berg, T., 1981, Scand. J. Clin. Lab. Invest., 41:63.

Patsch, J.R., Gotto, A.M., Jr., Eisenberg, S., and Olivecrona, T., 1978, Proc. Natl. Acad. Sci. USA, 75:4519.

Patsch, J.R., Karlin, J.B., Scott, L., Smith, L.C., and Gotto, A.M., Jr., 1982, submitted for publication.

Pattnaik, N.M., Montes, A., Hughes, L.B., and Zilversmit, D.B., 1978, Biochim. Biophys. Acta, 530:428.

Pattnaik, N.M., and Zilversmit, D.B., 1979, J. Biol. Chem., 254:2782.

Polz, E., Kostner, G.M., and Holasek, A., 1979, Hoppe-Seyler's Z. Physiol. Chem., 360:1061.

Pownall, H.J., Gotto, A.M., Jr., and Sparrow, J.T., 1982, submitted for publication.

Pownall, H.J., Hickson, D., and Gotto, A.M., Jr., 1981a, J. Biol. Chem., 256:9849.

Pownall, H.J., Hu, A., Albers, J.J., Gotto, A.M., Jr., and Sparrow, J.T., 1980, Proc. Natl. Acad. Sci. USA, 77:3154.

Pownall, H.J., Massey, J.B., Kusserow, S.K., and Gotto, A.M., Jr., 1978a, Biochemistry, 17:1183.

Pownall, H.J., Massey, J.B., Kusserow, S.K., and Gotto, A.M., Jr., 1979, Biochemistry, 18:574.

Pownall, H.J., Morrisett, J.D., and Gotto, A.M., Jr., J. Lipid Res. 18:14.

Pownall, H.J., Pao, Q., Hickson, D., Sparrow, J.T., Kusserow, S.K., and Massey, J.B., 1981b, Biochemistry, 20:6630.

Pownall, H.J., Pao, Q., Rhode, M., and Gotto, A.M., Jr., 1978b, Biochem. Biophys. Res. Comm., 85:408.

Pownall, H.J., and Smith, L.C., 1973, J. Am. Chem. Soc., 95:3136.

Quinn, P.J., 1981, Prog. Biophys. Molec. Biol., 38:1.

Rajaram, O.V., White, G.H., and Barter, P.J., 1980, Biochim. Biophys. Acta. 617:383.

Redgrave, T.G., and Small, D.M., 1979, J. Clin. Invest., 64:162.

Robinson, D.W., 1970, Compr. Biochem., 18:51.

Roseman, M.A., and Thompson, T.E., 1980, Biochemistry, 19:439.

Scanu, A.M., and Landsberger, F.R., eds., 1980, Ann. N.Y. Acad. Sci., 348.

Schaefer, E.J., Eisenberg, S., and Levy, R.I., 1978, J. Lipid. Res., 19:667.

Scow, R.O., Blanchette-Mackie, E.J., and Smith, L.C., 1976, Circ. Res., 39:149.

Segrest, J.P., Jackson, R.L., Morrisett, J.D., and Gotto, A.M., Jr., 1974, FEBS Lett., 38:247.

Sengupta, P., Sackman, E., Kuhnle, W., and Scholz, J.P., 1976, Biochim. Biophys. Acta, 436:869.

Shen, B.W., Scanu, A.M., and Kezdy, F.J., 1977, Proc. Natl. Acad. Sci. USA, 74:837.

Sherrill, B.C., and Dietschy, J.M., 1978, J. Biol. Chem., 253:1859.

Sherrill, B.C., Innerarity, T.L., and Mahley, R.W., 1980, J. Biol. Chem., 255:1804.

Sigurdsson, G., Noel, S.P., and Havel, R.J., 1979, J. Lipid Res., 20:316.

Sklar, L.A., Craig, I.F., and Pownall, H.J., 1981, J. Biol. Chem., 256:4286.

Smith, L.C., and Doody, M.C., 1981, in: "Polynuclear Aromatic Hydrocarbons: Chemistry and Biological Effects", M. Cooke and A.J. Dennis, eds., Battelle Press, Columbus, Ohio, p. 615.

Smith, L.C., Pownall, H.J., and Gotto, A.M., Jr., 1978, Ann. Rev. Biochem., 47:751.

Smith, L.C., and Scow, R.O., 1979, Prog. Biochem. Pharm., 15:65.

Smith, L.C., Voyta, J.C., Catapano, A.L., Kinnunen, P.K.J., Gotto,
    A.M., Jr., and Sparrow, J.T., 1980, Ann. N.Y. Acad. Sci.,
    248:213.
Smith, L.C., Voyta, J.C., Kinnunen, P.K.J., Gotto, A.M., Jr., and
    Sparrow, J.T., 1982a, Biophys. J., 37:174.
Smith, L.C., Voyta, J.C., Kinnunen, P.K.J., Gotto, A.M., Jr., and
    Sparrow, J.T., 1982b, J. Am. Oil Chem. Soc., in press.
Smith, R., Dawson, J.R., and Tanford, C., 1972, J. Biol. Chem.,
    247:3376.
Soutar, A.K., Garner, C.W., Baker, H.N., Sparrow, J.T., Jackson,
    R.L., Gotto, A.M., Jr., and L.C. Smith, 1975, Biochemistry,
    14:3057.
Sparrow, J.T., and Gotto, A.M., Jr., 1978, Circ., 58:11.
Steele, J.C.H., Jr., and Reynolds, J.A., 1979a, J. Biol. Chem.,
    254:1633.
Steele, J.C.H., Jr., and Reynolds, J.A., 1979b, J. Biol. Chem.,
    254:1639.
Steinberg, D., 1981, in: "Lipoproteins, Atherosclerosis and Coro-
    nary Heart Disease" , N.E. Miller and B. Lewis, eds., Elsevier,
    Amsterdam, p. 31.
Stone, W.L., and Reynolds, J.A., 1975, J. Biol. Chem., 250:8045.
Tall, A.R., and Small, D.M., 1980, Adv. Lipid. Res., 17:1.
Utermann, G., Jaeschke, M., and Mangel, J., 1975, FEBS Lett.,
    56:352.
Van't Hooft, F.M., Hardman, D.A., Kane, J.P., and Havel, R.J.,
    1982, Proc. Natl. Acad. Sci. USA, 79:179.
Verdery, R.B., 1981, Can. J. Biochem., 59:586.
Vitello, L.B., and Scanu, A.M., 1976, Biochemistry, 15:1161.
Voyta, J.C., Kinnunen, P.K.J., and Smith, L.C., 1980, unpublished
    observations.
Wandel, M., Norum, K.R., Berg, T., and Ose, L., 1981, Scand. J.
    Gastroent., 16:71.
Weisgraber, K.H., Bersot, T.P., and Mahley, R.W., 1978, Biochem.
    Biophys. Res. Comm., 85:287.
Weisgraber, K.H., Bersot, T.P., Mahley, R.W., Franceschini, G.,
    and Sirtori, C.R., 1980, J. Clin. Invest., 66:901.
Windler, E., Chao, Y., and Havel, R.J., 1980, J. Biol. Chem.,
    255:5475.
Witztum, J., and Schonfeld, G., 1979, Diabetes, 28:326.
Wolinsky, H., 1980, Cir. Res., 47:301.
Zannis, V.I., and Breslow, J.L., 1982, Mol. Cell. Biochem., 42:3.
Zilversmit, D.B., 1980, Circ. Res., 60:473.

Zilversmit, D.B., Hughes, L.B., Balmer, J., 1975, Biochim. Bio-
    phys. Acta, 409:393.

CRYSTALLOGRAPHIC STUDIES OF THE PROTEIN BIOSYNTHESIS SYSTEM

Anders Liljas and Marie Leijonmarck

Institute of Molecular Biology
Uppsala University
Uppsala, Sweden

INTRODUCTION

The protein biosynthesis system provides a very good illustration of the difficulties involved in studying the structure and function of a supramolecular assembly. The complexity of the system and the lack of appropriate experimental tools have forced those in the field to use methods which are less than ideally suited for the problems, as well as to develop new ones.

Among the achievements that have been made during the last few years are the complete sequence determination of all the ribosomal components from Escherichia coli; the determination of the shape of the ribosomal particles and the location of many ribosomal components; and the gradual understanding of the secondary structures of the ribosomal RNA-molecules. Many areas nevertheless remain poorly understood: the quaternary structure of the whole ribosome; the tertiary structure of its components; and the cooperative interactions of these components during protein biosynthesis.

The aim of this paper is to discuss the importance, the feasibility and achievements of crystallographic work on the protein synthesis system. The diffraction analysis of crystals is unique in its ability to accurately determine quaternary and tertiary structures. Additionally such an analysis can provide us with a more precise and detailed perception of the functional interactions. In

245

particular we will deal with the work on the ribosomal particles,
ribosomal components and with some of the proteins (or soluble
factors) that transiently interact with the ribosome during specific
phases of protein synthesis.

A part of the material presented at the summer school is described in a more extensive review (Liljas, 1982).

The Pieces of the Puzzle

A great number of molecules participate in protein biosynthesis. Among these are the tRNA molecules and the tRNA-synthetases
that charge the tRNA:s with appropriate amino acids. The crystallographic work on these is a separate subject and will not be reviewed
here. Around ten protein molecules, or factors, are involved in
specific functions and bind to the ribosome only during some of the
cycles that describe the synthesis of a protein. These are the ini-
tiation, elongation and termination factors. The bacterial ribosome
has three RNA molecules (5S, 16S and 23S RNA) and around fifty different protein molecules (numbered S1-S21 and L1-L34) organized into
two ribosomal subunits (30S and 50S). All the ribosomal components,
except one, are present in only one copy per ribosome (Hardy, 1975).
The deviating component is the protein L7/L12, present in four
copies per ribosome (Hardy, 1975; Subramanian, 1975).

It remains a difficult task to determine whether a protein is
a true ribosomal component and in such a case what its function
might be. Homologous proteins from other species enforce the observations (Matheson et al., 1980; Fahnestock et al.,1981). The finding
that certain mutant bacteria can lack one of a number of proteins from
their ribosomes might indicate that a cooperative system such as
the ribosome still can function with one of its components absent
(Dabbs, 1977, 1979). If one wants to study structure-function relationship for ribosomal components by crystallography one has to
be fairly selective in order to avoid spurious components of the
system as well as components for which no function is known. Furthermore, great preparational care has to be taken: 1. To maintain high
activity of the ribosomes (Jelenc, 1980); 2. To prevent conformational damage and loss of components (Stöffler et al., 1980a, 1980b);
and 3. To avoid denaturating steps that may prevent crystallization

of isolated ribosomal proteins (Liljas and Kurland, 1976; Dijk and Littlechild, 1979).

Crystallography of Elongation Factors

Microcrystals have been obtained of several of the factors involved in protein synthesis, but until recently only elongation factor Tu (EF-Tu) has given crystals good enough for diffraction analysis (Sneden et al., 1973). Table 1. gives a summary of the different types of crystals obtained for this protein and clearly illustrates some of the problems involved.

The best crystals are obtained when EF-Tu is proteolytically degraded either spontaneously or with trypsin (Sneden et al., 1973; Gast et al., 1976, 1977; Jurnak et al., 1977; Morikava et al., 1978).

Most of the crystal forms obtained after such treatment have two molecules per asymmetric unit. This can be due to the fact that

Table 1. Crystal Forms of Elongation Factors

Factor	Precipitant	Space group	Cell dimensions ($\overset{\circ}{A}$) a	b	c	Max res ($\overset{\circ}{A}$)	Molecules per a.u.	Fragmentation	Ref
EF-Tu·GDP	Isopropanol	$C222_1$	98.2	100.1	160.6	2.8	2	Spontaneous	a
	MPD	(pseudo $P4_12_12$	70.2	70.2	160.6		1)		b
	PEG 6000								c
"	$(NH_4)_2SO_4$ citrate	$P2_12_12_1$	62	67	245		2	No	b
"	PEG 6000	$P3_1$	80	80	326		8	No	b
"	"	$P3_1$	80	80	328		4	No	b
"	"	P6	80	80	329		2	No	b
"	"	$P3_12\ 1$	80	80	161	2.8	1	No	b
"	"	$P6_22\ 2$	80	80	161		0.6	No	b
"	"	$P3_11\ 2$	80	80	161		1	No	b
"	"	$P2_12_12_1$	144	93	69	2.5	2	Trypsin	d
"	PEG 4000	$P4_32_12$	70.4	70.4	161.4	2.8	1	Trypsin	e,f
EF-Tu·EF-Ts	PEG 6000	$P2_12_12_1$	81.7	110.5	206.0	4.5	2 Tu·Ts		g
"	"	$P2_12_12_1$	74.3	108.7	198.5	3.0	2 Tu·Ts		g

References to Table 1.
a Sneden et al. (1973)   e Jurnak et al. (1977)
b Leberman et al. (1976)   f Jurnak et al. (1980)
c Morikava et al. (1978)   g Leberman et al. (1981)
d Gast et al. (1977)

the crystal packing is very much improved by having two molecules
with slightly different conformations in the asymmetric unit
(Morikava et al., 1978). Since the crystals of undigested EF-Tu
have different space groups and cell dimensions, it is evident that
the cleavage either removes a peptide that is unsuitable for a
certain type of packing, or permits the protein to adopt a different conformation. The packing of intact EF-Tu is very complicated. A whole range of unit cells are obtained and the number of
molecules per asymmetric unit can be quite variable (Leberman et
al., 1976).

The main problem with this material is that the positions in
the asymmetric units seem to be only half occupied, and the most
favourable crystals are difficult to obtain reproducibly. For these
reasons the work has been focused on determining the structure of
proteolytically degraded EF-Tu.

The structure determination of EF-Tu has been carried out to
different resolution by three groups (Kabsch et al., 1977; Jurnak
et al., 1980; Rubin et al., 1981). The higher resolution model
(Rubin et al., 1981) seems to be in agreement with the low resolution models. The molecule has the dimensions 75 Å x 50 Å x 45 Å
and is composed of three domains (Morikava et al., 1978). The N-
terminal peptide produced by trypsin or spontaneous cleavage still
remains bound to the main part of the molecule, but a 14 amino acid
long peptide is excised between Arg-44 and Arg-58 (Jurnak et al.,
1980; Jones et al., 1980; Laursen et al., 1981).

GDP and GTP bind as allosteric factors to EF-Tu (Kaziro, 1978).
The fragmented proteins seem to have a reduced affinity for GTP
(Jurnak et al., 1980). The crystallography has been carried out
for the GDP complex. In the structure determined at 2.9 Å resolution by Rubin et al. (1981) the peptide chain can be followed
only in domains I and II (Fig. 1). Domain III has different orientations for the two molecules in the crystallographic asymmetric
unit and seems to be less ordered. Domain I brings the GDP molecule
to a type of structure (Fig. 2) which frequently is involved in
binding nucleotides (Rossmann and Argos, 1981). This so called
"Rossmann" fold in EF-Tu is compared to the corresponding structure in lactate dehydrogenase (LDH) in Fig. 3.

The binding of GTP to EF-Tu induces a different conformation
in the protein which does not involve secondary structure changes

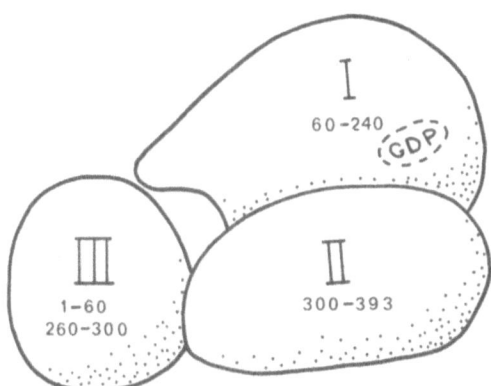

Fig. 1. The domain structure of EF-Tu (B.C.F. Clark, T. LaCour and
        J. Nyborg, private communication). The residues of each
        domain are indicated, as well as the location of the bound
        GDP molecule. Domain III obviously functions as a con-
        nection between domains I and II and has different orien-
        tations for the two molecules in the asymmetric unit
        (Morikava et al., 1978; Rubin et al., 1981).

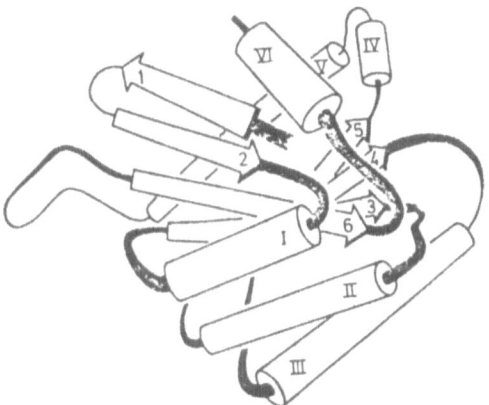

Fig. 2. The folding of domain I of EF-Tu (Rubin et al., 1981).
        Helices are represented by cylinders, and arrows indicate
        β-strands pointing from N-terminus towards C-terminus. The
        GDP molecule is bound between β-strands 3 and 4 and helices
        IV and VI (reproduced from Rubin et al. (1981), with per-
        mission).

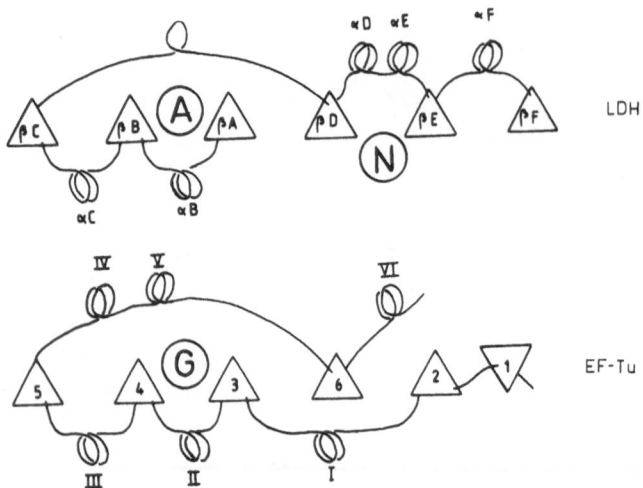

Fig. 3. The "Rossmann" fold of EF-Tu (Rubin et al., 1981) as compared to the NAD binding domain of lactate dehydrogenase
(LDH) (Rossmann et al., 1975). NAD is a dinucleotide and
consequently the nucleotide binding fold is repeated. A and
N represent the locations of the adenine and nicotinamide
moieties. EF-Tu binds GDP or GTP (indicated with a G) at
the standard location of the nucleotide binding fold. Only
one of these structures is formed in EF-Tu.

but probably domain reorientation (Kaziro, 1978; Liljas, 1982).
This conformation is both able to bind aminoacylated tRNA and bind
to the ribosome. Unfortunately the fragmented protein binds neither
GTP nor tRNA. However, it is known that the aminoacyl end of the
tRNA interacts with domain I of EF-Tu (Jonák et al., 1979; Duffy
et al., 1981). Thus His-66 can be crosslinked to N^{ε} -bromoacetyl-
lysyl-tRNA and chemical blocking of Cys-81 prevents tRNA from binding. tRNA and short fragments of its aminoacyl-acceptor end prevent
this cysteine from reacting. This interaction site for tRNA is
neither far from the assumed site for the γ-phosphate of GTP nor
far from the site of proteolytic cleavage (Arg-58) near the connection to domain III.

The reason for the inability to bind tRNA could be that the fourteen missing residues are involved in the binding, but it could also be that removing this part of the domain connection prevents the domains from adopting the proper relative orientation.

EF-Tu is released from the ribosome and the tRNA at the ribosomal A-site upon cleavage of the GTP molecule to GDP and inorganic phosphate (Kaziro, 1978; Weissbach, 1980). It has become evident that EF-Tu itself carries the GTPase activity. An antibiotic, kirromycin, can in the absence of ribosomes or tRNA induce this activity by binding to EF-Tu (Wolf et al., 1974). Furthermore in a kirromycin resistant mutant the factor has the ability to hydrolyze GTP without any other components added (Ivell et al., 1981). One must conclude that EF-Tu has one conformation that binds GTP and one which hydrolyses the nucleotide. The normal inducer of GTP hydrolysis must be some component of the ribosome. We will return to this topic at the end of this paper.

EF-Tu thus has several conformations, one of which has been crystallized. It seems clear that the other conformations cannot be deduced from the one that is presently analyzed since fragmented EF-Tu is inactive in the processes one would like to study. For a fuller picture of the functional states much more extensive crystallographic studies of EF-Tu are required. Along these lines a complex of EF-Tu with the elongation factor Ts has been crystallized (Leberman et al., 1981). When the structure is available it will show how EF-Ts, by changing the conformation of EF-Tu, enhances the rate of exchange of GDP for GTP (Kaziro, 1978; Weissbach, 1980).

There is yet another elongation factor, EF-G. Its function is to translocate the peptidyl-tRNA from the ribosomal A-site to the P-site after the peptidyl transfer. No crystal structure is available, but the N-terminal region of the amino acid sequence (Alakhov and Ovchinnikov, 1981) shows significant homology to the same region of EF-Tu (Laursen et al., 1981).

Some of the characteristics of EF-Tu are shared by EF-G. EF-G binds to the ribosome in complex with GTP, but dissociates after cleavage of the GTP molecule. It has been shown also for EF-G that the protein conformations with GTP and GDP are different (Kaziro, 1978). Furthermore most of the data availabe indicate that EF-Tu and EF-G bind to at least partly overlapping sites on the ribosome (Liljas, 1982).

In order to be able to discuss the details of the interaction between elongation factors and the ribosome we need to review the current models of the structure of the ribosome.

Crystallographic Studies of the Structure of Ribosomal Particles

Numerous attempts have been made to find or produce crystalline arrays of ribosomes for structure analysis. Table 2 shows a list of what has been reported in the literature. In only one of these cases were three dimensional crystals obtained (Yonath et al., 1980). It has, however, been shown practically that two-dimensional crystals can provide data for structure determination at relatively high resolution (Henderson and Unwin, 1975). Helical arrays give principally the same possibilities (Crowther and Klug, 1975).

Unfortunately, the work on prokaryotic ribosomes has not yet yielded any structural results, but the crystallizations of B.

Table 2. Crystallization of Ribosomal Particles

Species	Conditions for crystallisation	Arrangement of particles	Structural analysis	Ref.
Entamoeba invadens	In vivo. Chromatoid bodies	Helical. 600 Å wide Repeat distance 1290 Å	EM reconstruction at 107-129 Å resolution	a
Chick embryo	In vivo after incubation at low temperature	P4 crystalline monolayer 537 Å unit cell	EM reconstruction work (data at 60 Å resolution)	b,c
Lacerta sicula (lizard) oocytes	Ribosomal bodies formed during winter rest. In vivo Membrane bound	P422 double layer crystals 595 Å unit cell	EM reconstruction. 90 Å resolution	d,e
Human brain	In vivo in senile humans Membrane associated	Monolayers that can stack a=b=132 Å, γ =56°	EM reconstruction	f
E. coli	Incubation with vinblastine	Helical polysomes 600 Å wide Pitch = 180 Å	Inspection of EM	g
E. coli 30S	Ethanol, vapor diffusion	Helices or ribbons 320x180 Å Axial repeat 220 Å	Optical diffraction of EM (60 Å resolution possible)	h
B. stearo 50S	Organic solvents, vapor diffusion	Three dimensional crystals	EM reconstruction possible to $\sim$ 40 Å resolution. X-ray diffraction (powder) observed to $\sim$3.5 Å resolution	i
E. coli 50S				j

References to Table 2.
a Lake and Slayter (1972)   e Unwin (1977; 1979)   h Clark et al. (1979)
b Byers (1971)   f O'Brien et al. (1980)   i Yonath et al. (1980)
c Lake et al. (1974)   g Kingsbury and Voelz (1969) j Lake (1981)
d Taddei (1972)

stearothermophilus 50S subunits (Yonath et al., 1980) and E. coli
30S subunits (Clark et al., 1979; Lake, 1981) are very promising
for the future. For eukaryotes it is uncertain whether the material
from brains of senile humans can be ribosomal (O´Brien et al.,
1980). The cell dimensions are surprisingly small (Table 2). A more
careful analysis is needed to establish whether it could be a
smaller fragment of the ribosome.

The most thoroughly studied material comes from chick embryos
(Byers, 1971; Lake et al., 1974) and lizard oocytes (Unwin, 1977;
1979). The resolutions are 60 and 90 Å, respectively, too low for
any detailed interpretation. The lizard ribosomes show a number
of interesting features, one of which is a lower density in the
middle of the ribosomal particle between the two subunits (Unwin,
1977). This could be interpreted as a sizeable cavity between the
two subunits. Unfortunately very little is known about lizard oocyte
ribosomes in general. A stain collecting region between the subunits
in electron micrographs of prokaryotic ribosomes (Tischendorf et
al., 1975; Lake, 1976; Boublik et al., 1977) may correspond to this
feature in the eucaryotic ribosomes.

Electron Microscopy and Neutron Scattering Studies of Ribosomal Quaternary Structure

The methods that have provided most information about the
distribution of ribosomal components within the shape of the subunits are electron microscopy and neutron scattering.

After an initial period of disagreement the models of the subunits derived from electron microscopy are converging with respect
to each other (Kahan et al., 1981; Lake, 1981; Kastner et al.,
1981) and with respect to the 30S model derived from neutron scattering (Ramakrishnan et al., 1981). An attempt to combine the current information is presented in Fig. 4. The necessity of carefully
controlled experiments has considerably reduced the number of
localized proteins from a few years ago (Lake, 1978; Stöffler et
al., 1980a; 1980b). Another change in the models concerns the shape
of the proteins. They were previously described as very elongated,
but the present data suggests that only a few proteins are dis-
tinctly elongated (Kahan et al., 1981; Ramakrishnan et al., 1981).

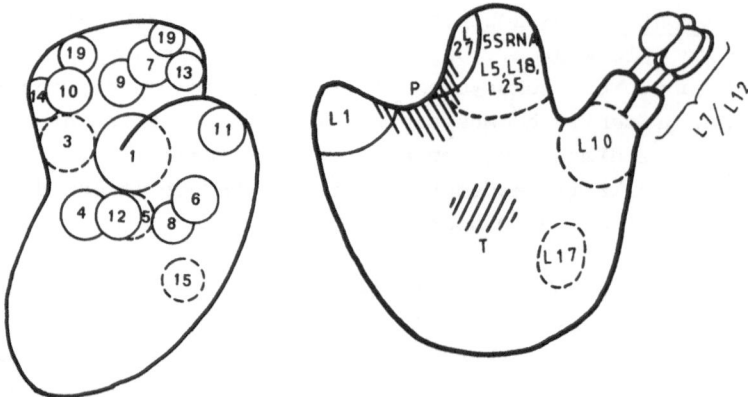

Fig. 4. The topology of the ribosomal subunits determined by immune
       electron microscopy (IEM) and neutron scattering as com-
       piled by Liljas (1982). A. Protein arrangement in the 30S
       particle. Proteins S3-S8, S10 and S11 have been located by
       both methods with good correspondence (Kahan et al., 1981;
       Lake, 1981; Ramakrishnan et al., 1981). The positions for
       proteins S1, S9, S15 have been determined only with neutron
       scattering (Ramakrishnan et al., 1981), and proteins S12-
       S14 and S19 only with immune electron microscopy (Kahan
       et al., 1981). B. The location of components and sites of
       the large ribosomal subunit determined with electron micro-
       scopy. P indicates the binding site for puromycin, which
       presumably indicates the peptidyl transfer site. T is the
       location for bound thiostrepton, an inhibitor of elongation
       factor binding (Stöffler and Stöffler-Meilicke, 1981). L1
       (Dabbs et al., 1981), L7/L12 (Strycharz et al., 1978;
       Marquis et al., 1981; Tokimatsu et al., 1981) L17 and L27
       (Lake, 1981) and the 3´ terminus of 5S RNA (Shatsky et al.,
       1980) have been localized using IEM. The proteins L5, L18
       and L25 bind to the 5S RNA (Garrett et al., 1981) and must
       be located in the immediate environment. L10 is the protein
       that connects the four L7/L12 molecules to the RNA and its
       position was estimated from this fact.

    The present models include more than half of the 30S proteins,
but only a few from the 50S subunit. The arrangement of the two

subunits within the 70S particle is even more poorly understood.
Three different models exist (Kastner et al., 1981; Lake, 1976;
Boublik et al., 1977). Data obtained by other methods has not
enabled us to choose between them (Liljas, 1982).

Crystallographic Studies of Ribosomal Protein

The possibility to determine high resolution crystal structures of ribosomes or subunits seems very remote at present. The
only way to approach the tertiary structure is then to crystallize
components or domains of the ribosome. Such experiments have during
the last few years gained momentum after the first successful
crystallization of a purified ribosomal component (Liljas and
Kurland, 1976; Liljas et al., 1978), and after the development of
techniques to purify ribosomal proteins in general without denatur-
ing them (Dijk and Littlechild, 1979).

In selecting ribosomal material from various species for
crystallization it is important to realize that proteins with no
relevance for protein synthesis can stick to the ribosomes. Furthermore, several ribosomal proteins have been found to be dispensable
in vivo (Dabbs, 1977; 1979). It would seem advisable to initially
concentrate on the ribosomal proteins that are clearly essential
in assembly of the subunits involved in some ribosomal function.

Table 3 gives a summary of crystallized ribosomal proteins.
The proteins that have been crystallized so far are 50S proteins
from E. coli and B. stearothermophilus. If one wants to crystallize
a majority of the ribosomal components a great number of different
species probably has to be tried, as has been the case for other
macromolecules (McPherson, 1976). Furthermore, fragmentation as
observed for L7/L12 (Liljas et al., 1978), or "nicking", as in the
case of EF-Tu (see above), might in some cases be important for
obtaining useful crystals. The only precipitants that have given
crystals of ribosomal proteins are ammonium sulfate and in one case
phosphate.

One characteristic feature of these crystals is that they
do not diffract to better than 3-3.5 $\overset{o}{A}$ resolution. The size of the
crystals would seem enough for quite good diffraction. The lack of
diffracting power could be due to a number of reasons. One is that

Table 3. Crystallized Ribosomal Proteins

Species	Protein	E. coli correspondence	Form	Cell dim. (Å)	Space group	Crystallized in	Molecules per a.u.	Limit of diffr. (Å)	Ref.
E. coli	L7/L12(1-36)	-		72x72x137	$P6_1$	1.2 M AS	4	3.5	a,b
E. coli	L7/L12(47-120)	-		55x55x43	$P4_32_12$	4 M AS	1	1.7	b
E. coli	L29	-		52x119x90	Monoclinic $\beta =71$	1.2 M AS		6	c
B. stearo	$(B-L8)\cdot(B-L13)_4$	$(L7/L12)_4\cdot(L10)$		110x110x244	$P4_22_2$	2.5 M AS	2(x5)	3.5	d
B. stearo	B-L17	?		135x38x49	$P2_12_12$	2.4 M AS	1	3	e
B. stearo	B-L10	L6(or L3)	I	?	?	1.5 M AS		?	c
			II	62x62x124	Hexagonal	2.4 M phosphate		4	c
B. stearo	B-L34	?	I	44x44x64	Ortorombic	3.2 M AS		3	c
			II	44x44x129	Tetragonal	3.2 M AS		?	c

References to Table 3.
a Liljas and Kurland (1976)   d Liljas and Newcomer (1981)
b Liljas et al. (1978)   e Appelt et al. 1979)
c Appelt et al. (1981)

the crystals might have a very low content of protein, but high
content of solvent. Such crystals could be obtained if the proteins
are unusually elongated with obvious difficulties for dense packing.
Another possibility is that the proteins have a fair amount of
flexibility with the net result that parts of the protein are not
identically oriented throughout the crystals. These parts would
then not contribute much to the diffraction.

The C-terminal fragment (CTF) of L7/L12 diffracts very well
in contrast to all other crystallized ribosomal proteins. This can
be explained by the almost globular shape of the stable domain while
the flexible part (see below) is less than 10% of the structure. In
addition the C-terminal region of L7/L12 is known to protrude from
the 50S particle (Marquis et al., 1981) and can probably be regarded as a globular soluble protein. Fig. 5 shows two precission
photographs of crystals of L7/L12 CTF.

Obviously the crystallization of complexes of ribosomal components will be a greater step towards the elucidation of ribosomal
tertiary structure. In one set of experiments it provides the
crystal structure of several components as well as the detailed
interaction. So far one pentameric protein complex has been crystallized from B. stearothermophilus ribosomes (Liljas and Newcomer,
1981). The proteins involved correspond to the E. coli proteins
L7/L12 and L10 which form a 4:1 complex in the ribosome and in
solution (Osterberg et al., 1977; Pettersson and Liljas, 1979;
Marquis and Fahnestock, 1980). Two precession photographs of these
crystals are shown in Fig. 6.

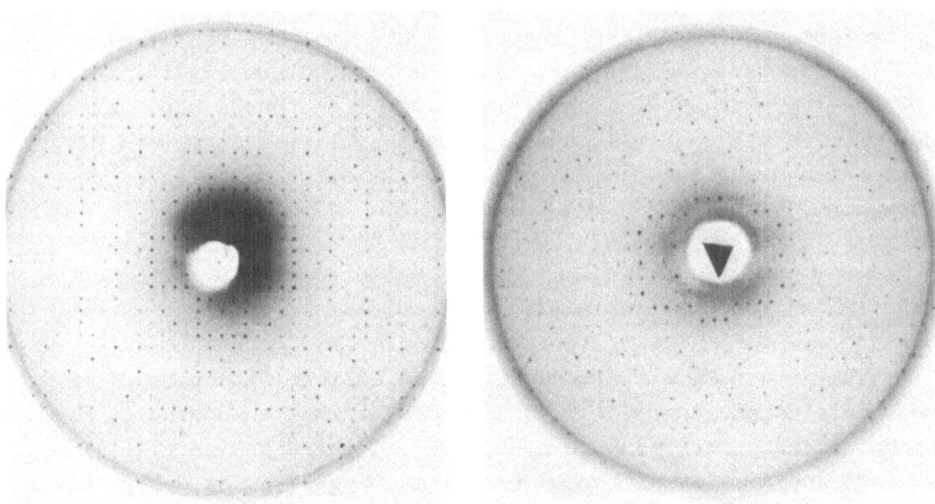

Fig. 5. 16° precession photographs of crystals of L7/L12 CTF with
the h-axis horizontal. A. The hk0 projection. B. The h0l
projection.

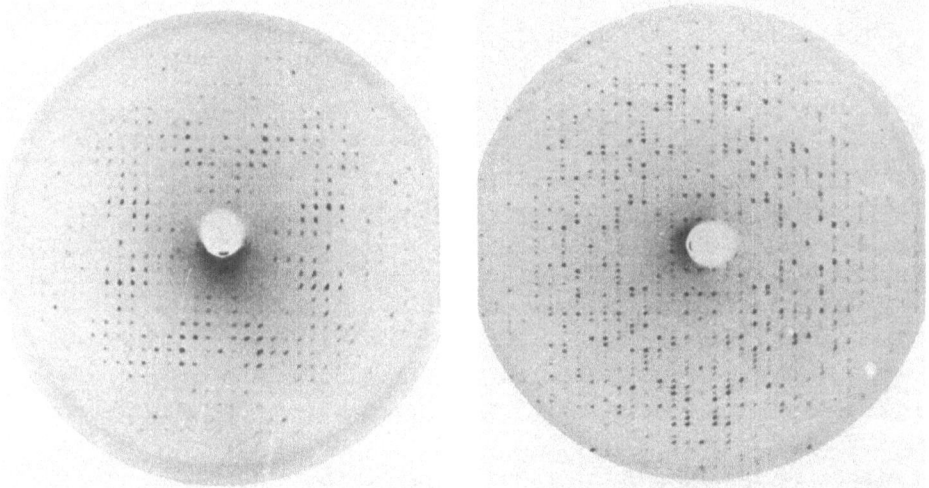

Fig. 6. 7° precession photographs of the protein complex (B-L13)$_4$·
(B-L8) from B. stearothermophilus. The complex corresponds
to (L7/L12)$_4$·(L10) in E. coli. The h-axis is horizontal.
A. The hk0 projection. B. the h0l projection.

Only one of the crystallized proteins has yet matured into a crystal structure. This is the L7/L12 CTF (Leijonmarck et al., 1980; Leijonmarck and Liljas, 1982). The structure was initially solved with phase angles determined from isomorphous heavy atom derivatives at 2.6 Å resolution. Subsequently the structure has been refined and the phase angles extended stepwise to 1.7 Å resolution.

The Structure of L7/L12

The structure of L7/L12 CTF has an ellipsoidal shape of approximate dimensions 35 Å x 25 Å x 20 Å. It has an unusually high amount of secondary structure: a layer of three antiparallel β - strands and a layer of three α -helices (Fig. 7). The design of this ribosomal component shows only features common to most soluble globular proteins. However, no similar arrangement of secondary structure elements has yet been found. The fragment represents a structure that must have considerable stability and be relatively independent of other ribosomal constituents. Despite the high amount of charged residues (about one third) this domain has a very hydrophobic interior. Almost all main chain oxygens and nitrogens are hydrogen bonded and all the side chains between the β -structure and the α -helix layer are aliphatic hydrocarbons.

This defined domain structure comprises residues 53-120, whereas the crystallized fragment contains an additional six residues starting from position 47. It has not been possible to locate these residues with any confidence in the electron density map. The reason is probably that this piece of chain is not an integral part of the chain is not an integral part of the domain, but forms a flexible link in the unfragmented protein to the N-terminal region.

The surface of CTF is rich in charged side chains with the exception of two regions which are hydrophobic. Such surface could be involved in protein-protein interactions. In fact one of these surfaces is involved in intramolecular contacts across a crystallographic two-fold axis. The fact that L7/L12 in the ribosome is present in four copies (Hardy, 1975; Subramanian, 1975), and normally occurs as dimers (Osterberg et al., 1976), makes this interaction interesting. It turns out that the main interaction across the twofold axis is a continuation of the β-sheet to a fourth short

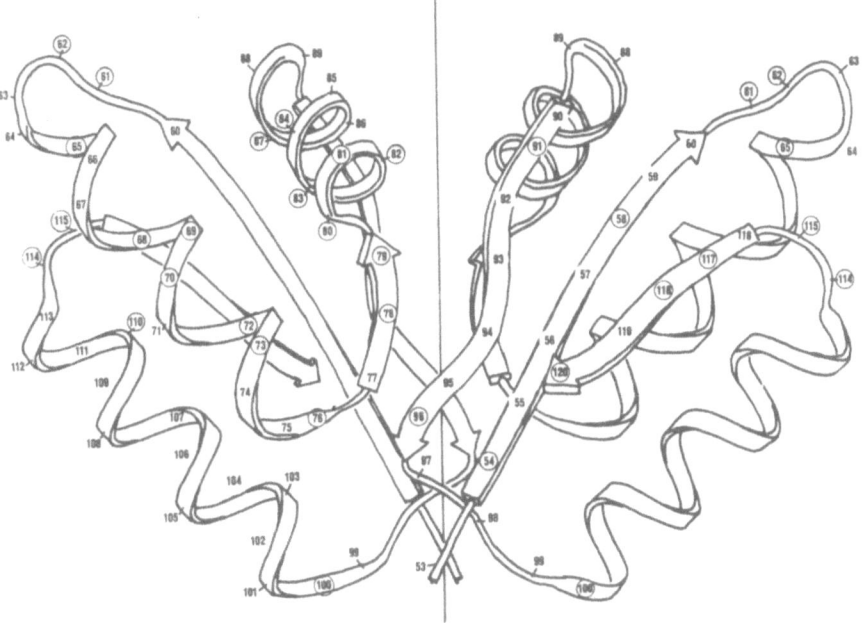

Fig. 7. The crystal structure of L7/L12 CTF. The spirals are α -
helices and the arrows represent the β -strands. The positions of the amino acid residues are indicated; the invariant ones in the five known amino acid sequences are en-
circled. The vertical line is a crystallographic two-fold
axis that might correspond to a molecular two-fold axis relating one CTF domain to another. The residues 77-79 form
a fourth strand as a continuation of the β -structure across
the dimer surface.

strand in the neighbouring molecule. Several oligomeric proteins
have such β-structures that extend from one subunit into another.
Some examples are concanavalin A (Reeke et al., 1975), prealbumin
(Blake et al., 1974) and alcohol dehydrogenase (Brändén et al.,
1973). The surface involved in the contact is of a size corresponding to what has been found for molecules known to exist as dimers
in solution (Leijonmarck and Liljas, 1982). Several of the residues
involved in the contact are highly conserved in the known sequences of L7/L12 (Fig. 8). Thus the circumstantial evidence suggests
that the dimer interaction in the CTF crystals is not merely an
effect of crystal packing, but an interaction of relevance for the

```
                    10        20        30        40
M.LYSO.   M N K E Q I L E A I K A M T V L E L N D L V K A I E E E F G V T A A A P V - V
B.STEARO.   . T . . . . I Q . V . N . . . . . . . . E . . . . . . . . . . . . . . . . .
C.PASTEUR   . S . . . . I Q . . . G . . . . . . . E . . . S . . . . . . . S . . . . .
B.SUBT.   A L . I . E . I A S V . E A . . . . . . . . . . . . . . . . . . . . . . A .
E.COLI   S I T . D . . I . . V A . . S . M D V V E . I S . M . . K . . . S . . . A . A .
V.COSTI.   S I T N . . . . D . . A . . S . M Q V V E . I E . M . . K . . . S . . . A .
NRCC 11227 A L T Q . D . I N . V A E . S . M . V A E . . S . M . . K . . . S . . . A . - .
A.GLACI.   A K L S T . D L . . Q F . G L . L I . . S E F . . . F . . T . E . G . . . . .
S.GRISE.   A K L S Q D D L . A Q F E E . . L I . . S E F . . . F . . K . D . . . A . A


                    50        60        70        80
M.LYSO.   A G G A A A - - - - A - - A E E K T E F D V V L A S A G A E K I K V I K V V R E I T G L G L
B.SUBT.   . . . . . . G G - - . - - . . . . - - . . . L I . . G . . S Q . . . . . . . . . . .
E.COLI   . A . P V - - - - E . - - . . . . . . . . I . K A . . . N . V A . . . A . . G A . . . . .
NRCC 1127 . - . P G - G G - E . E E . . . Q . . . N L . . T . . . E K . V N . . . . . . . . . . . .
S.GRISE.   . - . P . . G G A P . E E . . - Q D . . . . I . T G . . E K . . Q . . . . . . . . . . S . . . .


                    90        100       110       120
M.LYSO.   K E A K E V V D N A P K A - - L K E G V S K D E A E E I K A K L E E V G A S V E V K
B.SUBT.   . . . . . . L . . . T . . P L E V . . . I A . E . . . . L . . . . . . . . . . . .
E.COLI   . . . . . D L . E S A . A . - - . . . . . . . . D . A L . K A . . . A . . E . .
NRCC 11227 . . . . . A A . N G . . A T - - . . . . M . . Q N G N . A . T . . . . A . . . . L .
S.GRISE.   . . . . . D L . . G T . . P - - V L . K . A . E A . . K A A E S . K A A . . . . . . .
```

Fig. 8. A compilation of amino acid sequence data from the protein
       in bacterial ribosomes that corresponds to the protein
       L7/L12 in E. coli. The bulk of the data is taken from the
       review by Matheson et al. (1980). Additional data has been
       presented by Itoh (1981) and Itoh et al. (1981). The dots
       indicate residues identical to the ones in M. lysodeikti-
       cus, dashes represent deletions compared to one or several
       of the other sequences. The numbering used corresponds to
       the E. coli sequence (Terhorst et al., 1973) which has been
       known longest.

structure of L7/L12 in the ribosome. In such a case two possibilities
exist: the interaction observed can be relevant for the L7/L12
dimer or it can exist between C-terminal domains in different
dimers. The present information does not permit us to choose between these two possibilities.

    The L7/L12 dimer is strongly held together by interactions be-
tween the N-terminal regions (Gudkov and Behlke, 1978). The N-ter-
mini in addition bind to the protein L10 (van Agthoven et al.,

1975; Koteliansky et al., 1978; Gudkov et al., 1980). The C-terminal
region of L10 binds four copies of L7/L12, whereas the N-terminal
region of the same protein binds to the 23S RNA (Gudkov et al.,
1980). These findings can be summarized schematically (Fig. 9).
Here the L7/L12 molecules are oriented in a parallel fashion. This
is supported mainly by electron microscopic evidence (Marquis et
al., 1981; Tokimatsu et al., 1981). The elongated nature of both
the L7/L12 dimers and the pentameric complex $(L7/L12)_4 \cdot L10$ is supported both by X-ray scattering data (Osterberg et al., 1976;
1977), and data from electron microscopy (Boublik et al., 1976;
Strycharz et al., 1978; Kastner et al., 1981). The pentameric
protein domain composed of L7/L12 and L10 constitutes the thin stalk
protuberance extending from the 50S subunit. The C-terminal domains, whose structure has been determined, are located at the tip
of the stalk (Marquis et al., 1981). Other models for the arrangement of L7/L12 in the ribosome and in the pentameric complex have
been proposed (Boublik et al., 1976; Behlke and Gudkov, 1980).

Another feature of the L7/L12 CTF structure must be mentioned
at this point. A sulfate ion from the crystallization medium binds
at the other of the two hydrophobic surfaces of the domain (Leijonmarck et al., 1980). Lys-65 is the only charged residue in this
area and interacts with the sulfate. Binding of solvent ions in
this fashion has been observed to occur selectively at functional
sites (Arnone, 1972 ; Adams et al., 1973; Buehner et al., 1974;
Banner et al., 1975; Steitz et al., 1977). The L7/L12 residues involved in sulfate binding are invariant in the five sequences available. The hydrophobic surface with the bound ion thus has the
characteristics of a functional site. It is then somewhat surprising
that the suggested arrangement of L7/L12 in the ribosome places
this potential functional surface at the very end of the stalk
protuberance facing away from the rest of the ribosome.

The Binding Site for Elongation Factors

Several approaches have been tried in the effort to localize
the binding sites for molecules that interact with the ribosome,
such as tRNA:s and various factors. One striking observation is
that both subunits seem to be involved in most of these interactions.
Accordingly the binding sites for many of the transiently bound
molecules must be located at the subunit interface. This is true

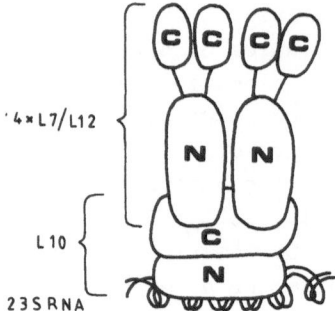

Fig. 9. One possible model for the $(L7/L12)_4 \cdot L10$ complex represented
         schematically. The dimers of L7/L12 are primarily held to-
         gether by their N-termini. As discussed in the text and by
         Liljas (1982), the L7/L12 molecules are probably oriented
         in a parallel fashion. Other models have been proposed
         (Behlke and Gudkov, 1980). The detailed involvement of
         L10 was determined by Gudkov et al. (1980).

for EF-Tu and EF-G, which furthermore require L7/L12 for their
proper function (see Matheson et al., 1980 and Weissbach, 1980, for
reviews). There is much evidence to suggest that these two factors
bind at partly overlapping sites on the ribosome (Liljas, 1982).
One interaction with the ribosome common to both these factors is
the induction of GTP hydrolysis.

Among those proteins observed to be near the binding sites
for elongation factors, two deserve particular attention: L11 and
L7/L12. L11 has been observed to crosslink to EF-G along with several other proteins (Maassen and Möller, 1981). It can also be
labelled by affinity probes attached to the guanine nucleotide bound
to EF-G (Maassen and Möller, 1974; 1978). Furthermore, the removal
of L11 seems to reduce the ribosomal binding as well as the GTPase
activity of EF-G (Highland and Howard, 1975; Schrier and Möller,
1975). However, mutants entirely lacking L11 are viable, but these
types of mutants in some bacterial species have reduced EF-G GTPase
activity (Stark and Cundliffe, 1979; Stöffler et al., 1980a; 1980b).
This finding argues against a fundamental role for L11.

L11 is in the vicinity of L7/L12 as observed by crosslinking
(Expert-Bezançon et al., 1976; Kenny and Traut, 1979) and binding

studied (Highland and Howard, 1975; Dijk et al., 1979; Behlke and
Gudkov, 1980). L11 seems to stimulate the binding of the
$(L7/L12)_4 \cdot L10$ complex to the 23S RNA.

The removal of L7/L12 from the ribosome has severe effects on
the binding of elongation factors and GTP hydrolysis (Hamel et al.,
1972; Highland and Howard, 1975; Schrier and Möller, 1975). These
observations were verified by Koteliansky et al. (1977) and Glick
(1977), but these authors concluded that since protein synthesis
was not dramatically reduced L7/L12 could not be essential. However,
it turns out that when the rate of in vitro protein synthesis is
optimized by using a buffer with a composition similar to the intracellular one and a system to regenerate the GTP (Jelenc and
Kurland, 1979), the removal of L7/L12 has dramatic consequences
(Pettersson and Kurland, 1980): the rate and fidelity of protein
synthesis are severely reduced. These results strongly indicate
that in vitro protein synthesis experiments can be very misleading
if they are not performed at conditions that approach the situation
in vivo (Kurland, 1982). Furthermore, one can conclude that L7/L12
is an essential component of the ribosome.

The elongation factor activity that is most affected by removal of L7/L12 is the GTPase activity. L7/L12 could be part of the
ribosomal system that triggers the GTP hydrolysis of the factors.
Even if there is very little information we may assume that there
is a direct interaction between L7/L12 and the factors. The lack
of affinity labelling of L7/L12 from the bound GTP might indicate
that this protein does not interact with the GTP end of the elongation factor (Fig. 1). We have, however, learned from the kirromycin resistant mutants of EF-Tu (see above) that GTPase activity
can be induced from regions far from the bound nucleotide (Duisterwinkel et al., 1981). Thus an interaction between L7/L12 and elongation factors at some region distant from the nucleotide binding
site could induce the domain rearrangement of the factor that is
required for GTP hydrolysis. There is yet no information about where
such a region could be on EF-Tu.

Some indications have been obtained for which region of L7/L12
can interact with the elongation factors. The modification of Arg-73
leads to reduced binding and GTP hydrolysis for EF-G (Koteliansky
et al., 1978). The modification of two unidentified lysines of
L7/L12 totally blocks EF-G GTPase and poly-Phe synthesis (Lee et

al., 1981). Modification of Lys-51, Lys-59 (Maassen et al., 1981) and Lys-120 (Lee et al., 1981) has no effect on ribosomal functions. Even though there are several invariant lysines (Fig. 8) which possibly could be the essential ones, Lys-65 with the bound sulfate deserves special attention. If this part of L7/L12 is directly involved with the elongation factors when bound to the ribosome, significant conformational changes are clearly required to bring it from its location at the tip of the stalk protuberance to cooperate with factors bound at the subunit interface. The existence of such conformational changes as well as a detailed analysis of the interaction of L7/L12 with EF-Tu and EF-G would greatly advance our understanding of one of the intricate interactions between ribosomes and transiently bound soluble proteins during translation of the genetic material into protein.

CONCLUDING REMARKS

The crystallographic studies of the protein biosynthesis system have just begun. Nevertheless the results can already be used for attempts to make models of the functional interactions in the ribosome. Clearly these models can stimulate specific biochemical experiments for further elaboration or disproval of the models. Similarly, it is obvious that the crystallographic investigations have to include a number of molecular conformations in order to provide an understanding of the functional interactions. Efforts along these lines will give us more detailed and more definite pictures of this system than most other methods can give. Even if the resolution obtained is lower than many crystallographers would want the present results show that this field is approachable and should become quite rewarding.

ACKNOWLEDGEMENTS

We are grateful to Prof. C.G. Kurland, Drs.A. Jones and M. Newcomer for stimulating interactions and assistance. Prof. B.F.C. Clark and Drs.J. Nyborg and L. LaCour are thanked for permission to use unpublished material. We have appreciated the art-work by Mr. S. Lövgren.

These studies have been supported by the Swedish Natural Science Research Council, The Knut and Alice Wallenberg Foundation and Magnus Bergwall´s Foundation.

REFERENCES

Adams, M.J., Liljas, A., and Rossmann, M.G., 1973, Functional anion binding sites in dogfish M_4 lactate dehydrogenase, J. Mol. Biol., 76:519.

Alakhov, Yu., B. and Ovchinnikov, Yu. A., 1981, Study of the structure and structural - functional properties of the elongation factor G from E. coli, Poster abstract from Seventh EMBO Annual Symposium.

Appelt, K., Dijk, J., and Epp, D., 1979, The crystallization of protein BL17 from the 50S ribosomal subunit of Bacillus stearothermophilus, FEBS Letters, 103:66.

Appelt, K., Dijk, J., Reinhardt, R., Sanhuesa, S., White, S.W., Wilson, K.S., and Yonath, A., 1981, The crystallization of ribosomal proteins from the 50S subunit of the Escherichia coli and Bacillus stearothermophilus ribosome, J. Biol. Chem., 256:11787.

Arnone, A., 1972, X-ray diffraction study of binding of 2,3-diphosphoglycerate to human deoxyhaemoglobin, Nature, 237:146.

Banner, D.W., Bloomer, A.C., Petsko, G.A., Phillips, D.C., Pogson, C.I., Wilson, I.A., Corran, P.H., Furth, A.J., Milman, J.D., Offord, R.E., Priddle, J.D., and Waley, S.G., 1975, Structure of chicken muscle triose phosphate isomerase determined crystallographically at 2.5 Å resolution using amino acid sequence data, Nature, 255:609.

Behlke, J., and Gudkov, A.T., 1980, Interaction between the ribosomal proteins L7/L12 and L10 as L7/L12-L10 and L11 from Escherichia coli, Studia Biophysica, 81:169.

Blake, C.C.F., Geisow, M.J., Swan, I.D.A., Rerat, C., and Rerat, B., 1974, Structure of human plasma prealbumin at 2.5 Å resolution. A preliminary report on the polypeptide chain conformation, quaternary structure and thyroxine binding, J. Mol. Biol., 88:1.

Boublik, M., Hellmann, W., and Kleinschmidt, A.K., 1977, Size and structure of E. coli ribosomes by electron microscopy, Cytobiologie, 14:293.

Boublik, M., Hellmann, W., and Roth, H.E., 1976, Localization of ribosomal protein L7/L12 in the 50S subunit of Escherichia coli ribosome by electron microscopy, J. Mol. Biol., 107:479.

Brändén, C.-I., Eklund, H., Nordström, B., Boiwe, T., Söderlund, G., Zeppezauer, E., Ohlsson, I., and Åkesson, Å., 1973, Structure of liver alcohol dehydrogenase at 2.9 Å resolution, Proc. Nat. Acad. Sci. USA, 70:293.

Buehner, M., Ford, G.C., Moras, D., Olsen, K.W., and Rossmann, M.G., 1974, Three-dimensional structure of D-glyceraldehyde-3 phosphate dehydrogenase, J. Mol. Biol., 90:25.

Byers, B., 1971, Chick embryo ribosome crystals: analysis of bonding and functional activity in vitro, Proc. Nat. Acad. Sci. USA, 68:440.

Clark,M.W., Hammons, M., Langer, M., and Lake, J.A., 1979, Helical arrays of Escherichia coli small ribosomal subunits produced in vitro, J. Mol. Biol., 135:507.

Crowther, R.A., and Klug, A., 1975, Structural analysis of macromolecular assemblies by image reconstruction from electron micrographs, Ann. Rev. Biochem., 44:161.

Dabbs, E.R., 1977, A spectomycin dependent mutant of Escherichia coli, Mol. Gen. Genet., 151:261.

Dabbs, E.R., 1979, Selection for Escherichia coli mutants with proteins missing from the ribosome, J. Bact., 140:734.

Dabbs, E.R., Ehrlich, R., Hasenbank, R., Schroeter, B.-H., Stöffler-Meilicke, M., and Stöffler, G., 1981, Mutants of Escherichia coli lacking ribosomal protein L1, J. Mol. Biol., 149:553.

Dijk, J., Garrett, R.A., and Müller, R., 1979, Studies on the binding of the ribosomal protein complex L7/L12-L10 and protein L11 to the 5´-one third of 23S RNA: a functional centre of the 50S subunit, Nucl. Acids Res., 6:2717.

Dijk, J. and Littlechild, J., 1979, Purification of ribosomal proteins from Escherichia coli under nondenaturing conditions, Methods in Enzymology, 59:481.

Duffy, L.K., Gerber, L., Johnson, A.E., and Miller, D.L., 1981, Identification of a histidine residue near the aminoacyl transfer ribonucleic acid binding site of elongation factor Tu, Biochemistry, 20:4663.

Duisterwinkel, F.J., De Graaf, J.M., Kraal, B., and Bosch, L., 1981, A kirromycin resistant elongation factor EF-Tu from Escherichia coli contains a threonine instead of an alanine residue in position 375, FEBS Lett., 131:89.

Expert-Bezançon, A., Barritault, D., Milet, M., and Hayes, D.H., 1976, Close proximity of Escherichia coli 50S subunit proteins L7/L12 and L10 and L11, J. Mol. Biol., 108:781.

Fahnestock, S.R., Strycharz, W.A., and Marquis, D.M., 1981, Immunochemical evidence of homologies among 50S ribosomal proteins of Bacillus stearothermophilus and Escherichia coli, J. Biol. Chem., 256:10111.

Garrett, R.A., Douthwaite, S., and Noller, H.F., 1981, Structure and role of 5S RNA-protein complexes in protein biosynthesis, TIBS, 6:137.

Gast, W.H., Kabsch, W., Wittinghofer, A., and Lebermann, R., 1977, Crystals of a large tryptic peptide (fragment A) of elongation factor EF-Tu from Escherichia coli, FEBS Letters, 74:88.

Gast, W.H., Lebermann, R., Schulz, G.E., and Wittinghofer, A., 1976, Crystals of partially trypsin-digested elongation factor Tu, J. Mol. Biol., 106:943.

Glick, B.R., 1977, The role of Escherichia coli ribosomal proteins L7 and L12 in peptide chain propagation, FEBS Lett., 73:1.

Gudkov, A.T. and Behlke, J., 1978, The N-terminal sequence protein of L7/L12 is responsible for its dimerization, Eur. J. Biochem., 90:309.

Gudkov, A.T., Tumanova, L.G., Gongadze, G.M., and Bushnev, V.N., 1980, Role of different regions of ribosomal proteins L7 and L10 in their complex formation and in the interaction with the ribosomal 50S subunit, FEBS Lett., 109:34.

Hamel, E., Koka, M., and Nakamoto, T., 1972, Requirement of an Escherichia coli 50S ribosomal protein component for effective interaction of the ribosome with T and G factors and with guanosine triphosphate, J. Biol. Chem., 247:805.

Hardy, S.J.S., 1975, The stoichiometry of the ribosomal proteins of Escherichia coli, Mol. Gen. Genet., 140:253.

Henderson, R., and Unwin, P.N.T., 1975, Three-dimensional model of purple membrane obtained by electron microscopy, Nature, 257:28.

Highland, J.H., and Howard, G.A., 1975, Assembly of ribosomal proteins L7, L10, L11 and L12 on the 50S subunit of Escherichia coli, J. Biol. Chem., 250:831.

Itoh, T., 1981, Primary structure of an acidic ribosomal protein from Micrococcus lysodeikticus, FEBS Lett., 127:67.

Itoh, T., Sugiyama, M., and Higo, K., 1981, The primary structure of an acidic ribosomal protein from Streptomyces griseus, Biochim. Biophys. Acta, in press.

Ivell, R., Fasano, O., Crechet, J.B., and Parmeggiani, A., 1981, Characterization of a kirromycin-resistant elongation factor Tu from Escherichia coli, Biochemistry, 20:1355.

Jelenc, P.C., 1980, Rapid purification of highly active ribosomes from Escherichia coli, Anal. Biochem., 105:369.

Jelenc, P.C., and Kurland, C.G., 1979, Nucleotide triphosphate regeneration decreases the frequency of translation errors, Proc. Nat. Acad. Sci. USA, 76:3174.

Jonák, J., Rychlík, I., Smrt, J., and Holý, A., 1979, The binding site for the 3´-terminus of aminoacyl-tRNA in the molecule of elongation factor Tu from Escherichia coli, FEBS Lett., 98:329.

Jones, M.D., Petersen, T.E., Nielsen, K.M., Magnusson, S., Sottrup-Jensen, L., Gausing, K., and Clark, B.F.C., 1980, The complete amino-acid sequence of elongation factor Tu from Escherichia coli, Eur. J. Biochem., 108:507.

Jurnak, F., McPherson, A., Wang, A.H.J., Rich, A., 1980, Biochemical and structural studies of the tetragonal crystalline modification of the Escherichia coli elongation factor Tu, J. Biol. Chem., 255:6751.

Jurnak, F., Rich, A., and Miller, D., 1977, Preliminary X-ray diffraction data for tetragonal crystals of trypsinized Escherichia coli elongation factor, J. Mol. Biol., 115:103.

Kabsch, W., Gast, W.H., Schulz, G.E., Lebermann, R., 1977, Low resolution structure of partially trypsin-degraded polypeptide elongation factor, EF-Tu, from Escherichia coli, J. Mol. Biol., 117:999.

Kahan, L., Winkelmann, D.A., and Lake, J.A., 1981, Ribosomal proteins S3, S6, S8 and S10 of Escherichia coli localized on the external surface of the small subunit by immune electronmicroscopy, J. Mol. Biol., 145:193.

Kastner, B., Stöffler-Meilicke, M., and Stöffler, G., 1981, Arrangement of the subunits in the ribosome of Escherichia coli: demonstration by immunoelectron microscopy, Proc. Nat. Acad. Sci. USA, 78:6652.

Kaziro, Y., 1978, The role of guanosine 5´-triphosphate in polypeptide chain elongation, Biochim. Biophys. Acta, 505:95.

Kenny, J.W., and Traut, R.R., 1979, Identification of fifteen neighbouring protein pairs in the Escherichia coli 50S ribosomal subunit crosslinked with 2-iminothiolane, J. Mol. Biol., 127:243.

Kingsbury, E.W., and Voelz, H., 1969, Induction of helical arrays of ribosomes by vinblastine sulfate in Escherichia coli, Science, 165-768.

Koteliansky, V.E., Domogatsky, S.P., and Gudkov, A.T., 1978,

Dimer state of protein L7/L12 and EF-G dependent reactions on ribosomes, Eur. J. Biochem., 90:319.

Koteliansky, V.E., Domogatsky, S.P., Gudkov, A.T., and Spirin, A.S., 1977, Elongation factor-dependent reactions on ribosomes deprived of proteins L7 and L12, FEBS Lett., 73:6.

Kurland, C.G., 1982, Translational accuracy in vitro, in press.

Lake, J.A., 1976, Ribosome structure determined by electron microscopy of Escherichia coli small subunits, large subunits and monomeric ribosomes, J. Mol. Biol., 105:131.

Lake, J.A., 1978, Electron microscopy of specific proteins: three-dimensional mapping of ribosomal proteins using antibody labels, in: "Advanced Techniques in Biological Electron Microscopy II", J.K. Koehler, ed., Springer Verlag, Berlin, Heidelberg, p. 173.

Lake, J.A., 1981, The ribosome, Scientific American, 245:56.

Lake, J.A., Nonomura, Y., and Sabatini, D.D., 1974, Ribosome structure as studied by electron microscopy, in: "Ribosomes", M. Nomura, A. Tissiere, P. Lengyel, eds., Cold Spring Harbor Laboratory, Long Island, N.Y., p. 543.

Lake, J.A., and Slayter, H.S., 1972, Three-dimensional structure of the chromatoid body helix of Entamoeba invadens, J. Mol. Biol., 66:271.

Laursen, R.A., L'Italien, J.J., Nagarkatti, S., and Miller, D.L., 1981, The amino acid sequence of elongation factor Tu of Escherichia coli. The complete sequence, J. Biol. Chem., 256:8102.

Labermann, R., Schulz, G.E., and Suck, D., 1981, Crystallization and preliminary X-ray diffraction data of the EF-Tu·EF-Ts (EF-T) complex of Escherichia coli, FEBS Lett., 124:279.

Lebermann, R., Wittinghofer, A., and Schulz, G.E., 1976, Polymorphism in crystalline elongation factor Tu·GDP from Escherichia coli, J. Mol. Biol., 106:951.

Lee, C.C., Cantor, C.R., and Wittmann-Liebold, B., 1981, The number of copies of ribosome-bound proteins L7 and L12 required for protein synthesis activity, J. Biol. Chem., 256:41.

Leijonmarck, M., Eriksson, S., and Liljas, A., 1980, Crystal structure of a ribosomal component at 2.6 Å resolution, Nature, 286:824.

Leijonmarck, M., and Liljas, A., 1982, Manuscript in preparation.

Liljas, A., 1982, Structural studies of ribosome, Progr. Biophys. Mol. Biol., in press.

Liljas, A., Eriksson, S., Donner, D., and Kurland, C.G., 1978, Iso-

lation and crystallization of stable domains of the protein
L7/L12 from Escherichia coli ribosomes, FEBS Letters, 88:300.

Liljas, A., and Kurland, C.G., 1976, Crystallization of ribosomal
protein L7/L12 from Escherichia coli, FEBS Lett., 71:130.

Liljas, A., and Newcomer, M.E., 1981, Purification and crystallization of a protein complex from Bacillus stearothermophilus
ribosomes, J. Mol. Biol., 153, in press.

Maassen, J.A. and Möller, W., 1974, Identification by photo-af-
finity labelling of the proteins in Escherichia coli ribosomes involved in elongation factor G-dependent GDP binding,
Proc. Nat. Acat. Sci. USA, 71:1277.

Maassen, J.A., and Möller, W., 1978, Elongation factor G-dependent
binding of a photoreactive GTP analogue of Escherichia coli
ribosomes result in labelling of protein L11, J. Biol. Chem.,
253:2777.

Maassen, J.A., and Möller, W., 1981, Photochemical cross-linking
of elongation factor G to 70S ribosomes from Escherichia coli
by 4-(6-formyl-3-azidophenoxy)-butyrimidate, Eur. J. Biochem.,
115:279.

Maassen, J.A., Shop, E.N., and Möller, W., 1981, Structural analysis
of ribosomal protein L7/L12 by the heterobifunctional cross-
linker 4-(6-formyl-3-azidophenoxy)-butyrimidate, Biochemistry,
20:1020.

Marquis, D.M., and Fahnestock, S.R., 1980, Stoichiometry and structure of a complex of acidic ribosomal proteins, J. Mol. Biol.,
142:161.

Marquis, D.M., Fahnestock, S.R., Henderson, E., Woo, D., Schwinge,
S., Clark, M.W., and Lake, J.A., 1981, The L7/L12 stalk, a
conserved feature of the procaryotic ribosome, is attached
to the large subunit through its N-terminus, J. Mol. Biol.,
150:121.

Matheson, A.T., Möller, W., Amons, R., and Yaguchi, M., 1980, Com-
parative studies on the structure of ribosomal proteins, with
emphasis on the alanine-rich acidic ribosomal, "A" protein,
in: "Ribosome Structure, Function and Genetics", G. Chambliss,
G.R. Craven, J. Davies, K. Davis, L. Kahan, and M. Nomura, eds.,
University Park Press, Baltimore, p. 297.

McPherson, A., Jr., 1976, The growth and preliminary investigation
of protein and nucleic acid crystals for X-ray diffraction
analysis, Methods of Biochemical Analysis, 23:249.

Morikava, K., La Cour, T.F.M., Nyborg, J., Rasmussen, K.M., Miller,
D.L., and Clark, B.F.C., 1978, High resolution X-ray crystal-

lographic analysis of a modified form of the elongation factor Tu: Guanosine Diphosphate complex, J. Mol. Biol., 125:325.

O'Brien, L., Shelly, K., Towfighi, J., and McPherson, A., 1980, Crystalline ribosomes are present in brains from senile humans, Proc. Nat. Acad. Sci. USA, 77:2260.

Osterberg, R., Sjöberg, B., Liljas, A., and Pettersson, I., 1976, Small-angle X-ray scattering and crosslinking study of the proteins L7/L12 from Escherichia coli ribosomes, FEBS Lett., 66:48.

Osterberg, R., Sjöberg, B., Pettersson, I., Liljas, A., and Kurland, C.G., 1977, Small-angle X-ray scattering study of the protein complex of L7/L12 and L10 from Escherichia coli ribosomes, FEBS Lett., 73:22.

Pettersson, I., and Kurland, C.G., 1980, Ribosomal protein L7/L12 is required for optimal translation, Proc. Nat. Acad. Sci. USA, 77:4007.

Pettersson, I., and Liljas, A., 1979, The stoichiometry and reconstitution of a stable protein complex from Escherichia coli ribosomes, FEBS Lett., 98:139.

Ramakrishnan, V.R., Yabuki, S., Sillers, I.-Y., Schindler, D.G., Engelman, D.M., and Moore, P.B., 1981, On the position of S6, S11 and S15 in the 30S ribosomal subunit of E. coli, J. Mol. Biol., in press.

Reeke, G.N., Becker, J.W., and Edelman, G.M., 1975, The covalent and threedimensional structure of concanavalin A. IV. Atomic coordinates, hydrogen bonding and quaternary structure, J. Biol. Chem., 250:1525.

Rossmann, M.G., and Argos, P., 1981, Protein folding, Ann. Rev. Biochem., 50:497.

Rossmann, M.G., Liljas, A., Brändén, C.I., and Banaszak, L.J., 1975, Evolutionary and structural relationships among dehydrogenases, in: "The Enzymes", P.D. Boyer, ed., Academic Press, London, LL:61.

Rubin, J.R., Morikawa, K., Nyborg, J., La Cour, T.F.M., Clark, B. F.C., and Miller, D.L., 1981, Structural features of the GDP binding site of elongation factor Tu from Escherichia coli as determined by X-ray diffraction, FEBS Lett., 129:177.

Schrier, P.I., and Möller, W., 1975, The involvement of 50S ribosomal protein L11 in the EF-G dependent GTP hydrolysis of E. coli ribosomes, FEBS Lett., 54:130.

Shatsky, I.N., Estafieva, A.G., Bystrova, T.F., Bogdanov, A.A., and Vasiliev, V.D., 1980, Topography of RNA in the ribosome:

Location of the 3´-end of 5S RNA on the central protuberance of the 50S subunit, FEBS Lett., 121:97.

Sneden, D., Miller, D.L., Kim, S.H., and Rich, A., 1973, Preliminary X-ray analysis of the crystalline complex between polypeptide chain elongation factor Tu, and GDP, Nature, 241:530.

Stark, M.J.R. and Cundliffe, E., 1979, On the biological role of ribosomal protein BM-L11 of Bacillus megaterium homologous with Escherichia coli ribosomal protein L11, J. Mol. Biol., 134:767.

Stöffler, G., Bald, R., Kastner, B., Lührmann, R., Stöffler-Meilicke, M., and Tischendorf, G., 1980a, Structural organization of the Escherichia coli ribosome and localization of functional domain, in: "Ribosomes. Structure, Function and Genetics", G. Chambliss, G.R. Craven, J. Davies, K. Davis, L. Kahan, and M. Nomura, eds., University Park Press, Baltimore.

Stöffler, G., Cundliffe, E., Stöffler-Meilicke, M., and Dabbs, E. R., 1980b, Mutants of Escherichia coli lacking ribosomal protein L11, J. Biol. Chem., 255:10517.

Stöffler, G., and Stöffler-Meilicke, M., 1981, Structural organization of the Escherichia coli ribosomes and localization of functional domains, in: "International Cell Biology 1980-1981", H.G. Schweiger, ed., Springer, New York, p. 93.

Steitz, T.A., Anderson, W.F., Fletterick, R.J., and Anderson, C.M., 1977, High resolution crystal structures of yeast hexokinase complexes with substrates, activators and inhibitors. Evidence for an allosteric control site, J. Biol. Chem., 252:4494.

Strycharz, W.A., Nomura, M., and Lake, J.A., 1978, Ribosomal proteins L7/L12 localized at a single region of the large subunit by immune electron microscopy, J. Mol. Biol., 126:123.

Subramanian, A.R., 1975, Copies of proteins L7 and L12 and heterogeneity of the large subunit of Escherichia coli ribosome, J. Mol. Biol., 95:1.

Taddei, C., 1972, Ribosome arrangement during oogenesis of Lacerta sicula Raf., Exp. Cell. Res., 70:285.

Terhorst, C., Möller, W., Laursen, R., Wittmann-Liebold, B., 1973, The primary structure of an acidic protein from 50S ribosomes of Escherichia coli which is involved in GTP hydrolysis dependent on elongation factors G and T, Eur. J. Biochem., 34:138.

Tischendorf, G.W., Zeichhardt, H., and Stöffler, G., 1975, Architecture of the Escherichia coli ribosome as determined by immune electron microscopy, Proc. Nat. Acad. Sci. USA, 72:4820.

Tokimatsu, H., Strycharz, W.A., and Dahlberg, A.E., 1981, Gel electrophoretic studies on ribosomal proteins L7/L12 and the Escherichia coli 50S subunit, J. Mol. Biol., 152:397.

Unwin, N., 1977, Three-dimensional model of membrane-bound ribosomes obtained by electron microscopy, Nature, 269:118.

Unwin, N., 1979, Attachment of ribosome crystal to intracellular membranes, J. Mol. Biol., 132:69.

Van Agthoven, A.J., Maassen, J.A., Schrier, P.I., and Möller, W., 1975, Inhibition of EF-G dependent GTPase by an aminoterminal fragment of L7/L12, Biochem. Biophys. Res. Commun., 64:1184.

Weissbach, H., 1980, Soluble factors in protein synthesis, in: "Ribosomes. Structure, Function and Genetics", G. Chambliss, G.R. Craven, J. Davies, K. Davis, L. Kahan, and M. Nomura, eds., University Park Press, Baltimore, p. 377.

Wolf, H., Chinali, G., and Parmeggiani, A., 1974, Kirromycin, an inhibition of protein biosynthesis that acts on elongation factor Tu, Proc. Nat. Acad. Sci. USA, 71:4910.

Yonath, A.E., Müssig, J., Tesche, B., Lorenz, S., Erdmann, V.A., and Wittmann, H.G., 1980, Crystallization of the large ribosomal subunits from Bacillus stearothermophilus, Biochem. Internat., 1:428.

HEMŐGLOBIN OXYGEN BINDING, ERYTHROCYTE SHAPE TRANSFORMATIONS, AND
MODELING OF CELL DIFFERENTIATION AS EXAMPLES OF THEORETICAL APPROACHES
IN STUDYING THE STRUCTURE-FUNCTION RELATIONSHIP IN BIOLOGICAL SYSTEMS

Saša Svetina

Institute of Biophysics
Medical Faculty and "J. Stefan" Institute
E. Kardelj University
Ljubljana, Yugoslavia

GENERAL INTRODUCTION

Any attempt to express functional behavior of a biological
system in terms of the physical and chemical properties of its constituents must involve relating together different levels of complexity of a system´s structure. At different levels of the system,
the structures are described by different observable quantities determinable by different experimental approaches. The understanding
of the macroscopic behavior of a certain structure, including its
function, on the basis of its microscopic properties can be in
many instances achieved simply by correlating the observations obtained at different levels of study. However, in many other instances such an understanding is possible only on the basis of
quantitative relationships which are the result of an appropriate
theory. Theoretical approaches in biology play the same role as
theoretical approaches in physics and chemistry, such as statistical
mechanics where the thermodynamic properties of the system are
determined on the basis of properties of atoms and molecules and
their interactions, or quantum chemistry where properties of atoms
and molecules are determined from interactions between the constituents of the atom. Biological systems comprise a large number
of different levels of organization, from atoms and molecules to
macromolecules, solutions, supramolecular structures, cells, cell
populations, tissues, organs, organisms, etc. Relating system pro-

perties at two such levels must take into consideration specific
features in the description of a system at a given level and it is
expected that the corresponding theory reflects these features.
Therefore many different theoretical approaches need to be developed
in order to cover all possible relationships. These lectures are
aimed at giving an example of three different theoretical approaches, each relating a different pair of levels of structural
complexity in biological system. The relations between the macromolecular structure and the thermodynamic behavior of the system
of a macromolecule and its ligands will be dealt with in the
lecture on the relationship between hemoglobin structure and its
ligand binding behavior. Relationship between the cell shape and
the properties of the cell membrane will be discussed in the lecture
on interaction of charged membranes with surrounding ions and ion
induced shape and volume changes of red blood cell ghosts. The
properties of the cell population will be related to the properties
of a chemical system at the level of a single cell in the lecture
on modeling the molecular basis of cell differentiation. Although
the topics discussed in these lectures are all connected with the
system of oxygen delivery in vertebrate organisms, they represent
rather different aspects of this system and therefore the material
is organized in such a way that each lecture is contained in a
separate chapter. Chapters are independent and selfsufficient as
regards the nomenclature. References are also listed separately for
each chapter.

1. THE RELATIONSHIP BETWEEN HEMOGLOBIN STRUCTURE
   AND ITS LIGAND BINDING BEHAVIOR

Introduction

      The problem treated in this lecture concerns the possibility
of relating the thermodynamic behavior of the system of hemoglobin
and its ligands to the structure of the hemoglobin molecule in a
quantitative manner. In the first part of the lecture the basic
information about the system will be presented. Then the necessary
theoretical background will be introduced by showing the thermodynamic formalism appropriate to the description of hemoglobin
ligands binding curves. In a separate section a critical account
will be given on models which have been introduced to relate structure and function in hemoglobin. Finally, it will be shown how the

generalized Adair equation can be used to interrelate in quantitative terms the oxygen binding curves of a class of chemically
modified and mutant hemoglobins.

Basic Information About the System of Hemoglobin and its Ligands

Hemoglobin (Antonini and Brunori, 1971; Baldwin, 1975) is a
protein whose primary function is to transport oxygen from the
lungs to tissues and also to facilitate the transport of carbon
dioxide from tissues to the lungs. It is a tetramer composed of two
α and two β subunits. Each of these four subunits can bind one
oxygen molecule. The macroscopic property of the hemoglobin-oxygen
system which can be related to the physiological function of hemoglobin is the oxygen binding curve. In Fig. 1.1a the average
fraction of occupied oxygen binding sites (y) as a function of
partial oxygen pressure (p) is shown for a tetramer hemoglobin as
well as for a separated subunit. In contrast to the hyperbolic
binding curve in the case of a monomer the oxygen binding curve in
the case of a tetramer is of a sigmoid shape. The sigmoid shape
reflects the cooperativity in the binding of the four oxygen molecules to the hemoglobin molecule: the more oxygen molecules influence each others´ binding the steeper the curve. An appropriate
way to visualize this steepness is the Hill plot in which the
logarithm of the ratio of occupied to free oxygen binding sites is
plotted against the logarithm of partial oxygen pressure (Fig.
1.1b). The value of the slope at the straight middle part of the
curve obtained is called the Hill parameter (n_H). The Hill parameter can not exceed the number of ligand binding sites on the
molecule. Therefore in the case of hemoglobin it must be less than
4. If there were no cooperativity, the value of the Hill parameter
would be 1. The actual value of the Hill parameter for normal human
hemoglobin (HbA) is about 3. In evaluating physiological behavior
it is usually sufficient to characterize the corresponding oxygen
binding curves by the values of partial oxygen pressure at which
half of the binding sites are occupied ($p_{1/2}$) and by the values of the Hill
parameter. Values of these parameters obtained for some chemically modified and some mutant hemoglobins are presented as an example in Table 1.

Hemoglobin is a regulatory protein which can be seen from the
fact that the oxygen binding curve can be changed due to some effectors. Two examples will be presented. The first is the shift of

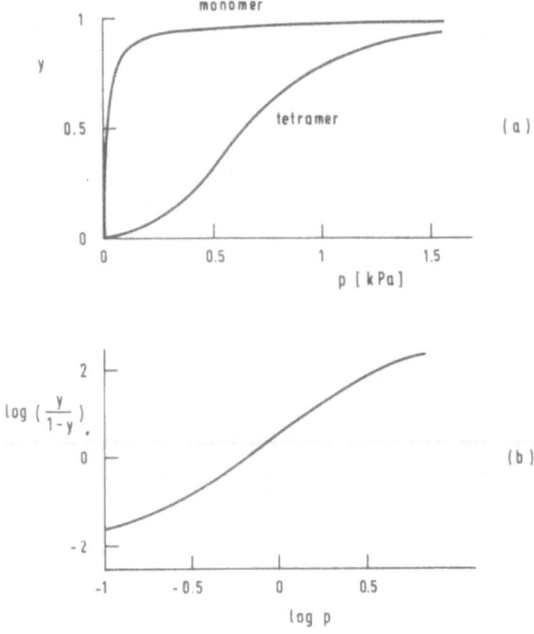

Fig. 1.1. (a) Fraction of occupied oxygen binding sites in hemoglobin and in a separated subunit as a function of partial
oxygen pressure. The hemoglobin oxygen binding curve is
calculated from the Adair Eq. (1.5) with the values of
the parameters $S_1 = 0.22$ kPa^{-1}, $S_2 = 0.10$ kPa^{-2}, $S_3 =$
0.12 kPa^{-3} and $S_4 = 6.69$ kPa^{-4} calculated from the data
given by Imai and Yonetani (1975) for HbA at pH 7.4 and
0.1 mole/l Cl$^-$. The binding in the monomer case is calculated from the single site binding curve with the value
of the binding constant 55.7 kPa^{-1}. (b) Hill plot of the
same data on oxygen binding to hemoglobin.

the oxygen binding curve on changes in solution pH (Kilmartin and
Rossi-Bernardi, 1973). In the physiological pH range the oxygen
binding curve is shifted to the right if pH is decreased. This shift
has a twofold functional significance. It enhances the release of
oxygen in tissues and it is related to the uptake of hydrogen ions
by hemoglobin, thus increasing the capacity of the blood to carry
carbon dioxide. Another example is the shift of the oxygen binding
curve caused by the increased concentration in erythrocytes of 2,3-

Table 1. Half Saturation Oxygen Pressure and Hill Parameter
of Some Chemically Modified and Mutant Hemoglobins
(from Data Collected by Baldwin (1975) and Data of
Coleman (1977)). Half Saturation Oxygen Pressure is
Given Relative to the Value of the Normal Hemoglobin
(HbA).

Hemoglobin	$p_{1/2}/p_{1/2}^{HbA}$	n_H
HbA	1	3.0
des His	0.25	2.6
des Arg	0.11	2.0
des Arg-Tyr	0.04	1.1
NES Hb	0.4	2.4
A-I	0.47	2.7
A-II	0.18	2.0
Sawara	0.44	2.8
J. Capetown	0.40	2.2
Tarrant	0.10	1.8
Chesapeake	0.19	1.4
Hiroshima	0.32	2.2
Rainer	0.04	1.2

diphosphoglycerate (DPG) (Benesch and Benesch, 1974). This shift
plays an important role in adaptation mechanisms in cases of external perturbations like oxygen deficiency. Both hydrogen ions
(Perutz et al., 1980) and DPG (Arnone, 1972), the effector of the
oxygen binding curve, are known to bind the hemoglobin molecule.
Their binding sites are different from the binding sites of oxygen
molecules. They affect oxygen binding allosterically.

A large number of structural investigations were performed in
order to clarify the behavior of hemoglobin. The most significant
were certainly the X-ray determinations of the molecule's atomic
structure. An important outcome of the structural investigations
was the differences observed in the structures of oxygenated (oxy)
and deoxygenated (deoxy) hemoglobin molecules (Perutz, 1976). These
differences were observed at tertiary as well as at quaternary
levels. It also became obvious that the cooperativity in oxygen

binding is not a consequence of the direct interaction among oxygen molecules bound to hemoglobin but is rather a consequence of
the effects mediated through the whole protein structure. Structural investigations revealed structural differences among chemically
modified and mutant hemoglobins. The effect of hydrogen ions and
DPG on the oxygen binding curve has become understood on the basis
of structural investigations because it was found that these effectors bind differently to oxy and deoxy structures of the hemoglobin molecule.

Thermodynamic Description of Hemoglobin Binding Curves

As a prerequisite to the determination of the quantitative
structure-function relationship it is necessary to define the parameters with which the system of hemoglobin and its ligands is described at the thermodynamic level. The ligand binding curve ex-
presses the thermodynamic properties of the system at equilibrium
and therefore this discussion is limited to the thermodynamic
equilibrium.

Structural investigations of hemoglobin have suggested that
the molecule has different structures or different conformations.
Each conformation can be considered as a separate state of the
system. These are conformational states. The molecule can also be
differently ligated. We may therefore also introduce ligational
states of the system. In general it is hoped that the total number
of states of such systems is small. In thermodynamic equilibrium
there is a certain probability that a given molecule is in a given
state. The values of these probabilities depend on the relative
free energy values which can be ascribed to the states of the system.
According to the formalisms of statistical thermodynamics, the probability that a molecule is in the i-th state is

$$f_i = \frac{e^{-F_i/kT} \lambda^{m_i}}{g} = \frac{S_i \lambda^{m_i}}{g}, \quad S_i = e^{-F_i/kT}, \quad i = 1,2,\ldots,N. \quad (1.1)$$

Here F_i is the free energy of the i-th state and k and T have the
usual meaning. As indicated in Eq. (1.1) the state can be characterized by the value of the parameter S_i instead of the free energy
value F_i. λ is the absolute ligand activity at the binding site.
In equilibrium it is proportional to the ligand concentration. m_i

is the number of occupied ligand binding sites at the i-th state
and N the number of states. g, known as the binding polynomial
(Wyman, 1972), is an analog of the grand partition function of the
system (Svetina, 1971) and is defined as

$$g = \sum_{i=1}^{N} e^{-F_i/kT} \lambda^{m_i} = \sum_{i=1}^{N} S_i \lambda^{m_i}. \tag{1.2}$$

If probabilities f_i are known the binding curve can be calculated
from the expression

$$y = \sum_{i=1}^{N} m_i f_i = \frac{1}{N} \frac{\partial \ln g}{\partial \ln \lambda}. \tag{1.3}$$

Differences between the free energy values F_i and the free energy
value of a selected state can be considered as parameters of the
system. With N states there are then N-1 parameters. The values of
these parameters should in principle be derivable from the thermodynamic data of the system under consideration.

The Adair equation (AE) (Adair, 1925) for the binding of oxygen to hemoglobin is derived as an example. It is obtained if only
ligational states are taken into consideration and if it is assumed
that all four oxygen binding sites are equivalent. The number of
ligational states is in this case $2^4 = 16$. However, due to the assumed equivalency of oxygen binding sites there are only five differently ligated hemoglobin molecules and consequently only four
parameters, S_i, are needed. The binding polynomial for the binding
of oxygen to hemoglobin is then

$$g = 1 + 4S_1 \lambda + 6S_2 \lambda^2 + 4S_3 \lambda^3 + S_4 \lambda^4 \tag{1.4}$$

and the resulting binding curve is according to Eq. (1.3)

$$y = \frac{1}{4} \frac{\partial \ln g}{\partial \ln \lambda} = \frac{S_1 \lambda + 3S_2 \lambda^2 + 3S_3 \lambda^3 + S_4 \lambda^4}{g}. \tag{1.5}$$

The numerical factors appearing in expression (1.4) show the number
of equivalent states.

AE was originally derived from considering the equilibrium of
the four consecutive chemical reactions

$$Hb(O_2)_{i-1} + O_2 \rightleftharpoons Hb(O_2)_i, \quad K_i = \frac{\left[Hb(O_2)_i\right]}{\left[Hb(O_2)_{i-1}\right]\left[O_2\right]} . \qquad (1.6)$$

The binding constants K_i (Adair constants) are related to the parameters S_i in the following way:

$$4S_i = K_1$$
$$6S_2 = K_1K_2$$
$$4S_3 = K_1K_2K_3 \qquad\qquad (1.7)$$
$$S_4 = K_1K_2K_3K_4 .$$

It is instructive to visualize how the values of probabilities f_i change in the course of hemoglobin oxygenation. In Fig. 1.2 values of m_if_i are shown as a function of y for a chosen set of values of S_i.

It is straightforward to generalize the procedure described in order to write down the binding polynomial for the case in which it is taken into consideration that the binding of oxygen to the α and β subunits is different (Karabeg-Musemić and Svetina, 1979). If the symmetry properties of the hemoglobin molecule are taken into account, the generalized binding polynomial is

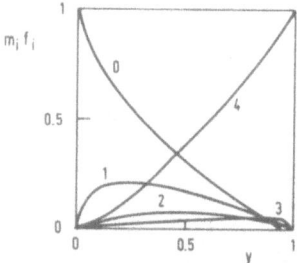

Fig. 1.2. Fractions of molecules of hemoglobin occupied by a different number of oxygen molecules (m_if_i) as a function of the fraction of occupied oxygen binding sites in hemoglobin. Curves are calculated by using parameters of the Adair equation given in Fig. 1.1.

$$g = 1 + 2S_{\alpha}\lambda_{\alpha} + 2S_{\beta}\lambda_{\beta} + S_{\alpha\alpha}\lambda_{\alpha}^2 + 2S_{\alpha 1\beta 1}\lambda_{\alpha}\lambda_{\beta} + 2S_{\alpha 1\beta 2}\lambda_{\alpha}\lambda_{\beta}$$

$$+ S_{\beta\beta}\lambda_{\beta}^2 + 2S_{\alpha\alpha\beta}\lambda_{\alpha}^2\lambda_{\beta} + 2S_{\alpha\beta\beta}\lambda_{\alpha}\lambda_{\beta}^2 + S_{\alpha\alpha\beta\beta}\lambda_{\alpha}^2\lambda_{\beta}^2 . \qquad (1.8)$$

λ_{α} and λ_{β} are absolute oxygen activities at the binding sites of α and β subunits, respectively, and in equilibrium they are proportional to the ligand concentration. The index i for the ligational states is replaced here by indices α, β, $\alpha\alpha$, etc. which explicitly show the ligation of each state. The fractional saturation of α and β subunits is defined as

$$y_{\alpha} = \frac{1}{2}\frac{\partial \ln g}{\partial \ln \lambda_{\alpha}} \quad \text{and} \quad y_{\beta} = \frac{1}{2}\frac{\partial \ln g}{\partial \ln \lambda_{\beta}} , \qquad (1.9)$$

respectively, and the binding curve, a generalized Adair equation (GAE), is

$$y = \frac{1}{2}(y_{\alpha} + y_{\beta}). \qquad (1.10)$$

By putting $\lambda_{\alpha} = \lambda_{\beta}$ the GAE reduces to the AE whereby the four parameters of the AE can be expressed in terms of the nine parameters of the GAE.

A proposal will be made here that in some specific cases the GAE represents a suitable way for the study of the structure-function relationship in the hemoglobin system. Before proceeding to put forward this proposal a short critical account on the modeling usually used in discussions relating the structure and function of this system will be presented.

Modeling of Structure-Function Relationship in Hemoglobin

The interest in the structure-function relationship in hemoglobin has been primarily focused on the molecular mechanism of the cooperative oxygen binding. The main relevant structural information related to this problem was the differences observed between the oxy and deoxy structures of hemoglobin. However, little is known about the structure of the intermediately oxygenated hemoglobin molecules. Several models have therefore been proposed in order to

predict the behavior of the hemoglobin system during intermediate
steps of oxygenation.

The existing models of hemoglobin cooperativity differ in
taking into consideration different ligational and conformational
states. Let us start this discussion with a rather general model
introduced by Herzfeld and Stanley (1974). They considered that
each subunit may be either ligated or unligated. They then assumed
that each subunit may be in either tense (t) or relaxed (r) conformation. Furthermore, at the quaternary level, the whole molecule
may be in either tense (T) or relaxed (R) conformation. The number
of states of such a model, then, is $2^4 \cdot 2^4 \cdot 2 = 512$. This model can
be simplified if it is assumed that tertiary conformational changes
occur in parallel with the ligand binding. In such a case the
number of states is $2^4 \cdot 2 = 32$. In a tetramer with four different
subunits ($\alpha\beta\gamma\delta$) all these states have, in general, different free
energies. In the case of hemoglobin which is a tetramer of the
$\alpha_2\beta_2$ type there is a degeneracy in some states so that the number
of different states is 20. For tetramers of the α_4 type the number
of states with different free energies is 10. In Table 2 the
number of states and parameters of this general model is compared
with the corresponding number in the models described below as well
as with the number in the Adair and generalized Adair schemes, the
latter being considered special cases of the general model in which
only ligational states are taken into consideration.

The earlier models of hemoglobin cooperativity can be derived
as special cases of the above general scheme. The model of Monod,
Wyman and Changeux (1965) (MWC, concerted or allosteric model)
comprises all 32 states. In this model subunits are considered to
be eqivalent so that the number of different states is reduced to
10. Nine parameters would be needed in order to quantify such a
model. However, in the MWC model the number of parameters is considerably reduced by making the specific assumption that the binding of the ligand either to T or R conformation of the molecule
is not cooperative. There are therefore 3 parameters of the model,
the allosteric constant L which is the ratio between the molecules
in T and R conformation in the absence of oxygen, the oxygen binding
constant for binding to the subunits in the T structure, and the
oxygen binding constant for binding to the R structure. The MWC
model has been later extended to account for differences in the
binding of oxygen to the α and β subunits (Ogata and McConnell,
1972). The model obtained comprises 5 parameters. Another type of

Table 2. Number of States (N) and Number of Parameters (N_p)
of Selected Models of Cooperative Oxygen Binding to
Hemoglobin.

Tetramer:	$\alpha\beta\gamma\delta$		$\alpha_2\beta_2$		α_4	
	N	N_p	N	N_p	N	N_p
Herzfeld and Stanley	32	31	20	19	10	9
AE					5	4
GAE			10	9		
MWC					10	3
Ogata and McConnell			20	5		
Pauling					5	2
KNF					5	3

hemoglobin cooperativity model was advanced by Koshland, Nemethy
and Filmer (1966) (KNF, sequential model). In these models it was
assumed that there are certain bonds between the hemoglobin subunits
in the deoxy conformation which break sequentially during the
process of oxygenation. Depending on the specific type of bonds
included, the models developed have two, three or four parameters.
An earlier model by Pauling (1935) is a special case of the KNF
model. Many other models of cooperativity exist. It should at least
be mentioned that the Perutz stereochemical model (Perutz, 1970)
has been mathematically described as a hybrid between the concerted and sequential models of cooperativity (Szabo and Karplus,
1972).

Different criteria can be introduced in order to evaluate the
applicability of a given model to describe hemoglobin behavior. A
possible criterion is the least number of model parameters which
allows a satisfactory description of experimental data. This criterion did not prove to be very useful because it was found out
that it is possible easily to fit an oxygen binding curve by means
of most of the models proposed. The oxygen binding curve by itself
does not seem to be a very sensitive function. A more stringent
criterion could be the ability of a model to describe changes in
the oxygen binding curves of structurally modified hemoglobins. According to this criterion a model is useful if it allows us to

relate the oxygen binding properties of the two hemoglobins by
changing the value of one parameter of the model in such a way that
the change of the value of this parameter is obvious on structural
grounds. This procedure can work properly only if the parameters
of the models have a true physical meaning. The inherent danger of
modeling is that the number of different conformational states in
a real system may be more than the number of parameters with which
the oxygen binding curve can be satisfactorily fitted. It can be
concluded that although simple models employing a small number of
parameters were very valuable because they provided basic insights
into possible mechanisms of cooperativity, they may not be used to
relate oxygen binding to different hemoglobins in a simple but
also quantitative manner.

The above critical evaluation of the modeling approach was
performed in order to justify the attempts to relate some structural information directly to the thermodynamic level of study of
the system (Svetina, 1973), specifically to the information at
the level of the GAE. The number of parameters in such a case is
higher, but less assumptions about the behavior of the system need
to be made.

Hemoglobin Structure-Function Relationship by the Use of
the Generalized Adair Equation

The problem of how the oxygen binding curves of the two structurally different hemoglobins are related to structural changes is
reformulated here into the problem of how the two corresponding
sets of parameters of the GAE are related to this structural change.
It will be shown how the parameters of the GAE can be interrelated
in the case where the structural changes are located within either
α or β subunits (Karabeg-Musemić and Svetina, 1979). In such a
case the effect of the structural change can be formally described
by the change of the absolute oxygen activity at the binding sites
of corresponding subunits. If binding to the α subunits is changed
the effect can be described by replacing λ_{α} in Eq. (1.8) with the
product $a\lambda_{\alpha}$ where a is the constant measuring the magnitude of
the effect of the structural change on the oxygen affinity of the α
subunit. Similarly, the effect of the structural change in the β
subunit can be accounted for by substituting $b\lambda_{\beta}$ for λ_{β} in
Eq. (1.8), where b is a constant.

It is clear that the oxygen binding curve of a structurally
modified hemoglobin which belongs to the treated class of modified
hemoglobins can be obtained by the use of the parameters of the
GAE of the normal hemoglobin and the appropriate value of either
the constant a or the constant b. The present disadvantage of this
approach lies in the lack of knowledge about the values of the parameters of the GAE. These values are difficult to determine because it is not possible to detect separately the intermediately
oxygenated hemoglobin molecules which would allow determination of
curves like those shown in Fig. 1.2. Nevertheless it is possible
to get some information about the parameters of the GAE and this
will be briefly described. There are some combinations of the probabilities f_i which can be determined and which give partial information about the desired parameters. One such combination is the
oxygen binding curve from which the four parameters of the AE can
be determined. Additional information can be obtained by measuring
y_α or y_β (see Eq. (1.9)) which was done in some cases with appropriately labeled hemoglobin subunits (Asakura, 1978). Some information about the parameters of the GAE can be obtained by ana-
lyzing families of oxygen binding curves such as those obtained
either at different pH values (Imai and Yonetani, 1975) or different concentrations of DPG (Imaizumi et al., 1979). It is known
from structural determinations that binding sites for DPG and hydrogen ions are distributed asymmetrically with regard to the two
kinds of hemoglobin subunits. The binding site for DPG is for instance located in between the two β -subunits (Arnone, 1972). Also
the ionizable groups which contribute to the Bohr effect are distributed unevenly on the two types of subunits (Perutz et al., 1980).
Therefore the corresponding families of oxygen binding curves reflect the differences in oxygen binding to the two subunits as
described by the GAE. Another possible method of obtaining information about the parameters of the GAE is the simultaneous analysis
of the two or more oxygen binding curves of the hemoglobins belonging to the above mentioned class of modified hemoglobins (Karabeg-
Musemić and Svetina, 1979). Each curve gives four values and the
two curves are related by only one additional parameter.

Recently a set of values of the parameters of the GAE has been
obtained from the analysis of the dependence of the oxygen binding
curve on pH and DPG (Šmigoc, 1981). The set obtained is characterized by certain approximations made in modeling the pH and DPG dependence of the GAE parameters. Nevertheless, it is used here to

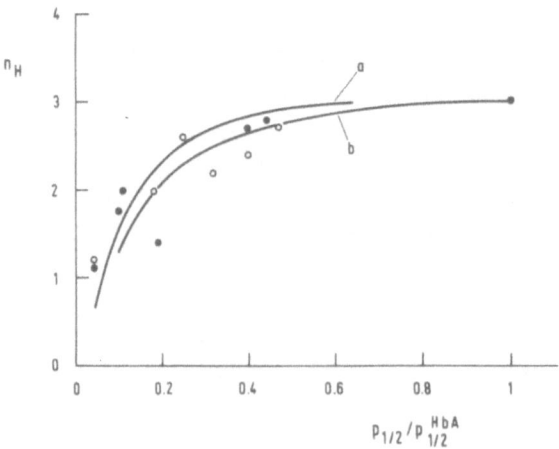

Fig. 1.3. Correlation between the Hill parameter and partial oxygen
pressure at half saturation for chemically modified and
mutant hemoglobins from Table 1. •: hemoglobins modified
at α subunits; o: hemoglobins modified at β subunits.
Curves represent calculated values obtained by varying
the parameters a (curve a) and the parameter b (curve b),
introduced as described in the text. Parameters used in
the calculation are four parameters of the AE as given
in Fig. 1.1, and five parameters of the GAE $S_\alpha = 0.6\, S_1$,
$S_{\alpha\alpha} = 0.6\, S_2$, $S_{\alpha1\beta1} = S_{\alpha1\beta2} = 1.2\, S_2$, and $S_{\alpha\alpha\beta} = 2\, S_3$ obtained (Šmigoc, 1981) from the analysis of the
dependence of the oxygen binding curve on pH (Imai and
Yonetani, 1975) and DPG (Imaizumi et al., 1979).

illustrate the change expected in the oxygen binding curves following structural modifications within subunits. In Fig. 1.3 is
shown the correlation between the half saturation oxygen pressure
and the Hill parameter for changes in parameter a (a) and for
changes in parameter b (b). In Fig. 1.3 are also given values for
hemoglobins from Table 1. These hemoglobins, of course, are not
supposed to be all of a type in which structural change would be
located within subunits. However, some of them, possibly the two
spin-labeled hemoglobins A-I and A-II (Coleman, 1977) may belong
to the discussed class of hemoglobins. Oxygen binding to the spin
labeled hemoglobin A-I is for instance obtained by taking b=5.0 and
oxygen binding to the spin labeled hemoglobin A-II by taking b = 48.

Conclusions

It remains here to discuss the relevance of the approach presented to studies of the structure-function relationship in hemoglobin. The final goal of the structure-function relationship is to obtain the macroscopic parameters of the system, for example oxygen binding constants, in terms of the properties of constituent atoms and interatomic interactions. However, hemoglobin is such a complex molecule that this may never be done. The purpose here was to point out that there may exist some parameters of the system which could serve as a possible link between atomic and macroscopic information about the system. Such parameters may be the parameters a and b described in the previous section. Provided that there are appropriate structurally different hemoglobins, the comparison of their oxygen binding curves should allow a reliable determination of the value of such a parameter. On the other hand there is the possibility that in the structural change within the subunit only a small portion of the molecule is involved, which would make it possible that the corresponding change in free energy could be calculated by some microscopic approaches.

2. THE INTERACTION OF CHARGED MEMBRANES WITH SURROUNDING IONS AND ION INDUCED SHAPE AND VOLUME CHANGES OF RED BLOOD CELL GHOSTS

An interpretation is presented of the ion induced shape and volume changes of red blood cell (RBC) ghosts, based firstly, on the dependence of the membrane surface area on the concentration of electrolyte solution and, secondly, on the dependence of the RBC ghost shape and volume on the difference in the areas of the two monolayers forming the RBC membrane. In the first part of the chapter the calculation of the membrane area changes is demonstrated, in which electrostatic interactions in the system and the elastic properties of the membrane are taken into consideration. In the second part the procedure for calculating the RBC ghost volume is described. In the third part both results are combined to obtain the dependence of the RBC ghost volume on the electrolyte concentration.

The free energy of the charged monolayer is expressed as the sum of elastic and electrostatic contributions:

$$F = F_{elastic} + F_{electrostatic} . \qquad (2.1)$$

Elastic energy can be approximated by the harmonic potential function (Evans and Hochmuth, 1978)

$$F_{elastic} = \frac{1}{2} K \cdot A_o (\frac{A}{A_o} - 1)^2 \qquad (2.2)$$

where A is the RBC membrane area, A_o the area with the minimum elastic energy, and K the elastic constant. Its value for the RBC membrane has been determined (Evans and Hochmuth, 1978).

The electrostatic free energy can be calculated by considering the charged membrane in the ionic environment as a Gouy-Chapman double layer system (McLaughlin, 1976). It is then assumed that charges are uniformly distributed on the surface of the membrane with the surface charge density $\sigma = e/A$, e being the total surface charge. Surface charges attract ions of opposite charge and repel ions of the same charge. Thus a double layer is formed. By a standard procedure the distribution of charges can be determined in a Poisson-Boltzmann approximation. The effective width of the double layer depends on the concentration of the electrolyte at some distance from the membrane and is given by the Debye length

$$\frac{1}{\varkappa} = (\frac{\varepsilon \varepsilon_o kT}{2ne_o^2})^{1/2} \qquad . \qquad (2.3)$$

Here ε_o, k and T have the usual meaning, e_o is unit charge, ε the dielectric constant and n the electrolyte concentration. The free energy of this system has been determined (Jähnig, 1976) and is

$$F_{electrostatic} = e \Psi_o - 4 \varepsilon \varepsilon_o \varkappa (\frac{kT}{e_o})^2 (ch \frac{e_o \Psi_o}{2kT} - 1)A \qquad (2.4)$$

where Ψ_o, the surface potential, is given by

$$\frac{kT}{e_o} 2 \varkappa \varepsilon \varepsilon_o sh \frac{e_o \Psi_o}{2kT} = \frac{e}{A} \qquad (2.5)$$

For a given value of surface charge and electolyte concentration the membrane area is determined by the minimum value of the total free energy, equation (2.1). At different electrolyte concentrations this minimum is found at different values of the membrane area. The dependence of the membrane area on the electrolyte con-

centration obtained is shown in Fig. 2.1. Membrane area is smaller at
higher electrolyte concentrations, the consequence of the enhanced
screening and therefore decreased repulsion between the surface
charges. It is to be noted that the surface charges affect the
membrane area very little which means that electrostatic effects
are a small perturbation to other factors determining the membrane
area.

The effect of the electrolyte concentration on the membrane
area which was obtained can be related to the RBC ghost shape
changes in view of the bilayer couples hypothesis (Sheetz and
Singer, 1974). According to this hypothesis the RBC shape changes
are caused by different changes in the areas of the two monolayers
forming the membrane. The two monolayers of RBC are differently
charged, the phosphatidylserine as the only charged phospholipid
component of the RBC membrane being located only on the inner part
of the membrane (Verkleij et al., 1973). According to the bilayer
couples hypothesis the two membrane monolayers can be, although in
close contact, dilated or compressed independently of each other.
The electrolyte therefore affects the area of the inner monolayer
and not that of the outer monolayer. Recently a quantitative method

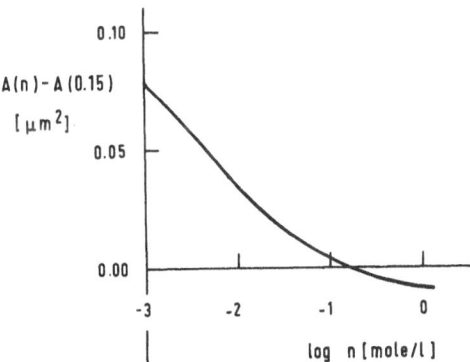

Fig. 2.1. Ion induced change in membrane area as a function of
electrolyte concentration as calculated by minimization
of Eq. (2.1.) Parameters are K = 0.45 N/m, e = $1.3 \cdot 10^{-12}$
As, A_o = 137 μm^2.

has been developed (Svetina et al., 1982) by which it is possible
to predict the RBC shape at a given difference between the two monolayer areas. A simple geometrical model for the family of axisym-
metrically shaped RBC´s has been introduced and the cell shape
determined by finding the shape with the minimum total bending
energy of the membrane whereby the cell volume (V), membrane area
and difference between the two membrane monolayer areas (Δ A) are
kept constant. With V = 90 μm^3, A = 137 μm^2 and with δ being the
mean distance between the two monolayers, the RBC was shown to be
a stomatocyte below $\Delta A/\delta$ = 83 μm, to be a discocyte in the interval 83 μm < $\Delta A/\delta$ < 89 μm, and that no axisymmetrical shape of
the proposed type can be realized above $\Delta A/\delta$ = 89 μm.

An attempt is made here to combine the effect of electrolytes
on the areas of membranes with the observation (Johnson et al.,
1980) that RBC ghosts, in response to changes in electrolyte concentration, exhibit some of the characteristic shapes seen in the
intact RBC´s, which are accompanied by volume changes. It is assumed that the bilayer couples hypothesis can also be applied in
the case of RBC ghosts. The method for the determination of shapes
has to be slightly modified because the membrane of these ghosts is
permeable to cations and therefore the ghost volume does not depend
on electrolyte concentration according to the osmotic laws. In the
modified procedure the minimum total membrane bending energy is
sought for various axisymmetrical shapes by keeping constant only
the membrane area and the difference in areas of the two monolayers
constituting the membrane. The result of this minimization procedure is the shape as in the case of RBC. However, the volume corresponding to a given shape can also be determined. Figure 2.2
shows the calculated RBC ghost volume as a function of $\Delta A/\delta$. It
is assumed that in the course of the hemolysis process the cell is
of spherical shape. The increase of $\Delta A/\delta$ causes the ghost to
shrink.

From the dependence of the membrane area on the electrolyte
concentration obtained (Fig. 2.1) and applying the result obtained
only to the inner monolayer and from the results obtained for the
dependence of the cell volume on the difference of the monolayer
areas (Fig. 2.2) it is then possible to determine the dependence of
the ghost volume on the electrolyte concentration. The result obtained is compared in Fig. 2.3 with the results of Johnson et al.
(1980). In this calculation the only parameters were the RBC mem-

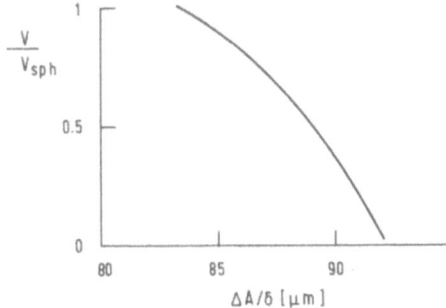

Fig. 2.2. RBC ghost volume relative to the volume of spherical ghost
(V$_{sph}$) with the membrane area 137 μm^2 as a function of
$\Delta A/\delta$, calculated by the method of Svetina et al. (1982)
modified as described in the text.

brane elastic constant, the width of the membrane area and the
surface charge. The value of δ was chosen to be 5 nm and the value
for the surface charge was found such that the curve calculated fitted the experimental points. The value obtained, $1.3 \cdot 10^{-12}$ As, can

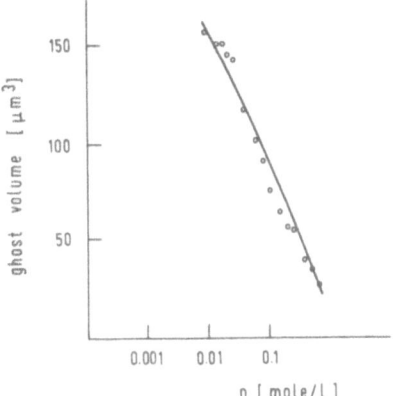

Fig. 2.3. Ghost volume as a function of electrolyte concentration
as obtained from the results shown in Fig. 2.1 and Fig.
2.2 (solid line) by taking δ = 5 nm and assuming that
hemolysis occurs at n = 0.07 mole/l. Points are measured
values obtained from Fig. 1 of Johnson et al. (1980).

be compared with the value $4.5 \cdot 10^{-12}$ As, an estimate for RBC surface
charge obtained from information about the amount of phosphatidylserine in the RBC membrane (Verkleij et al., 1973). The two values
are of the same order of magnitude. In view of approximations made
in calculating electrostatic free energy and in calculating shape
by assuming for it a simple geometrical model, the result obtained
points to the interaction between the charged membrane and the surrounding electrolyte as a probable mechanism for the RBC ghost
volume changes observed. It should be pointed out that the treatment presented is suited to the experiment of Johnson et al. (1980)
with RBC´s pretreated with heat where no transition to the echino-
cytic shapes occurred while increasing the electrolyte concentration.

The analysis presented shows the relationship between the molecular level of the RBC system, which involves the interaction between the membrane charges and charges of the surrounding electrolyte, and the macroscopic level of the system, which is the shape
of the cell. It is important to note that this relationship could
be obtained, firstly, by realizing that RBC shape changes can be
interpreted on the basis of differences in the areas of the two
sides of the RBC membrane (Sheetz and Singer, 1974) and, secondly,
by the possibility of quantifying this interpretation (Svetina et
al., 1982). The example presented clearly shows the role of theoretical approaches in studying biological systems. They either
refute or strengthen the qualitative idea about the mechanism at
work. Moreover, they bring up new questions. A possible question
posed by the analysis presented is, for instance, why the real RBC
membrane surface charge density and that used in the calculation
differ. Attempts to resolve questions of this type may potentially
increase our knowledge about the system treated and may also stimulate studies on simpler model systems such as phospholipid membranes
with the aim of evaluating the appropriateness of approximations
like that used here for the determination of the electrostatic free
energy.

3. ON MODELING THE MOLECULAR BASIS OF CELL DIFFERENTIATION

The objective of this chapter is to show and discuss a possible link between chemical processes in a cell and the phenomenon
of cell differentiation. In the first part an example of a chemical

system exhibiting two stable steady states will be described. In the second part of the chapter it will be shown how the kinetic properties of this system have to be invoked in order for it to be used for the interpretation of commitment phenomenon observed in the differentiation of murine erythroleukemia (MEL) cells. In particular a theoretical treatment will be described which allows us to relate the cell population properties of the differentiating system, such as the commitment kinetics, to the properties of the proposed underlying chemical processes at the level of a cell.

The process of differentiation involves the transition of a cell from one possible state with regard to a set of genes transcribed into another such possible state. The idea that such a transition might be described by a simple biochemical system is not new. Monod and Jacob (1961) proposed several possible chemical schemes and subsequently corresponding mathematical representations have also been developed (Griffith, 1968; Edelstein, 1972). More recently chemical systems of this kind have been employed in discussing a possible molecular basis for positional differentiation in morphogenesis (Edelstein, 1972; Lewis et al., 1977; Meinhardt and Gierer, 1980). A simple chemical model, to be demonstrated in the first part of this chapter is in its essential properties equivalent to the models described in the works cited above.

The essential feature of any chemical system underlying the differentiation is that it must exhibit the property of having multiple stable steady states. The system to be described here is a two state model. It is schematically presented in Fig. 3.1. The three chemical processes linked in a closed loop are the synthesis of certain substance P which is the product of gene G, regulation of this synthesis by the repressor R, and the modulation of the repressor action by the substance P. Let us first give the expression describing the regulation of the synthesis taking place in a cell. The binding of repressor R to the site of synthesis G can be considered as a chemical reaction

$$R + G \rightleftharpoons RG . \hspace{4cm} (3.1)$$

Here we shall take into consideration the possibility that there is only single binding site for the repressor in a cell. This is not essential for showing the existence of the two stable steady states of the system described. However, it is included here because of further developments of the model described later. Specific features of chemical reactions with the finite number of one of the reactants have been described previously (Berg and Blomberg, 1977). The time average of the rate of synthesis is in the case of a single binding site proportional to the fraction of time during which the repressor is not bound. This fraction will be denoted by $\bar{\theta}$ where $0 < \bar{\theta} < 1$. The value of $\bar{\theta}$ in general depends on the probability per unit time that the repressor which is bound is released and the probability per unit time that the repressor binds to the free binding site. The first probability is a constant (k_b), and the second is proportional to the concentration of the repressor ($k_a[R]$; k_a is a constant, $[R]$ is the repressor concentration). The fraction $\bar{\theta}$ can be determined by making the assumption that on the average, the number of transitions from free to bound and from bound to free state of the repressor binding site must be equal. Therefore

$$k_b(1 - \bar{\theta}) = k_a[R]\bar{\theta}, \tag{3.2}$$

and it follows that

$$\bar{\theta} = \frac{1}{1 + k_a[R]/k_b}. \tag{3.3}$$

The second feature of the system shown in Fig. 3.1 is the kinetics of accumulation in a cell of the substance P, the product of the synthetic process discussed. If first order kinetics is assumed, the rate at which the concentration of P changes (in order to simplify the procedure it is assumed here that the cell volume is constant) is given by

$$\frac{dP}{dt} = A\bar{\theta} - \gamma P, \tag{3.4}$$

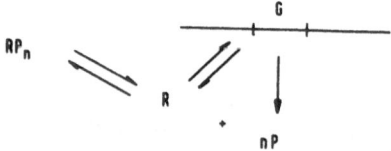

Fig. 3.1. Schematic representation of the chemical model exhibiting two stable steady states. Symbols used are described in the text.

where A is the maximum rate of synthesis of P and γ is the decay constant.

The third feature introduced is the effect of the substance P on its own synthesis by modulating the action of the repressor. It is specifically assumed here that n molecules of P cooperatively bind a repressor molecule:

$$R + nP \rightleftharpoons P_n R . \tag{3.5}$$

It is assumed that only those molecules of the repressor which are free of P can bind to the regulative binding site. In such a way the substance P determines the pool of repressor molecules available for their repressive action. For the chemical reaction (3.5) in thermal equilibrium, the concentration of free repressors in terms of the total repressor concentration $[R]_T$ is

$$[R] = [R]_T \frac{P_{1/2}^n}{P_{1/2}^n + P^n} , \tag{3.6}$$

where $P_{1/2}$ is the concentration of P at which one half of the repressors are free.

The steady state of the system is defined by the time constancy of P:

$$\frac{dP}{dt} = 0 . \tag{3.7}$$

The meaning of this equation is that the rate of synthesis equals the rate of degradation of P. The rate of synthesis of P depends on P and the corresponding dependency is obtained if the expression for $[R]$, equation (3.6), is inserted into Eq. (3.3). The curve obtained is shown in Fig. 3.2. The rate of degradation is in this model a linear function of P and is also shown in Fig. 3.2. With a chosen set of parameter values there are three points where the rate of synthesis and the rate of degradation are equal. These points represent two stable steady states, one with a low synthesis of P (P_L) and one with a high synthesis of P (P_H). The middle point (P_M) is the metastable state. The stability properties are obvious from examining the sign of the rate of accumulation of P which is

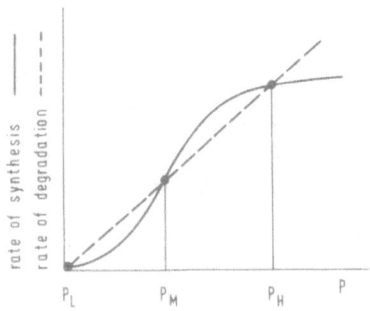

Fig. 3.2. Rate of synthesis and rate of degradation of the substance
         P as a function of P. Points show the steady states of the
         system. Symbols used are as described in the text.

$$\frac{dP}{dt} < 0 \qquad\qquad P_L < P < P_M ,$$

$$\frac{dP}{dt} > 0 \qquad\qquad P_M < P < P_H , \qquad\qquad (3.8)$$

$$\frac{dP}{dt} < 0 \qquad\qquad P_H < P .$$

It is to be noted that the two curves presented in Fig. 3.2
cross at three points if at least one of them has the inflection
point. This property is realized in the model presented by the cooperative binding of the product P to the repressor, and by the assumption that only a free repressor can bind to the regulative site.

The steady state aspects of the two-state chemical system described can be considered as a satisfactory basis for describing
the essential feature of the differentiating systems, that is the
ability of the system to be stable under two different chemical
regimes. The question remains, however, what causes the system to
make a transition from one to the other of the two stable steady
states. Before approaching this question it is desirable to analyze
in terms of the model the behavior of real systems in which differentiation has been studied.

A particularly useful system for studies of cell differentiation was found to be murine erythroleukemia cells (Marks and
Rifkind, 1978). These proerythroblasts, transformed by the Friend
virus complex, can be permanently grown in a cell culture. Exposure
to dimethylsulphoxide or a variety of other inducers stimulates MEL
cells to make a transition to the more advanced stages of erythroid
maturation. An analysis of the differentiated state of individual
cells, based on cloning cells in a semi-solid medium, has revealed
(Housman et al., 1978) that an irreversible commitment to differentiation occurs as a discrete event at a given time after the application of the inducer. Removal of the inducer prior to this event
returns a cell to its initial state, whereas its removal after this
event leaves it in the differentiated state.

It is tempting to relate properties of the differentiation of
MEL cells to the two state chemical system described. It is possible to assume that the non-differentiated state is the stable
steady state denoted by P_L. The action of the inducer can be understood as changing the value of one or more of the parameters determining the behavior of the system. The induction can be in
principle understood on the basis of such changes of parameters
that the rate of synthesis is larger than the rate of degradation
at all values of P. This would cause this substance to accumulate.
If after a certain time its value exceeded the value of P_M, the
system would, even in the absence of the inducer, remain stable at
the state P_H, which could be considered the committed state.

The above picture of the commitment process provides an interpretation about the lag period of the commitment kinetics shown in
Fig. 3.3. The lag period can be related to the time needed for the
substance P to reach the value P_M. However, it does not provide a
satisfactory explanation of the variability of commitment times
shown also in Fig. 3.3. As will be demonstrated in the following,
the commitment kinetics can be accounted for by the two state chemical model described if certain assumptions are made about its kinetic behavior. It has recently been shown (Svetina, 1981) that the
commitment kinetics can be understood on the basis of a chemical
system which is a simplification of the system shown in Fig. 3.1,
such that the effect of the substance P on the repressor is abol-
ished, if it is assumed that the characteristic time for the degradation and accumulation of substance P is of the same order of
magnitude as are characteristic times of the binding and release of
the repressor. In the mathematical representation of this model Eq.

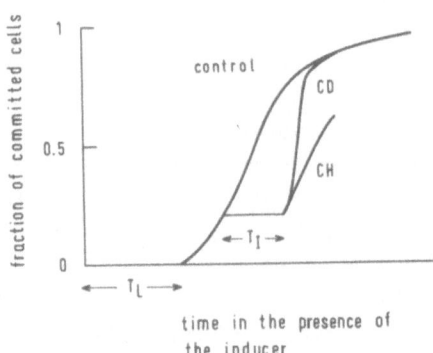

Fig. 3.3. Commitment kinetics of MEL cells drawn schematically from
        the data of Housman et al. (1978). Fraction of committed cells
        as a function of the time the cells spend in the presence
        of the inducer. T_L is the lag period. Also shown schema-
        tically is the effect on commitment kinetics of the inhi-
        biting concentrations of cycloheximide (CH) (Levenson and
        Housman, 1979) and cordycepin (CD) (Levenson et al., 1979).
        T_I is the time during which the inhibitor is applied.

(3.4.) has to be replaced by two equations, one for the allowed
synthesis in which $\bar{\theta}$ is replaced by unity, and one for the re-
pressed synthesis by taking zero instead of $\bar{\theta}$. In this model the
commitment was assumed to occur after the substance P reached a
certain critical value. During this time the repressor is released
from and bound to the regulative site several times and in a random
manner, and therefore the time needed to reach the critical level
P is different for each cell.

    Commitment kinetics is one property of the differentiating MEL
cells with the commitment event being another. This becomes evident from the analysis of the experiments where the induction process has been perturbed by the drug cordycepin (CD) (Levenson et
al., 1979). This drug applied in a suitable concentration completely blocked the commitment. After its removal the number of
committed cells rapidly increased and in a relatively short time
reached the control commitment curve. This behavior should be distinguished from the behavior due to the temporary presence of
another drug, the inhibitor of the protein synthesis cycloheximide
(CH) (Levenson and Housman, 1979) (Fig. 3.3), where after the re-

moval of the inhibitor the number of committed cells increased as
if the whole commitment curve were shifted to the right for the
duration of the presence of the drug. The cordycepin experiment is
indicative that the process determining the commitment kinetics was
continued despite the presence of CD and that there exists a consecutive process which is affected by this drug. The hypothesis is
made here that this second process is connected with the effect of
the substance P on the repressor. This hypothesis is appealing because in view of the properties of such closed loops discussed
above it also provides the molecular basis for the critical level
of substance P, this feature being rather artificially introduced
in the model describing the commitment kinetics.

The requirement that the chemical basis for the commitment
event would also account for the variability of commitment times
lead to the development of the model which is essentially a generalization of the two-state model described by Eqs. (3.1) to (3.6).
The generalization concerns the kinetic behavior of the model,
whereby the characteristic times of the regulative part of the
system are of the same order of magnitude as is that of the synthetic part of the system. The two-state model is a special case
of this generalization reached if the regulative characteristic
times are much shorter than the synthetic characteristic time. The
generalization brings about some new properties of the system and
in the following will be described first some steady state aspects
of the model and then the application of the model to simulate the
induction process.

The behavior of the generalized two-state model is studied by
the simulation computer experiment described in the caption of Fig.
3.4. The method used for obtaining the behavior of a cell on the
basis of the properties of the chemical system in a cell is essentially the same as developed earlier (Svetina, 1981). The computer simulation, run for a time interval much longer than the
characteristic times of the system, shows that there are actually
no real steady states. The system randomly switches in between a
high level state P_H and a low level state P_L. It is important that
in a given time most of the cells are either in a low or in a high
level state which means that the states can be considered quasi-
stable, whereby there is a probability per unit time that a cell
makes a transition to another state. As exemplified in Fig. 3.4,
the proportion of time a cell spends in a given state depends on
the value of model parameters such as k_b.

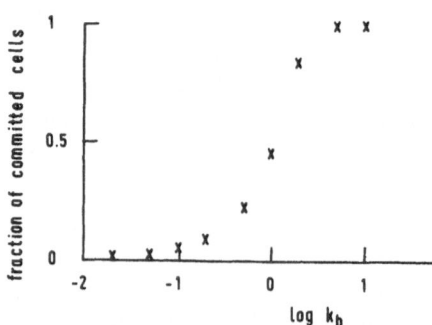

Fig. 3.4. Fraction of committed cells as a function of the parameter k_b predicted by the model shown in Fig. 3.1,
generalized in such a way that the repressor action is
modulated by the substance P^* which is synthesized in the
presence of the substance P. It is specifically assumed
that the effect of P on the synthesis of P^* is cooperative
with the Hill parameter 4, that the maximum rate of the
synthesis of P^* is A^*, and that P^* decays with the decay
constant γ^*. It is also assumed that four molecules of
P^* bind the repressor so that n = 4 (see Eq. (3.5)).
Parameters used are $k_a [R]_T / \gamma = 5$, $\gamma^*/\gamma = 5$, $\gamma P_{1/2}/A =$
0.8, and $\gamma^* P^*_{1/2}/A^* = 0.1$. $P_{1/2}$ is the concentration of
P at which the rate of synthesis of P^* is half the maximum. The meaning of $P^*_{1/2}$ is as seen from Eq. (3.6). The
method of calculation is analogous to that described by
Svetina (1981). Initially each cell has the values
$\gamma P_{1/2}/A = \gamma^* P^*_{1/2}/A^* = 0.5$. At time $10/\gamma$ the system is
scored for the values of $\gamma P/A$ and $\gamma^* P^*/A^*$. If $\gamma P/A >$
0.9 the cell is considered to be committed. Each point
is obtained from scoring 100 cells.

In view of the properties of the system described it is now
possible to give another possible interpretation of the induction
of the differentiation of MEL cells. The basic idea is again that
the values of parameters are such that in the absence of the inducer most cells are in a P_L state. The inducer causes changes of
values of one or several parameters of the model such that according

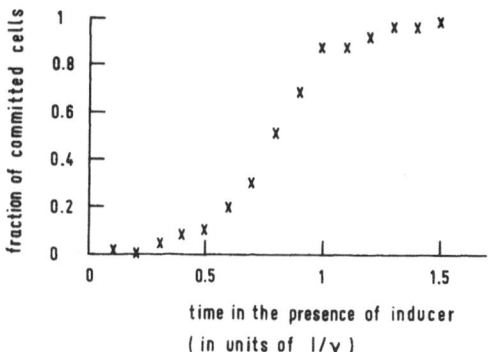

Fig. 3.5. Calculated commitment kinetics. Fraction of committed
cells as a function of the time the cells spend in the
presence of the inducer. It is assumed that the inducer
applied at t = 0 changes the value of the constant k_b
from 0.005 γ to 5 γ and that initially P = P* = 0. The
time course of P and P* is calculated as indicated in
the caption to Fig. 3.4. At the time given on the abscissa
the value of k_b is returned to the initial value. Cells
which have γ P/A > 0.9 at time 5/γ after the removal
of the inducer are considered to be committed. Each point
is obtained from scoring 100 cells.

to Fig. 3.4 the state P_H would be favourable. Figure 3.5 shows an
example of the commitment curve obtained if it is assumed that the
inducer changes the value of the parameter k_b. It should be pointed
out that with the chosen values of parameters the lifetime for a
cell in the P_H state is such that cells which reached this state
can really be considered irreversibly committed.

The main conclusion of the analysis presented is that at least
some selected properties of the induction of MEL cells are consistent with a simple two-state chemical model, provided that the model
is generalized to include some specific kinetic features. It is to
be noted that the ideas presented will remain hypothetical in nature
until the substances can be identified behaving kinetically as proposed in this model. However, this modeling may still prove useful
particularly in designing possible tests in the corresponding identification procedures.

GENERAL DISCUSSION

In the above three chapters were presented three different theoretical approaches applicable to three different systems. It seems to be appropriate to point out some common features and differences between them in relation to the general problem of the role of theoretical studies of biological systems. The important common problem in theoretical studies relating different structural levels of a biological system is to define in a suitable way the function of the system. The function must be identified with a measurable property of the system, such as the oxygen binding curve to hemoglobin, erythrocyte shape, and the commitment kinetics in the examples given. It seems then to be valuable in establishing quantitative relationships to identify the parameters of the systems which link the microscopic and macroscopic levels, such as the parameters a and b introduced in the case of the oxygen binding to hemoglobin and the difference in the monolayer areas ΔA discussed in the second chapter. Such parameters should be determinable by microscopic theories but should also be determinable from data obtained at the macroscopic level. It is at this point that a theory is needed such as the determination of the erythrocyte shape and the determination of the commitment curve in differentiating systems as described above. As exemplified the theories are different in their structure in considering specific features of the system under study. It is also to be noted that the aim of a theory is different at various levels of study. At lower levels of study such as hemoglobin and erythrocyte some kind of information is available at both levels to be related. The aim of a theory is to link this information in a quantitative manner. At higher levels of a biological system such as the level of cell differentiation there are many more unknowns. If knowledge is insufficient at some levels of study models may be introduced such as that discussed in the third chapter. These models may give better insight into possible mechanisms underlying the macroscopic property of the system studied. The aim of a theory such as the presented one relating cell chemistry and cell population kinetics is to test the consistency of the microscopic model with the macroscopic behavior of the system.

REFERENCES

Chapter 1

Adair, G., 1925, The haemoglobin system. VI. The oxygen dissociation curve of haemoglobin, J. Biol. Chem., 63:529.

Antonini, E.,and Brunori, M., 1971, "Hemoglobin and Myoglobin in Their Reactions with Ligands", North Holland, Amsterdam.

Arnone, A., 1972, X-ray diffraction study of binding of 2,3-diphosphoglycerate to human deoxyhemoglobin, Nature, 237:146.

Asakura, T., and Lau, P.-W., 1978, Sequence of oxygen binding by hemoglobin, Proc. Natl. Acad. Sci. USA, 75:5462.

Baldwin, J.M., 1975, Structure and function of hemoglobin, Progr. Biophys. Mol. Biol., 29:225.

Benesch, R.E.,and Benesch, R., 1974, The mechanism of interaction of red cell organic phosphates with hemoglobin, Adv. Protein Chem., 28:211.

Coleman, P.F., 1977, A study of conformational changes in two β -93 modified hemoglobin A´s using a triphosphate spin label, Biochemistry, 16:345.

Herzfeld, J.,and Stanley, H.E., 1974, A general approach to cooperativity and its application to the oxygen equilibrium of haemoglobin and its effectors, J. Mol. Biol., 82:231.

Imai, K.,and Yonetani, T., 1975, pH dependence of the Adair constants of human hemoglobin, J. Biol. Chem., 250:2227.

Imaizumi, K., Imai, K., and Tyuma, I., 1979, The linkage between the four-step binding of oxygen and the binding of heterotropic anionic ligands in hemoglobin, J. Biochem., 86:1829.

Karabeg-Musemić, R.,and Svetina, S., 1979, On the relationship between oxygen binding to normal and specifically modified hemoglobins, abstract S7-50, in: "FEBS Special Meeting on Enzymes", Dubrovnik-Cavtat.

Kilmartin, J.V.,and Rossi-Bernardi, L., 1973, Interaction of hemoglobin with hydrogen ions, carbon-dioxide, and organic phosphates, Physiol. Rev., 53:836.

Koshland, D.E., Jr., Nemethy, G., and Filmer, D., 1966, Comparison of experimental binding data and theoretical models in proteins containing subunits, Biochemistry, 5:365.

Monod, J., Wyman, J., and Changeux, J.-P., 1965, On the nature of allosteric transitions: a plausible model, J. Mol. Biol., 12:88.

Ogata, R.T.,and McConnell, H.M., 1972, States of hemoglobin in solution, Biochemistry, 11:4792.

Pauling, L., 1935, The oxygen equilibrium of hemoglobin and its
    structural interpretation, Proc. Natl. Acad. Sci. USA,
    21:186.
Perutz, M.F., 1970, Stereochemistry of cooperative effects in
    hemoglobin, Nature, 228:726.
Perutz, M.F., 1976, Structure and mechanism of haemoglobin, Br.
    Med. Bull., 32:195.
Perutz, M.F., Kilmartin, J.V., Nishikura, K., Fogg, J.H., Butler,
    P.J.G., and Rollema, H.S., 1980, Identification of residues
    contributing to the Bohr effect of human haemoglobin, J. Mol.
    Biol., 138:649.
Šmigoc, K., 1981, Generalized Adair-equation for binding of oxygen,
    hydrogen ions and 2,3-diphosphoglycerate to normal human
    hemoglobin, Mag. Thesis, University of Zagreb, in Slovene.
Svetina, S., 1971., Thermodynamic studies of mechanisms for co-
    operativity in proteins containing subunits, in: "Proc. First
    European Biophysics Congress", E. Broda, A. Locker and H.
    Springer-Lederer, eds., Verlag der Wiener Medizinischen Aka-
    demie, Vienna, p. 85.
Svetina, S., 1973, Thermodynamical studies of cooperative oxygen
    binding by hemoglobin, in: "Proc. VII. Internationales Sym-
    posium über Strukture und Funktion der Erytrozyten", Anhand-
    lungen der Akademie der Wissenschaften der DDR, Berlin,
    p. 135.
Szabo, A., and Karplus, M., 1972, Mathematical model for structure-
    function relations in haemoglobin, J. Mol. Biol., 72:163.
Wyman, J., 1972, On allosteric models, in: "Current Topics in
    Cellular Regulation", Academic Press, Inc., New York and
    London, p. 209.

Chapter 2

Evans, E.A., and Hochmuth, R.M., 1978, Mechanochemical properties
    of membranes, Curr. Top. Membr. Transp., 10:1.
Jähnig, F., 1976, Electrostatic free energy and shift of the phase
    transition for charged lipid membranes, Biophys. Chem.,
    4:309.
Johnson, R.M., Taylor, G., and Meyer, D.B., 1980, Shape and volume
    changes in erythrocyte ghosts and spectrin-acting networks,
    J. Cell. Biol., 86:371.

McLaughlin, S., 1976, Electrostatic potentials at membrane-solution interfaces, Curr. Top. Membr. Transp., 9:71.

Sheetz, M.P., and Singer, S.J., 1974, Biological membranes as bilayer couples. A molecular mechanism of drug-erythrocyte interactions, Proc. Natl. Acad. Sci. USA, 71:4457.

Svetina, S., Ottova-Leitmannova, A., and Glaser, R., 1982, Membrane bending energy in relation to bilayer couples concept of red blood cell shape transformations, J. Theor. Biol., 94:13.

Verkleij, A.J., Zwaal, R.F.A., Roelofsen, B., Comfurius, P., Kastelijn, D., and Van Deenen, L.L.M., 1973, The asymmetric distribution of phospholipids in the human red blood cell membrane, Biochim. Biophys. Acta, 323:178.

Chapter 3

Berg, O.G., and Blomberg, C., 1977, Mass action relations in vivo with application to the lac operon, J. Theor. Biol., 67:523.

Edelstein, B.B., 1972, The dynamics of cellular differentiation and associated pattern formation, J. Theor. Biol., 37:221.

Griffith, J.S., 1968, Mathematics of cellular control processes II. Positive feedback to one gene, J. Theor. Biol., 20:209.

Housman, D., Gusella, J., Geller, R., Levenson, R., and Weil, S., 1978, Differentiation of murine erythroleukemia cells: The central role of the commitment event, in: "Cold Spring Harbor Conferences on Cell Proliferation. Vol. 5. Differentiation of Normal and Neoplastic Hematopoietic Cells", B. Clarkson, P.A. Marks, and J.E. Till,eds., Cold Spring Harbor Press, Cold Spring Harbor, N.Y., p. 193.

Levenson, R., and Housman, D., 1979, Developmental program of murine erythroleukemia cells. Effect of the inhibition of protein synthesis, J. Cell. Biol., 82:715.

Levenson, R., Kernen, J., and Housman, D., 1979, Synchronization of MEL cell commitment with cordycepin, Cell, 18:1073.

Lewis, J., Slack, J.M.W., and Wolpert, L., 1977, Thresholds in development, J. Theor. Biol., 65:579.

Marks, P.A.,and Rifkind, R.A., 1978, Erythroleukemic differentiation, Ann. Rev. Biochem., 47:419.

Meinhardt, H.,and Gierer, A., 1980, Generation and regeneration of sequence of structures during morphogenesis, J. Theor. Biol. 85:429.

Monod, J.,and Jacob, F., 1961, General conclusions: Teleonomic
    mechanisms in cellular metabolism, growth, and differentiation,
    Cold Spring Harbor Symposia on Quantitative Biology, 26:389.
Svetina, S., 1981, Protein induction process and stochastic nature
    of cell commitment to proliferation and differentiation, J.
    Theor. Biol., 90:151.

EVOLUTION OF POLYNUCLEOTIDES

Peter Schuster

Institut für Theoretische Chemie und Strahlenchemie
Universität Wien
A 1090 Wien, Austria

INTRODUCTION

Selection and Darwinian evolution are phenomena which we commonly correlate to the existence of living organisms. When Charles Darwin about 130 years ago was working on his fundamental theory of biological evolution he had in mind highly developed organisms, plants and animals which he had studied for a lifetime. He built his universal concept on two basic principles the consequences of which he saw in nature. These principles are (1) the variability of multiplying organisms and (2) natural selection as a permanently operating optimization process. The quantity to be optimized is reproductive success under conditions dictated by the environment and the internal constraints of the organism.

The variability of nature has never been seriously called in question. It is so obvious to everybody who performs studies in the fields. With natural selection, however, Darwin himself and later on many other biologists had some troubles: the concept of "survival of the fittest" degenerates to a mere tautology when the degree of fitness cannot be measured independently of the probability of survival. In the early days one could only refer to the results achieved by animal breeders or farmers in nursery gardens as Darwin (1867) did when he discussed the action of selective forces.

309

More than one hundred years latter an enormous and impressive abundance of detailed information has become available. We know a lot on the mechanisms of inheritance and gene distribution through the marvelous and enlightening works in population genetics and molecular biology. Nevertheless the discussion on natural selection is still going on, sometimes carried by exceedingly superficial arguments, sometimes conducted on more profound grounds (Gould, 1978). Despite all the progress made, there seems to be still a need for deeper understanding of the mechanism through which natural selection works. It is not unnecessary, therefore, to search for simple enough systems which allow to study how biological evolution works. In this contribution we shall consider a particularly simple example. The entire system will contain a piece of viral RNA and a specific RNA replicase together with necessary low molecular weight materials. This simple system, nevertheless, will provide some insight into the molecular mechanisms of variation and selection.

Competition through self-replication is one important principle of biology. Nevertheless, sometimes we observe co-operation between replicating elements as well. Apparently, another principle must be in operation in those cases. The underlying problem can be analysed very generally and the resulting new form of organization has been called "hypercycle" (Eigen, 1971; Eigen and Schuster, 1979). We shall discuss this principle and its role in biological evolution as well as present day biology.

AUTOCATALYSIS, EXPONENTIAL GROWTH AND SELECTION

"Replication or multiplication of organisms living on large resources leads to exponential growth of population ...". This fact has been predicted long ago by the well-known British economist T.R. Malthus, who lived in the 18th and 19th century. He was also the first to recognize the tremendous nutrition problems created by an exponentially growing human population. Malthus' ideas had a certain influence on Darwin when he worked on his theory of evolution about fifty years later. What is unique about self-replication and exponential growth? In order to be able to see the underlying principle let us consider some exceedingly simple examples.

Growth Properties of a Single Autocatalyst

A simple, single step autocatalytic process leads to exponential growth when it is run under conditions far off equilibrium. For the reaction

$$I + A \; \underset{f_2}{\overset{f_1}{\rightleftharpoons}} \; 2I \qquad\qquad (1)$$

we obtain the kinetic equation (concentrations are denoted by $[I] = x$ and $[A] = a$):

$$\frac{dx}{dt} = \dot{x} = (f_1 a - f_2 x)x \;. \qquad\qquad (2)$$

Equation (2) can be integrated without difficulties. We obtain

$$x(t) = \frac{f_1(a_o + x_o)x_o \, \exp\{f_1(a_o + x_o)t\}}{f_1(a_o + x_o) - (f_1 + f_2)x_o(1 - \exp\{f_1(a_o + x_o)t\})} \;, \qquad (3)$$

wherein a_o and x_o stand for the concentrations of A and I at time $t = 0$. After long enough time $x(t)$ approaches the equilibrium value $\bar{x}$:

$$\lim_{t \to \infty} x(t) = \bar{x} = \frac{f_1}{f_1 + f_2}(a_o + x_o) \;. \qquad\qquad (3a)$$

Far off equilibrim which means $x_o \ll \bar{x}$ and t sufficiently small we observe an exponential growth law:

$$x_o \ll \bar{x}, \; t \text{ small}: \; x(t) \simeq x_o \, \exp\{f_1(a_o + x_o)t\} \qquad (3b)$$

Figure 1 shows a concrete numerical example. We observe an initial phase of exponential growth but then when the reaction is left alone it approaches equilibrium through a common relaxation process.

In order to keep the autocatalytic reaction within the interesting range of exponential growth we have to couple the system to an external reservoire or to other reactions which can be driven

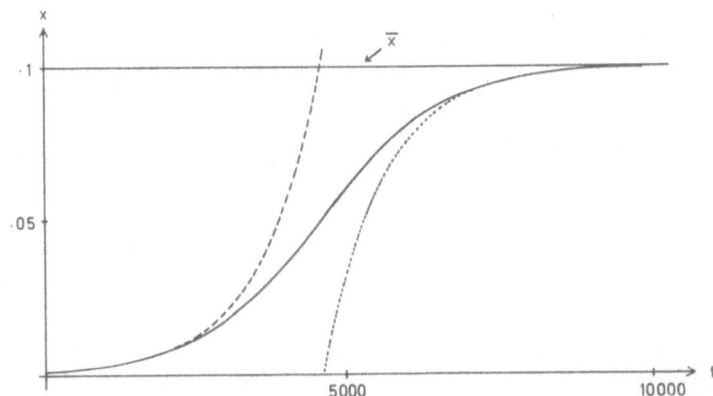

Fig. 1. The solution curve for an autocatalytic reaction far off
and close to thermodynamic equilibrium. Exponential growth
and relaxation towards equilibrium are shown as broken lines.
The rate constants chosen are $f_1 = 1$ $[t^{-1} \cdot c^{-1}]$ and $f_2 = 0.01$
$[t^{-1} \cdot c^{-1}]$ in arbitrary time $[t]$ and concentration $[c]$ units.

from outside. An easy to handle three-step mechanism contains a degradation and a recycling reaction together with the autocatalytic
process:

$$I + A \overset{f_1}{\underset{f_2}{\rightleftharpoons}} 2I \tag{4a}$$

$$I \overset{d_1}{\underset{d_2}{\rightleftharpoons}} B \tag{4b}$$

$$B \overset{g(E)}{\longrightarrow} A . \tag{4c}$$

Reaction (4) is described in mathematical terms by the differential
equation $[A] = a$, $[I] = x$, $[B] = b$:

$$\dot{a} = gb + f_2 x^2 - f_1 ax , \tag{4d}$$

$$\dot{x} = f_1 ax + d_2 b - f_2 x^2 - d_1 x , \tag{4e}$$

$$\dot{b} = d_1 x - (d_2 + g)b . \tag{4f}$$

The recycling reaction B $\longrightarrow$ A is controled by the environmental conditions. Because of the reaction (4c) we are dealing with an open
system which may approach a stationary state. Thermodynamic equilibrium can be reached only in the limiting case g $\longrightarrow$ 0. For g $\neq$ 0 we
find two stationary states $\overline{P}_1$ and $\overline{P}_2$. Note, that the total concentration of material available c$_o$ = a + x + b is a conserved quanti-
ty in reaction (4) since (a + x + b)$'$ = 0. Hence, c$_o$ can be considered as a parameter which is under external control. For short we
use vector notation: $\overline{P}$ = (a,x,b). Stationary states and concentra-
tions are indicated by bars, e.g. $\overline{P}$ = ($\overline{a}$,$\overline{x}$,$\overline{b}$). In particular we have

$$\overline{P}_1 = (c_o,0,0) \text{ and} \tag{5a}$$

$$\overline{P}_2 = \frac{1}{(d_2 + g) \left[f_1(d_1 + d_2 + g) + f_2(d_2 + g)\right]} \cdot$$

$$\cdot \{ f_2 c_o (d_2 + g)^2 + d_1 (d_1 + d_2 + g), \ f_1 c_o (d_2 + g)^2 - d_1(d_2 + g)g,$$

$$f_1 c_o d_1 (d_2 + g) - d_1^2 g \} \ .$$

At $\overline{P}_1$ only A is present, the concentrations of the autocatalyst and
its degradation product vanish ($\overline{x}$ = $\overline{b}$ = 0). $\overline{P}_1$ thus represents a
kind of "frozen" state at which no dynamical compensation exists.
$\overline{P}_2$ is the true reactive state. Here, forward and backward reactions
proceed at the same rates and all three compounds are present in
finite concentrations ($\overline{a}$ $\neq$ 0, $\overline{x}$ $\neq$ 0, $\overline{b}$ $\neq$ 0).

In order to learn more about the two states $\overline{P}_1$ and $\overline{P}_2$ we have
to perform stability analysis. In this particular example lineari-
zation of the differential equation (4) around the stationary states
yields the desired results already. The two eigenvalues of the Jacobian matrix - due to the conservation law a + x + b = c$_o$ we have
only two degrees of freedom in (4) - can be obtained by a rather
lengthy but straightforward calculation. The expressions obtained
thereby are rather clumsy and therefore we dispense here from all
details. The total concentration

$$c_o = c_{cr} = \frac{d_1 g}{f_1(d_2 + g)}$$

Table 1. Stability Analysis of the Two Stationary States of Equation (4d-f)

Range of Total Concentration	Stationary State	Eigenvalues of the Jacobian	Stability
$0 < c_o < c_{cr}$	$\bar{P}_1$	$\omega_1^{(1)} < 0$ $\omega_2^{(1)} < 0$	stable
	$\bar{P}_2$	$\omega_1^{(2)} > 0$ $\omega_2^{(2)} < 0$	unstable
$c_o = c_{cr} = \dfrac{d_1 g}{f_1(d_2+g)}$	$\bar{P}_1 = \bar{P}_2$	$\omega_1 = 0$ $\omega_2 = -\dfrac{d_1 d_2 + (d_2+g)^2}{d_2+g}$	bifurcation point
$c_o > c_{cr}$	$\bar{P}_1$	$\omega_1^{(1)} > 0$ $\omega_1^{(1)} < 0$	unstable
	$\bar{P}_2$	$\omega_1^{(2)} < 0$ $\omega_2^{(2)} < 0$	stable

is critical. As we can easily verify the two stationary points coincide: $\bar{P}_1(c_{cr}) = \bar{P}_2(c_{cr})$. The eigenvalues of the Jacobian are summarized in Table 1.

At low total concentration, i.e. at concentrations below the critical value $c_o < c_{cr}$, $\bar{P}_1$ is stable. Hence, the autocatalyst I and its degradation product B are absent at the steady state. At

$c_o = c_{cr}$ we observe a bifurcation of the simplest class: two stationary points exchange their properties. Note, that one eigenvalue
vanishes at the bifurcation point. This leads to a well known phenomenon called "critical slowing down". In the surrounding of the
critical point ω_1 is almost zero and the corresponding mode of reaction proceeds exceedingly slowly. Above c_{cr} the state $\bar{P}_2$ is
stable. The bifurcation diagram is shown schematically in Fig. 2.

The assumption that reactions (4a) and (4b) are run under conditions of practical irreversibility ($f_2 = 0$, $d_2 = 0$) which applies
well to most biochemical systems does not change the qualitative
features of the system. The only major change concerns the stationary concentration of A above the bifurcation point: $\bar{a}$ is independent of c_o in case of irreversibility; $\bar{a} = d_1/f_1$ for $c_o > c_{cr} = d_1/f_1$.
The autocatalytic reaction step thus consumes all A accessible ex-
cept the minimum quantity which is necessary to keep the reaction
going.

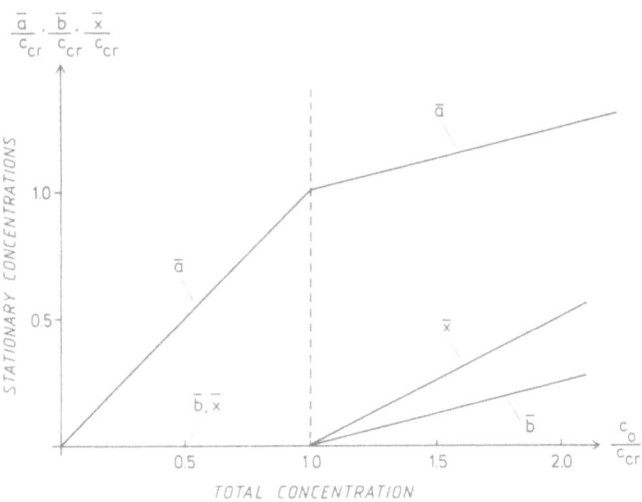

Fig. 2. Stable steady states and the bifurcation point of the three-
        step reaction (4). The numerical values of the rate con-
        stants chosen are: $f_1 = 4\ [t^{-1}\ c^{-1}]$, $f_2 = 2\ [t^{-1}\ c^{-1}]$,
        $d_1 = 3\ [t^{-1}\ ]$, $d_2 = 1\ [t^{-1}]$, and $g = 5\ [t^{-1}]$ in arbitrary
        time and concentration units. The critical concentration
        then is: $c_{cr} = 5/8\ [o]$.

Replication of synthetic homopolyribonucleotides has been studied by Friedemann Schneider and coworkers (Schneider et al., 1979; Heinrichs and Schneider, 1980). They studied RNA-synthesis at poly(A)-poly(U) templates by the unspecific RNA polymerase from E. coli in a stirred flow reactor. In general this reaction follows an overall autokatalytic kinetics. In the flow reactor the recycling process is replaced by an influx of A and an outflux of the material in the reactor. It is worth noticing that critical slowing down has been observed experimentally in this system.

Competition Between Two Autocatalysts

Next, we consider two autocatalysts growing on the same resource. We use the same recycling reaction as before in order to keep the system off equilibrium. For the sake of simplicity the reactions are assumed to be irreversible for practical purposes:

$$A + I_1 \xrightarrow{f_1} 2I_1 \ , \tag{6a}$$

$$A + I_2 \xrightarrow{f_2} 2I_2 \ , \tag{6b}$$

$$I_1 \xrightarrow{d_1} B \ , \tag{6c}$$

$$I_2 \xrightarrow{d_2} B \ , \tag{6d}$$

$$B \xrightarrow{g(E)} A \ . \tag{6e}$$

The reaction mechanisms (6) can be analysed in straightforward manner. We find three steady states and two bifurcation points (Fig. 3). Only one bifurcation is relevant since one steady state remains unstable for all values of the total concentration $c_0 = a + b + x_1 + x_2$. As we see from Fig. 3 the two autocatalysts compete and selection occurs. The autocatalyst which leads to the bifurcation at a lower critical concentration, $c_{cr}(I_i) = d_i/f_i$, $i = 1$ or 2, is the one that will be selected. Systems which have the potentiality to grow exponentially lead to selection. Thus, we observe one principle of Darwin's theory already with fairly simple chemical reactions.

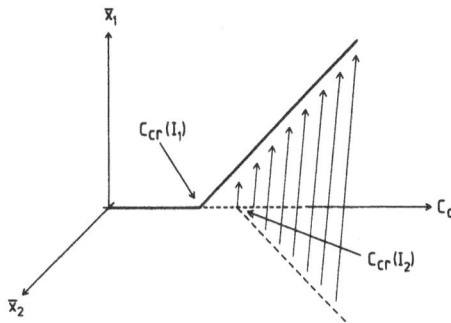

Fig. 3. Stationary states and bifurcation points in a system of
competing autocatalysts as described by the mechanism (6).
We observe three steady states: the zero state and two reactive states. Note, that the steady state which branches
off the zero state first (here at $c_{cr}(I_1) = d_1/f_1$) is always stable whereas the other reactive state remains un-
stable. The stable reactive state corresponds to $\bar{x}_2 = \overline{[I_2]}$
$= 0$. Hence, the autocatalysts compete and selection of I_1
takes place.

Enzyme Catalysed Replication

In this section we consider RNA replication as a characteristic
example of an autocatalytic reaction. RNA replication in most cases
is catalysed by a specific enzyme. The role of the enzyme in the
kinetics of replication is studied by means of a simple two step
mechanism:

$$I + E \xrightleftharpoons[k_2]{k_1} IE \ , \tag{7a}$$

$$IE + \sum_{\lambda=1}^{4} \nu_\lambda A_\lambda \xrightarrow{f'} IE + I \ . \tag{7b}$$

Herein we denote the RNA molecule by I, the enzyme by E and the
polynucleotide-enzyme complex by IE. The four energy rich monomers
which are to be incorporated into the newly synthesized RNA molecule
are symbolically characterized by A_i, $i = 1,\ldots,4$: $A_1 \equiv GTP$, $A_2 \equiv$

ATP, $A_3 \equiv$ CTP and $A_4 \equiv$ UTP. The following symbols are used for the individual concentrations:

$$[I] = x, \quad [E] = e \quad \text{and} \quad [IE] = y .$$

The conditions applied in serial transfer experiments come very close to the assumption of constant or buffered concentrations of the activated monomers: $[A_i] = a_i^o$, $i = 1,\ldots,4$. Therefore, they do not enter as variables into the kinetic equations. We can account for them by redefinition of the rate constant f:

$$f = f' \cdot F(a_1^o, a_2^o, a_3^o, a_4^o).$$

Thus, the system is described by the differential equation:

$$\dot{x} = (f + k_2) y - k_1 xe , \tag{7c}$$

$$\dot{y} = k_1 xe - k_2 y = -\dot{e} . \tag{7d}$$

The total enzyme concentration, $e_o = y + e$, is a constant in agreement with catalytic action. The total concentration of the template, $x_o = x + y$, is steadily growing since

$$\dot{x}_o = \dot{x} + \dot{y} = fy > 0 . \tag{7e}$$

Equation (7) can be integrated numerically. An example of the characteristic solution curve is shown in Fig. 4. We distinguish two phases of growth: exponential at low template concentration ($x_o \ll e_o$) and linear above the point of enzyme saturation ($x_o \simeq e_o$).

For the purpose of illustration we make a straightforward approximation: we assume quasiequilibrium of the complex IE (Gassner and Schuster, 1982) or, in other words, f is considered to be so small that reaction (7a) is practically at equilibrium. The equilibrium constant of (7a) is denoted by $K = k_1/k_2$. Then we find

$$\dot{x}_o = fy \simeq f \, \frac{x_o + e_o + K^{-1}}{2} \left(1 - \sqrt{1 - \frac{4 x_o e_o}{(x_o + e_o + K^{-1})^2}}\right) \simeq$$

$$f \, \frac{x_o e_o}{x_o + e_o + K^{-1}} . \tag{7f}$$

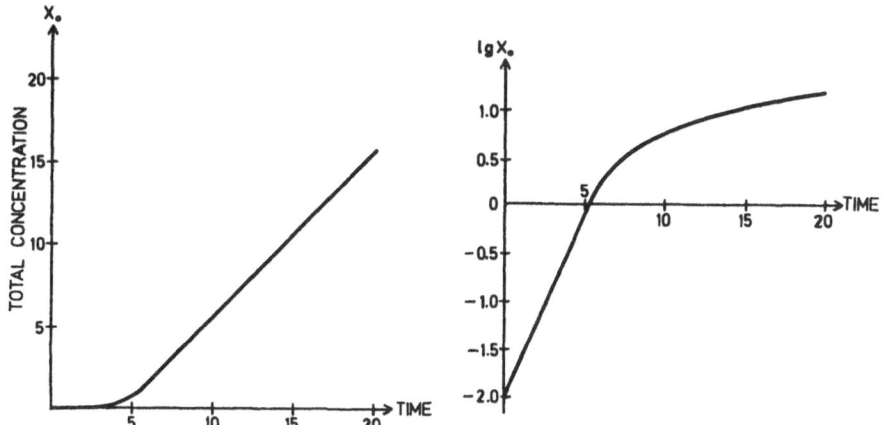

Fig. 4. Numerical integration of equation (7) as an example of
enzyme catalysed polynucleotide synthesis. The rate constants
and initial conditions applied are: $f = 1$ $[t^{-1}]$, $k_1 = 1000$
$[t^{-1} c^{-1}]$, $k_2 = 10$ $[t^{-1}]$, $x(0) = 0.01$ $[c]$, $y(0) = 0$ and
$e_0 = 1$ $[c]$ in arbitrary time and concentration units. Exponential growth observed at low concentrations of template
changes into linear growth at the point of saturation: $x \sim$
e_0. In order to be able to recognize both phases equally
well we present a linear and logarithmic plot.

The last term represents a very useful rational approximation to
the expression with the square-root above which is correct within
a factor of utmost two:

$$\frac{x_0 e_0}{x_0 + e_0 + K^{-1}} \leqslant \frac{x_0 + e_0 + K^{-1}}{2} \left(1 - \sqrt{1 - \frac{4x_0 e_0}{(x_0 + e_0 + K^{-1})^2}}\right) \leqslant$$

$$\frac{2x_0 e_0}{x_0 + e_0 + K^{-1}} \; .$$

Equation (7f) can be studied easily in case the equilibrium
constant fulfils the relation $K^{-1} > e_0$. Then we have the two asymp-
totic cases:

(1) Low concentrations of the template I, $x_0 \ll e_0$ and hence

$\dot{x}_o \sim fKe_o \cdot x_o$. We obtain exponential growth as we did in the numerical simulation. The mechanism (7) thus can be considered as a
"hidden" autocatalytic process.

(2) High concentration of the template I, $x > K^{-1} > e_o$ and hence $\dot{x}_o =$
fe_o = const. At the condition of "enzyme saturation" we find linear
growth as with the numerical simulation. The enzyme produces at
maximum rate since all enzyme is converted into the complex IE under these conditions.

The case $K^{-1} \leqslant e_o$ can be analysed similarly and leads to closely related results. It is less realistic, however, and we dispense
here from details.

It is straightforward to consider a system of different polynucleotides which are replicated by the same replicase. Again we
apply the quasiequilibrium approximation to the system

$$
\begin{aligned}
&I_1 + E \xrightleftharpoons{K_1} I_1E, \qquad I_1E \xrightarrow{f_1} I_1E + I_1 , \\
&I_2 + E \xrightleftharpoons{K_2} I_2E, \qquad I_2E \xrightarrow{f_2} I_2E + I_2 , \qquad\qquad (8)\\
&\quad\vdots \qquad\qquad\qquad\qquad\quad \vdots \\
&I_n + E \xrightleftharpoons{K_n} I_nE, \qquad I_nE \xrightarrow{f_n} I_nE + I_n .
\end{aligned}
$$

As in the previous example we assume that the concentrations of
activated monomers are buffered and appear implicitely in the rate
constants f_i, $i = 1,...,n$. Under these conditions the system follows different differential equations in the low and high concentration limit. In the first case we find

$$
\dot{x}_i^o = \frac{f_i K_i e_o}{1 + K_i e_o} x_i^o; \qquad i = 1,...,n . \qquad\qquad (8a)
$$

Equation (8a) leads to selection and exponential growth as we observed with mechanism (6). The quantity which is selected for, thus
is a function of f_i, K_i and e_o. In the particular case we mentioned
above, $K_i^{-1} > e_o$, the relevant quantity is $f_i K_i e_o$. In case of high
enzyme concentrations, $e_o > K_i^{-1}$, the system simply selects for the
largest rate constant f_i^o.

High concentration of templates led to linear growth in the example with one polynucleotide (7). For many polynucleotides we obtain

$$\dot{x}_i^o = \frac{f_i K_i e_o}{1 + \sum_j K_j x_j^o} x_i^o = f_i K_i e_o \cdot F(t) \cdot x_i^o; \quad i,j = 1,\ldots,n . \quad (8b)$$

Interestingly, this differential equation leads to selection as well. The reason for this behaviour is to be seen in the competition of different templates for the same enzyme. In order to be able to realize this we defined the common time dependent factor in all n equations as $F(t) = (1 + \sum_j K_j x_j^o)^{-1}$. How does the system behave after selection took place? Let us assume that I_m is the best adapted polynucleotide, i.e. the one with the largest value of $f_i K_i$: $f_m K_m = \max(f_i K_i; \ i = 1,\ldots,n)$, then all other templates will disappear after long enough time: $x_i \to 0$, $i \neq m$. The common factor then is of the form

$$\lim_{t \to \infty} F(t) = (1 + K_m x_m^o)^{-1} \simeq (K_m x_m^o)^{-1}$$

and the differential equation converges to that obtained for a single polynucleotide in the high concentration limit:

$$\dot{x}_m^o \cong f_m \cdot e_o = \text{constant.}$$

Thus, we observe again linear growth.

The catalytic action of an enzyme modifies the growth properties of an autocatalyst in the limit of high template concentrations: exponential growth changes into linear growth. The competitive behaviour of different autocatalysts, however, was not affected by the enzyme, only the quantity which is selected for may be different under different conditions.

SOME DATA ON SIMPLE RNA BACTERIOPHAGES

One of the simplest viruses known is the bacteriophage Qβ. It belongs to a class of single stranded RNA phages with very similar properties. The phage MS2 belongs to this family and is a close re-

lative of Qβ . We shall mention data of both organisms since not
all the information we need is available for a single virus. MS2
forms a virus particle consisting of one piece of plus strand RNA
about 3500 bases long. In the native virus particle the RNA is condensed onto a protein molecule called A-protein. This complex is
surrounded by a capsula built up from 200 coat protein molecules
(Eoyang and August, 1974).

The nucleotide sequence of MS2 has been determined (Fig. 5).
The genome consists of four genes which are read in different fra-
mes and which overlap partially (Fiers et al., 1976; Atkins et al.,
1979, Beremand and Blumenthal, 1979).

In nature Qβ and MS2 infect bacteria like <u>Escherichia coli</u>.
The life cycle of the virus consists of a regulated sequence of
processes (Fig. 6): (1) translation of viral RNA by the host´s machinery yielding the virus dependent subunit of the specific RNA
replicase, (2) RNA replication by the specific enzyme yielding
minus- and lateron plus-strands, (3) translation of viral RNA leading to A-protein, coat protein and lysis protein, (4) formation of
virus particles and (5) lysis of the bacterial cell.

The basic properties of viral RNA which are used in the life
cycle of the bacteriophage are the capabilities to act as template
for replication as well as to be recognized as messenger RNA by
the host translation system. In this contribution we shall be concerned mainly with replication because this process can be regarded
as a simple version of Darwinian evolution.

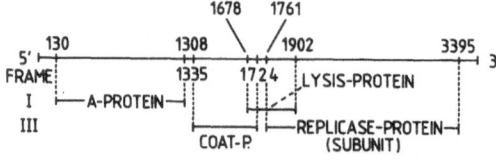

Fig. 5. The genome of the bacteriophage MS2.

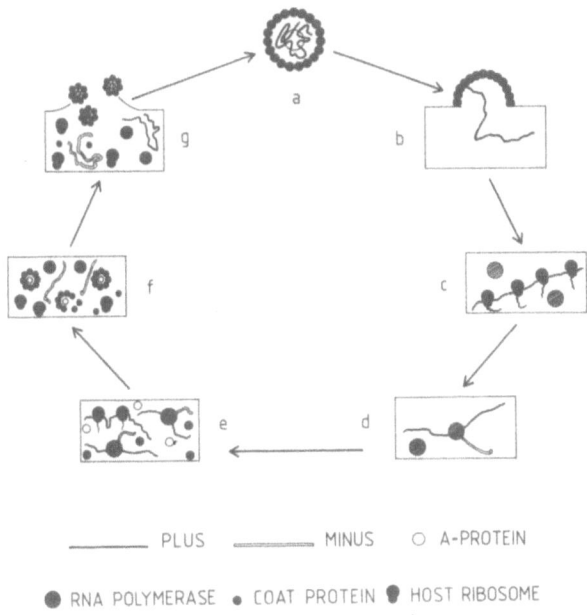

_____ PLUS   ═══════ MINUS   ○ A-PROTEIN

● RNA POLYMERASE   • COAT PROTEIN   ♥ HOST RIBOSOME

Fig. 6. The life cycle of a simple bacteriophage (Qβ or MS2) in
        a bacterial cell (E. coli).

        The process of RNA replication is shown schematically in Fig.
7. The enzyme processes template and newly synthesized strand in
such a way that the formation of (plus-minus) double strands is
avoided. The experimentally determined rate of RNA chain elongation
is highly variable (Mills et al., 1978). The replicase stop tempo-
rarily at certain "pause sites" in order to allow formation or re-
formation of secondary structures. In particular, hairpin loops are
formed in the product strand and reformed in the template strand.
Thereby, the formation of RNA double strands is avoided. This is
very important for efficient RNA synthesis. Double strands have to
melt at least in part before they are recognized as templates by
the enzyme. Biebricher et al. (1982) found a polynucleotide which
exists in two defined secondary structures. The more stable configuration, presumably a hairpin with a long double stranded region,
is very unefficient in replication.

        The replicating enzymes, e.g. Qβ replicase, are characterized
well biochemically. Qβ replicase is a tetramer. One subunit is de-

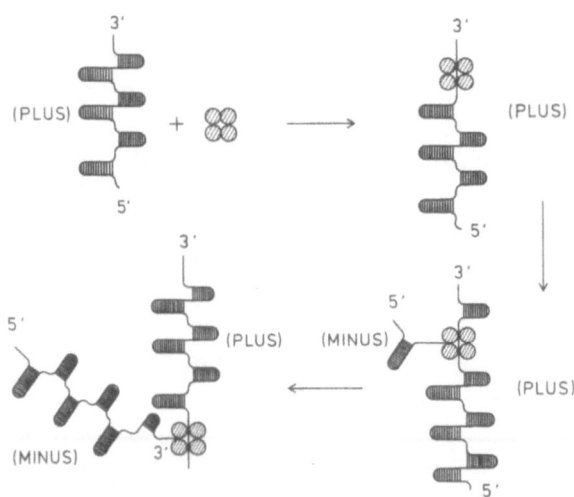

Fig. 7. Plus-minus replication of viral RNA by a specific enzyme.
The enzyme processes the template in such a way that formation of double strands is avoided.

pendent on viral RNA, i.e. it is coded on the RNA of the bacteriophage. The other three subunits are factors of ribosomal protein
synthesis of the host, namely the translation factors EF-Tu and EF-Ts
as well as the ribosomal protein S_1.

EXPERIMENTAL STUDIES ON RNA REPLICATION

     The first attempts to study viral RNA replication systemati-
cally in vitro were undertaken by Sol Spiegelman (1971) and his
group. In particular, they did serial transfer experiments on $Q\beta$ -
RNA replication by its specific replicase (Fig. 8). The most ex-
citing result they got was a stepwise increase in the rate of replication accompanied by changes in the other properties of the
RNA, infectiousness for E. coli cells or the molecular weight. The
faster replicating molecules are smaller, i.e. of lower molecular
weight, and cannot infect bacterial cells as the intact plus
strand does when it is injected (Fig. 9). Apparently, part of the
original RNA molecule was lost during replication and competition
between different polynucleotides led to selection of the faster
replicating (and no longer infectious) molecule.

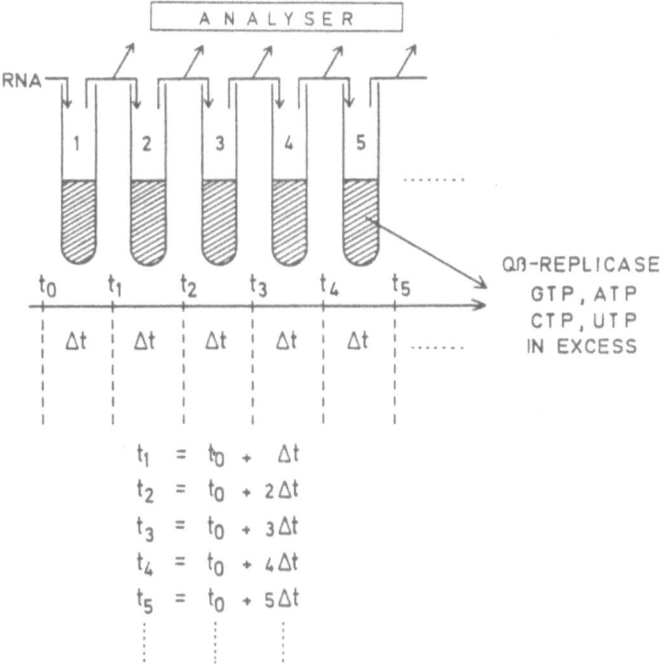

$$t_1 = t_0 + \Delta t$$
$$t_2 = t_0 + 2\Delta t$$
$$t_3 = t_0 + 3\Delta t$$
$$t_4 = t_0 + 4\Delta t$$
$$t_5 = t_0 + 5\Delta t$$

Fig. 8. The technique of serial transfer experiments (Spiegelman
        1971). RNA, in particular RNA of the bacteriophage Qβ ,
        grows in a medium which contains the enzyme Qβ replicase
        as well as the four nucleoside triphosphates ATP, UTP, GTP
        and CTP in excess. After a time Δt a sample is taken out
        of the test tube. Part of it is analysed, part of it is
        transferred into a new medium. The procedure is repeated
        after time intervals of Δt.

        Careful kinetic studies done in the laboratory of Manfred
Eigen (Küppers and Sumper, 1975; Sumper and Luce, 1975; Biebricher
et al., 1981) revealed several other interesting details on the
mechanism of Q β-RNA replication: (1) the specific enzyme recognizes
an RNA molecule only when it carries a certain recognition site
which consists of at least two CCC-triplets, (2) template-free enzyme
synthesizes RNA de novo when nucleoside triphosphates are available and (3) RNA-synthesis at excess nucleoside triphosphate concentrations and constant amount of enzyme shows three phases of
growth (Fig. 10) in the sequence exponential, linear and parabolic.

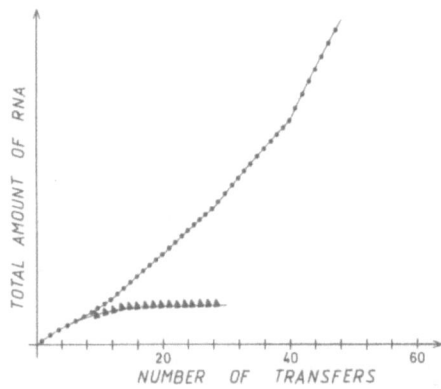

Fig. 9. Selection in serial transfer experiments (Spiegelman, 1971).
RNA replication in the test tube is initiated by addition
of Qβ RNA. Two quantities are recorded: the number of infectious particles (▲) and the amount of nucleoside triphosphate incorporated into RNA as a measure of polynucleotide synthesis (●). The rate of RNA synthesis increases
spontaneously in jumps. The percentage of infectious polynucleotides, however, becomes smaller and smaller. New RNA
molecules were formed which grow faster but are no longer
infectious. Since serial transfer selects for the excess
productivity (f-d) only, infectiousness is lost readily.
Several spontaneous, stepwise increases in the rate of RNA
synthesis are observed until an optimal value is reached.

Exponential growth is observed at excess enzyme concentration, i.e.
at low concentration of polynucleotides. It changes to linear growth
when the enzyme is saturated by template. The parabolic phase at
high concentrations of polynucleotides is consistent with product
inhibition. All data can be interpreted by means of the many-step
mechanism shown in Fig. 11. There is a pronounced similarity between the curves shown in Figs. 4 and 10. The simple two step mechanism (7) catches already some of the important features. The differences, namely the sharp turn of the curve at the point of enzyme
saturation and the flattening due to product inhibition at very
high template concentrations provide information on the particulars
of the mechanism shown in Fig. 11.

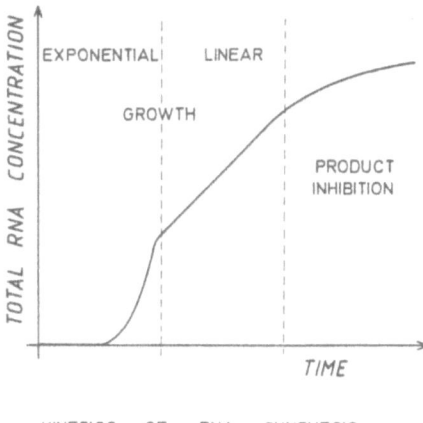

KINETICS   OF   RNA — SYNTHESIS

Fig. 10. Typical course of RNA synthesis at excess concentrations
of activated monomers and constant total concentration of
enzyme. Three phases of growth can be distinguished: exponential, linear and parabolic. Note, that there is an
abrupt flattening of the curve around the point of enzyme
saturation where exponential growth changes into linear.

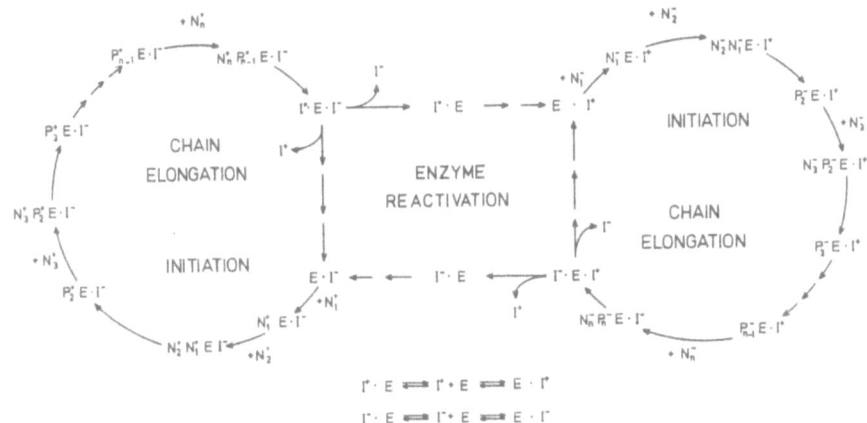

Fig. 11. A many step mechanism of RNA replication by Qβ replicase
according to Biebricher et al. (1981). E stands for the
enzyme, I^+ and I^- are plus and minus strands of the RNA,
$N_1^{(\pm)}$, $N_2^{(\pm)}$,...,$N_n^{(\pm)}$ are the appropriate nucleoside triphosphates for the corresponding positions and $P_2^{(\pm)}$,
$P_3^{(\pm)}$,...,$P_{n-1}^{(\pm)}$, finally, represent the growing polynucleotide chain bound to the enzyme.

OPTIMIZATION IN DARWINIAN SYSTEMS

Selection in the Evolution Reactor

Selection can be studied appropriately and in great detail
under conditions which allow a straightforward mathematical analysis (Eigen and Schuster, 1979; Schuster et al., 1980). A convenient system, more elaborate than serial transfer experiments or
stirred flow reactors, consists of an evolution reactor (Fig. 12)
which allows to control the concentrations of all important reactants or products. In the simplest case we keep these concentrations
constant as well as the sum of all concentrations of polynucleotides. Then, the relative concentrations of autocatalysts are the
only remaining variables.

Under the condition of constant organization, $\Sigma x_i = c_o = $ constant, a dynamical analysis of the processes taking place in the
reactor is straightforward. Chemical kinetics leads to a system of
differential equations:

$$\frac{dx_i}{dt} = \dot{x}_i = \Gamma_i(x_1 \ldots x_n) - \frac{x_i}{c_o} \phi_o; \quad i = 1,2 \ldots n \; . \tag{9}$$

Here we use c_o for the total concentration: $c_o = \Sigma x_i$. The growth
function Γ_i describes the kinetics of replication for the compound
I_i, ϕ_o is the constraint imposed on the system, in particular, the
dilution flux which leads to a constant total concentration in the
evolution reactor. In general, the growth function determines the
macroscopic phenomena we observe, like co-existence, selection or
co-operation. When mass action kinetics is applied the mathematical
analysis is straightforward (Eigen and Schuster, 1979; Schuster et
al., 1980; Hofbauer et al., 1981). The results are summarized in
Table 2.

Linear growth functions, the cases which are of particular interest for Darwinian behaviour lead to the differential equation

$$\dot{x}_i = (f_i - d_i - \frac{1}{c_o} \phi_o)x_i, \quad i = 1, \ldots, n \; . \tag{10a}$$

The rate constants f_i and d_i refer to formation and decomposition
reactions as defined before (see reactions (6) and (7)). Equation

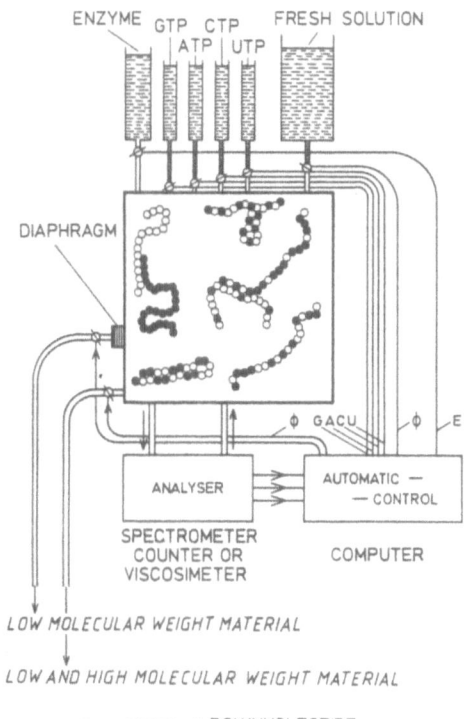

Fig. 12. The evolution reactor. This kind of flow reactor consists
of a reaction vessel which allows for temperature and pressure control. Its walls are impermeable to polynucleotides.
Energy rich material is poured from the environment into
the reactor. The degradation products are removed steadily.
Material transport is adjusted in such a way that the concentration of monomers is constant in the reactor. A dilution flux ϕ is installed in order to remove the excess
of polynucleotides produced by multiplication. Thus the
sum of concentrations $[I_1] + [I_2] + \ldots + [I_n] = \sum_i x_i = c$
may be controlled by the flux ϕ. Under "constant organization" ϕ is adjusted such that the concentration $c = c_o$
is constant. The regulation of ϕ requires internal control
which may be achieved by analysis of the solution and data
processing by a computer as indicated above.

Table 2. Polynomial Expansion of the Growth Function Γ_i.

$\Gamma_i \;=$	$k^{(i)}$	$+$	$\sum_j k_j^{(i)} x_j$	$+$	$\sum_{j,k} k_{jk}^{(i)} x_j x_k$
Reaction	Spontaneous formation of I_i		Template induced synthesis of I_i		Template induced and catalysed synthesis of I_i
Mass action kinetics	$\sum_\lambda \nu_\lambda^{(i)} A_\lambda \xrightarrow{k(i)} I_i$		$I_i + \sum_\lambda \nu_\lambda^{(i)} A_\lambda \xrightarrow{k_i^{(i)}} 2I_i$ $I_j + \sum_\lambda \nu_\lambda^{(i)} A_\lambda \xrightarrow{k_j^{(i)}} I_i + I_j$ (Mutation)		$I_i + I_k + \sum_\lambda \nu_\lambda^{(i)} A_\lambda \xrightarrow{k_{ik}^{(i)}} 2I_i + I_k$ $I_j + I_k + \sum_\lambda \nu_\lambda^{(i)} A_\lambda \xrightarrow{k_{jk}^{(i)}} I_i + I_j + I_k$ (Mutation)
Phenomenon	Co-existence		Selection and formation of a quasispecies "Darwinian" evolution		Co-operation of competitors Hypercyclic organization "Once for ever" decisions

$(10a)^{\bigstar}$ can be rewritten in more convenient form

$$\dot{x}_i = (E_i - \bar{E})x_i, \quad i = 1,\ldots,n \, . \qquad (10b)$$

Herein, we denote the excess production on template I_i by $E_i = f_i - d_i$ and its mean value by

$$\bar{E} = \frac{1}{c_o} \sum_i (f_i - d_i)x_i \, . \qquad (10c)$$

Equations (10b) inevitably lead to competition and selection and thus to Darwinian behaviour as it is shown schematically in Fig. 13. The appearance of a more efficient mutant leads to selection of this variant. During this evolutionary process the excess production E_i, turns out to be the crucial quantity, which determines the poly-nucleotide that is selected: I_m, $E_m = \max (E_i; \quad i = 1,\ldots,n)$. The mean excess production $\bar{E}$ is a very important quantity as well: it grows monotonously during the evolution in the reactor. In mathematical terms $\bar{E}$ is an Lyapounov function, in physical terms it represents the quantity which is optimized during the evolutionary process:

$$\lim_{t \to \infty} \bar{E} = \max (E_i; \quad i = 1,\ldots,n) = E_m \, . \qquad (10d)$$

Mutation and its Consequences for Replicating Systems

Variability of organisms caused by the imperfection of reproduction represents the other leg upon which Darwin's theory stands. Variations are introduced already at the level of viral RNA replication. Point mutations are one major source of variability. They are caused by mismatching of bases during replication. Systematic studies concerning the nature and the frequencies of point mutations were performed on the bacteriophage Qβ by Charles Weissmann and his coworkers (Batschelet et al., 1976; Domingo et al., 1976 and 1978). We shall consider the consequences of point mutations on replicating ensembles of polynucleotides later on.

$^{\bigstar}$As in reaction (7) we define the formation rate constant such that it is of dimension $[t^{-1}]$: $f_i = f_i' F(a_1^o, a_2^o, a_3^o, a_4^o)$.

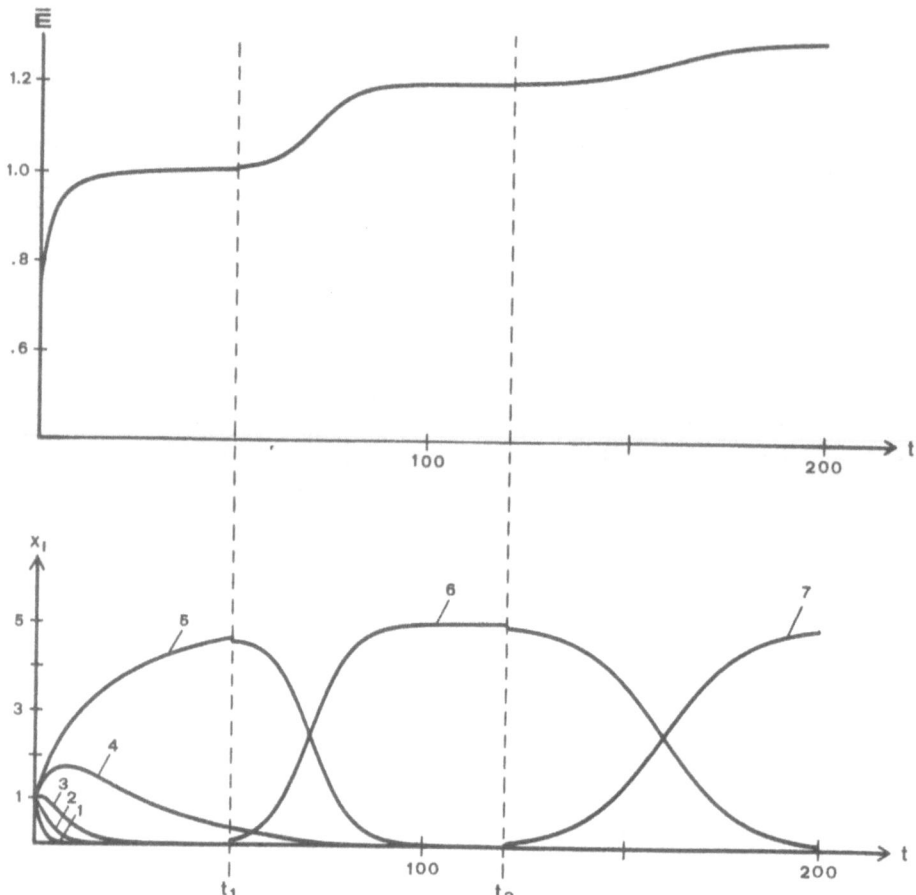

Fig. 13. Selection in an ensemble of polynucleotides under the
constraint of constant total concentration ($c = c_o$). We
present solution curves of the differential equation (9)
with $\Gamma_i (f_i - d_i)$, $x_i = E_i x_i$, i.e. $\dot{x}_i (E_i - \bar{E})$,
$i = 1,2...,7$ with $E_1 = 0.3$, $E_2 = 0.65$, $E_3 = 0.82$, $E_4 =$
0.96, $E_5 = 1.01$, $E_6 = 1.2$ and $E_7 = 1.3$ in arbitrary
time units and $\bar{E} = \sum_i E_i x_i / x_i$. The initial conditions are:
$x_1(0) = x_2(0) = \ldots = x_5(0) = 1$ and $x_6(0) = x_7(0) = 0$. The
mean excess productivity $\bar{E}$ starts from an initial value
of $\bar{E} = 0.748$ and increases steadily as the population be-
comes homogeneous: $\bar{E} \to 1.01$ and $x_1, x_2, x_3, x_4 \to 0$ as $x_5 \to c_o =$
5. One easily recognizes the cause for the increase of $\bar{E}$:
The less efficiently growing polynucleotides are elimina-
ted. They disappear in the same sequence as their excess

Another source of variations observed with Qβ replicase consists in major changes in sequence and length of the RNA. Most probably deletions or duplications of whole segments occur at certain
sites on the polynucleotide. The most complete studies on the replication process performed so far (Biebricher et al., 1981) were
undertaken in order to clarify an interesting catalytic property
of the replication enzyme first reported by Sumper and Luce (1975)
in absence of any detectable amount of longer RNA templates[x], Qβ
replicase synthesizes RNA de novo . For given external conditions
there are the conditions under which the serial transfer experiments
are conducted (temperature, concentration of reactants and enzymes,
ionic strength etc.) the de novo products have defined chain
lengths and replication properties. A change in the environment,
e.g. an increase in ionic strength or the addition of ethidium bromide, leads to increased resistence of the variants against these
agents. In case of ionic strength the increase in salt resistence
is accompanied by a stepwise increase of the RNA´s molecular weight.
Three well defined variants, 86, 113 and 220 bp long have been isolated and characterized (Biebricher et al., 1982). A hint on the
nature of the mechanism of this stepwise change in the chain lengths
can be seen in the discontinuous dynamics of the polymerization
process mentioned before. The replicating enzyme stops at several
"pause sites" and idles around for some while before polymerization

---

[x]The experiments were carried out under cloning conditions which
   allow to find single molecules of RNA.

---

productivity E_i increases, namely 1,2...5. At $t = t_1 = 50$
we observe a fluctuation $x_6 = 0 \rightarrow x_6 = \delta$ the appearance
of I_6 as a favourable mutant leads to further increase in
$\bar{E}$ which now approaches the value $\bar{E} = 1.2$ as I_5 disappears
when the population becomes homogeneous again: $x_1 = \ldots =$
$x_5 = 0$, $x_6 = 1$. The same story happens when a more efficient mutant appears at $t = t_2 = 120$. Now I_6 is replaced by
I_7 and $\bar{E}$ approaches the temporary optimum $\bar{E} = 1.3$. This
example illustrates well the nature of the selection process and the role of the mean excess productivity $\bar{E}$ as the
quantity to be optimized.

is continued. Insertions and deletions, eventually are the results
of irregular continuation of the polymerization process at these
"pause sites".

Spiegelman´s serial transfer experiments and the optimization
of de novo products have never been carried out under identical
conditions. Nevertheless, both series of experiments led to similar
adaption products coming either from larger or from smaller chain
lengths. What we observe with this replicating polynucleotides is
Darwinian evolution in its most simple form: an interplay between
variation through mutations and selection through competition of
self-reproducing molecules.

Replication with errors can be studied by means of equation
(9). We put

$$\Gamma_i = (f_i Q_i - d_i)x_i + \sum_{j \neq i} w_{ij}x_i; \quad i,j = 1,\ldots,n \qquad (11)$$

and analyze the corresponding differential equation. Herein, f_i and
d_i are the rate constants of nucleotide synthesis and degradation
as before, Q_i is the quality factor of the replication process, i.e.
the probability to obtain an error-free copy, w_{ij} finally is the
probability to obtain the sequence I_i as an error copy of I_j multi-
plied by the corresponding rate constants. The problem has been
formulated and studied extensively in the literature (Eigen, 1971;
Eigen and Schuster, 1979, Thompson and McBride, 1974, Jones et al.,
1975, Svetina and Schuster, 1982).

Depending on the accuracy of replication we distinguish two
cases: (1) the replication process is accurate enough to sustain a
stable ensemble of polynucleotide sequences. This ensemble consists
of a master sequence and a distribution of frequently occuring mutants. Such an ensemble has been called "quasispecies" in order to
indicate some analogy to the notion of a species in biology. Apart
from cases of kinetic degeneracy[*] the master sequence selected I_m,
is that with the maximum selective value w_{mm}:

---

[*]The situation is somewhat more complicated in case we are dealing
with two or more different sequences which have the same or almost
the same selective value.

$$I_m : w_{mm} = \max \{ w_{ii} : (i = 1, \dots, n) \} ; \quad w_{ii} = f_i Q_i - d_i. \quad (12)$$

Case (1), therefore, is in complete analogy to the previously treated dynamics of error-free replication; the only important difference
is the fact that a quasispecies is selected instead of a single sequence. (2) There is no stable distribution of sequences in case
the accuracy of replication is too low. New sequences appear whereas
those present at the moment will disappear after a few generations.

The two cases discussed above are separated by a precisely defined error threshold which corresponds to a critical accuracy of
the replication process (Q_{min}). Selection of a quasispecies with
the polynucleotide I_m as master sequence occurs if

$$Q_m > Q_{min}^{(m)} = \sigma_m^{-1} = \frac{d_m + \bar{E}_{-m}}{f_m}. \quad (13)$$

Herein we define a superiority parameter σ_m for the master sequence.
The "mean but the best" excess production $\bar{E}_{-m}$ reads in explicit form
($E_i = f_i - d_i$)

$$\bar{E}_{-m} = \frac{\sum_{i \neq m} x_m E_m}{\sum_{i \neq m} x_m} .$$

Let us consider now polynucleotide replication in more detail. We
can define a mean single digit accuracy $\bar{q}$ which is a measure of
the mean probability for the correct incorporation of a single base
into the growing chain. The accuracy of replication for the whole
sequence is given by

$$Q = \bar{q}^{\nu} \quad (14)$$

where ν is the degree of polymerization. Here we neglect deletions
and insertions. In contrast to this result in serial transfer experiments these variations play a minor role only in the in vivo
reproduction of RNA viruses. In serial transfer experiments the
rate of replication is the exclusive target of selection. The RNA
in in vivo systems has to fulfil other requirements as well like
to be translatable into functional proteins, and almost all deletions and insertions will be lethal.

Equation (13) has been derived by zero-order perturbation
theory. Nevertheless, it leads to excellent agreement with the exact
results when ν exceeds a certain limit ($\nu > 10$). In Fig. 14 we compare the exact solution with the expression from perturbation theory
for $\nu = 50$ (Svetina and Schuster, 1982). This figure shows the percentage of master sequence present in the quasispecies. This percentage decreases with decreasing $\bar{q}$ and becomes very small near the
critical accuracy $Q \simeq Q_{min}$. The minimum accuracy for the replication of the whole polynucleotide chain (Q_{min}) can be converted
into a maximum chain length (ν_{max}) for a given mechanism of replication. Combination of equations (13) and (14) leads to the inequa-
lity

$$\nu < \nu_{max} = - \frac{\ln \sigma}{\ln \bar{q}} \simeq \frac{\ln \sigma}{1 - \bar{q}} \quad . \tag{15}$$

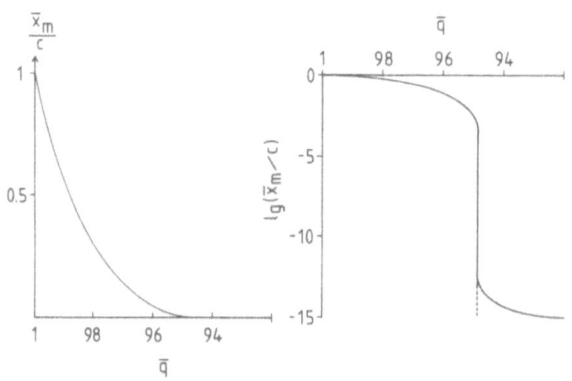

Fig. 14. Percentage of the master copy in a stable quasispecies.
We present the fraction x_m/c as a function of $\bar{q}$ for polynucleotides with a chain length of $\nu = 50$. Note that the
results of perturbation theory (broken line) coincide
almost perfectly with the exact solution (Svetina and
Schuster, 1982). The only exception being the range below
the critical accuracy (Q_{min}). Here, perturbation theory
predicts zero concentration whereas a very small fraction
of the master copy is still present due to the statistics
of erroneous replication.

The mean single digit accuracy of replication has been determined experimentally for three cases: enzyme-free template induced
replication (Lohrmann et al., 1980), enzyme-catalysed RNA replication in simple bacteriophages (Domingo et al., 1976) and DNA replication in procaryotes (Kunkel and Loeb, 1979; Kunkel et al.,
1979). We can use equation (15) to estimate the maximum chain length
of polynucleotides which are compatible with the fidelities of given
mechanisms of replication (Table 3).

Without the help of enzymes replicating RNA sequences seem to
be limited to about the size of t-RNA like molecules. The fidelity
of viral RNA replication with specific replicases allows to repro-
duce molecules up to a few thousands bases long. These are the actual sizes of RNA molecules found with simple bacteriophages and
with most animal viruses. Nature, thus approaches the error-limit
rather closely. Two experimental findings support this interpretation: (1) The virus populations studied in detail so far resemble
a quasispecies growing closely to the limit. Cloning experiments
have revealed substantial nucleotide sequence heterogeneity of the
RNA in natural populations of the bacteriphage $Q\beta$ (Domingo et al.,
1978) of foot and mouth disease virus (Domingo et al., 1980) and
of influenza virus (Ortin et al., 1980; Fields and Winter, 1981).
(2) Simple bacteriophages, the complete nucleotide sequences of
which are known at present, have overlapping genes in their genomes.
This, certainly, leads to a restricted adaptability of the proteins
encoded. The existence of a maximum chain length close to that observed provides an explanation: restricted adaptability is still
better than the loss of stability in replication.

DNA replication involves a highly developed enzymatic machinery
which uses a proof reading mechanism in order to increase the single
digit accuracy. Thereby, genomes as long as some million base pairs
become possible. In procaryotes nature again seems to operate close
to the error limit. Bacteria, nevertheless, have retained a certain
flexibility to adjust mutation rates to the evolutionary needs of
the moment. They do not seem to replicate as closely to the accuracy
limit as viruses do.

Replication leads to competition and, ultimately, to selection
of the most efficiently replicating species. The combination of
selection and mutation allows to optimize structures and properties
in the sense of Darwin's theory. The accuracy of the replication

Table 3. Single Digit Accuracy and Error Threshold in Some Chemical and Biological Systems.

Single Digit Accuracy $\bar{q}_m$	Error Rate per Digit $1-\bar{q}_m$	Superiority σ_m	Maximum Digit Content ν_{max}	Biological Examples
0.95	$5\cdot10^{-2}$	2	14	Enzyme-free RNA replication
		20	60	
		200	106	t-RNA precursors, $\nu = 80$
0.9995	$5\cdot10^{-4}$	2	1386	Single stranded RNA, replication
		20	5991	via specific replicases
		200	10597	Phage Qβ, $\nu = 4500$
0.999999	$1\cdot10^{-6}$	2	$0.7\cdot10^6$	DNA replication via polymerases
		20	$3.0\cdot10^6$	and proofreading
		200	$5.3\cdot10^6$	E. coli, $\nu = 4\cdot10^6$

process sets a limit to the size of the structures that can be optimized. Biological evolution, at least in the early stages of procaryotic development appears to be governed by these limits of polynucleotide replication.

CO-OPERATION OF SELFREPLICATING ELEMENTS

Apart from competitive ensembles of selfreplicating elements we observe co-operation in nature as well. Just to give some examples we mention the co-operative interaction between independent genes in some RNA viruses like influenza, the controled interaction of genes in organisms, various types of symbiotic co-operation and the interaction between cells in higher organisms. The main issue of this section thus centers around the question: how can we find a form of organization which allows several selfreplicating units to be present at the same time and in controled relative amounts. This control has to be exerted internally by means of an appropriate mechanism in order to make the system sufficiently resistent against variations in the environment.

At first we consider the mathematical problem. The analysis is simplest when we apply mass action kinetics. We have to study second order terms in order to search for co-operation since first order terms as applied in equation (11) lead always to competition and selection. Second order terms in most general form can be expressed by a growth function (see Table 2)

$$\Gamma_i = x_i \left(\sum_j k_{ij} x_j \right) .$$ (16)

For the sake of simplicity we omitted terms which result from mutations. The differential equation obtained by insertion of (16) into (9) has been studied in great detail (Hofbauer et al., 1980; Schuster et al., 1980). Co-operation between selfreplicating elements is observed in systems in which these elements are connected by a positive feedback loop of catalytic actions. In other words we require a closed loop of catalytic enhancement in order to stabilize the system against competition. Such a closed loop has been called an elementary hypercycle (Eigen and Schuster, 1979). In the context with the rate constants in equation (16) we find that some off-diagonal elements of the matrix of catalytic coefficients $K = \{ k_{ij} \}$, have to dominate all other elements including the diagonal terms. For simplicity we put all terms except the dominating ones equal zero:

$$K = \begin{pmatrix} 0 & 0 & \cdots\cdots\cdots & 0 & k_{1n} \\ k_{21} & 0 & \cdots\cdots\cdots & 0 & 0 \\ 0 & k_{32} & \cdots\cdots\cdots & 0 & 0 \\ \vdots & & & & \\ 0 & 0 & \cdots\cdots\cdots & k_{n,n-1} & 0 \end{pmatrix}. \tag{17}$$

The differential equation corresponding to the elementary hypercycle, schematically shown in (17), is of the form ($k_{i,i-1} \equiv \gamma_i$,
all indices are understood modulo n: n + 1 = 1, 0 = n):

$$\dot{x}_i = x_i (\gamma_i x_{i-1} - \frac{1}{c_o} \phi), \quad i = 1,\ldots,n . \tag{18}$$

The detailed mathematical analysis of equation (18) has been given
previously (Schuster et al., 1978; Eigen and Schuster, 1979; Schuster
et al., 1979). A general proof was presented that the elements of
equation (18) co-operate; no selection occurs. The dynamics of higher
dimensional elementary hypercycles ($n \geqslant 5$) is of a certain interest.
The individual concentrations oscillate in regular manner, controled by a stable limit cycle.

    The theorem of co-operation can be brought into more general
form. Mass action kinetics facilitates the analysis but is not a
necessary requirement for co-operation: the algebraic form of the
catalytic action was found to be unimportant; co-operation occurs
also in systems where the rate constant γ_i is replaced by some
continuous function F_i with a positive lower bound:

$$\Gamma_i = F_i x_i x_{i-1}, \quad i = 1,\ldots,n \tag{19}$$

with $0 < a_i \leqslant F_i \leqslant \infty$ (Hofbauer et al., 1981).

    Let us now consider the physics of catalytic interaction be-
tween individual polynucleotides. The most simple example for higher order catalytic action can be observed with primitive RNA bacteriophages. We consider the replication of $Q\beta$ in the host cell
(Fig. 15). Plus and minus strands act as templates in replication.

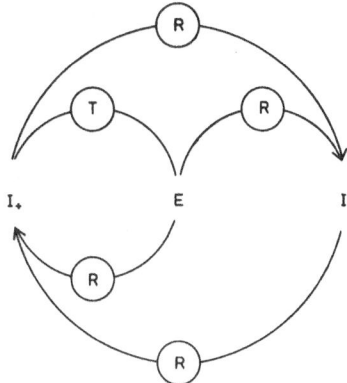

Fig. 15. RNA–phage infection of a bacterial cell as a simple hyper-
        cyclic process. Using the translation machinery (T) of the
        host cell the infectious plus strand (I_+) first instructs
        the synthesis of a protein subunit which associates with
        three host proteins to form a phage-specific RNA-replicase.
        This replicase complex (R) exclusively recognizes both
        phage-RNA strands, plus and minus, and replicates them.
        The result is a burst of phage-RNA production which follows
        a hyperbolic growth law.

Catalytic action on replication is the origin of the second factor
(x_j) in the individual terms of equation (16). RNA molecules are
lousy catalysts, otherwise it would be very hard to understand that
almost all catalysis in biochemistry is done by proteins. The role
of RNA (or DNA) in catalysis through the action of proteins is to
be seen in its function as template for translation. In case of Qβ
RNA the plus strand codes for one of the four subunits of Qβ –replicase. Positive catalytic feedback of RNA on RNA replication thus
occurs via translation.

    How to integrate several polynucleotides into a hypercycle?
Again we make use of the capability of RNA to instruct translation.
The conceptually simplest suggestion, but not necessarily that
which is easiest to realize physically introduces specificity into
RNA replication: the translation of the polynucleotides in the
hypercycle act as specific catalysts in the replication of other
RNA molecules (Fig. 16). The dynamics of such ensembles of poly-

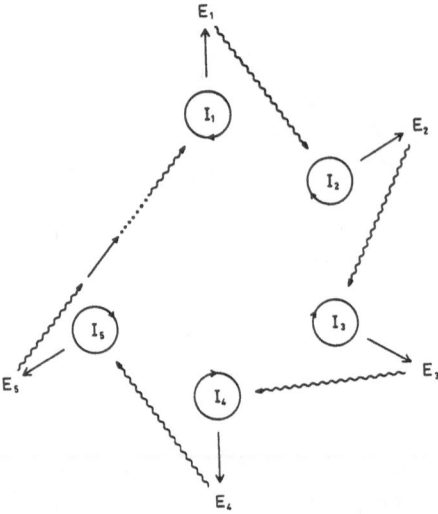

Fig. 16. Hypercycle with translation. The polynucleotides I_i re-
        present carriers of two kinds of instruction: (1) for their
        own replication due to their capability to act as template
        and (2) for the translation into proteins E_i which are
        the effective catalysts. The proteins act as specific re-
        plicases or as specific subunits in a multi-subunit enzyme.

nucleotides and proteins has been studied in some detail (Eigen
and Schuster, 1979; Eigen et al., 1980). The whole set of n polynucleotide and n protein molecules develops like an organized unit.
Relative concentrations are strictly regulated by internal control.

    How can hypercycles evolve after they were formed and passed
through an initial period? A system with hypercyclic organization
approaches an internal equilibrium or a state with oscillating concentrations. Thereafter we can consider the whole system as an integrated functional entity. In competition with the external world
such a hypercycle behaves like a single individuum growing with a
non-linear rate. In the simplest case we are dealing with quadratic
growth rates. Competition between m hypercycles then is determined
by the differential equation

$$\dot{C}_i = k_i C_i^2 - C_i \cdot \frac{\phi}{C} \; ; \quad i = 1, \ldots m \; . \tag{20}$$

This equation follows from (9) by putting $\Gamma_i = k_i C_i^2$. Herein we define the total concentration C_i of a given hypercycle H_i as the sum of the corresponding polynucleotide concentrations: $C_i = \Sigma I_j$ (I_j is a member of H_i) then the rate constant k_i may be expressed easily in terms of the rate constants γ_j:

$$k_i = (\Sigma_j \gamma_j^{-1})^{-1} \; . \tag{21}$$

The total concentration C ultimately is defined in analogy to the previous cases: $C = \Sigma C_i$.

Equation (20) is of some interest since it describes an example of restricted optimization. In an initial phase during which several hypercycles are present in comparable concentrations selection of the system with the largest value of k_i takes place. After one hypercycle won this initial competition the chance to replace the primarily selected system is practically zero for a new although more efficient hypercycle. Even large fluctuations in concentrations are not enough to defeat the established system (Fig. 17). There is no optimization of properties through adaption by successive replacement of less efficient competitors between hypercycles like that we observed with competitive polynucleotide sequences. One way how hypercycles can evolve is shown schematically in Fig. 18. Suitable mutations can be incorporated into the system when they fit properly into the network of catalytic reactions. Stepwise enlargement of the system through the incorporation of mutants thus is a highly conservative process: what has been established once is kept in the future. We may speak of "once for ever" decisions.

POLYNUCLEOTIDES AND PREBIOTIC EVOLUTION

Prebiotic evolution from unorganized molecules to individual protocells can be characterized by three important logical steps of molecular selforganization which, in principle, need not be separated by long intervals on the time axis. A sketch which makes it easier to follow the text is given in Fig. 19. These three steps are: (1) The origin of the first oligo- or polynucleotides introduces the capability of selfreplication and changes the dynamics of the system by making Darwinian selection possible at the mole-

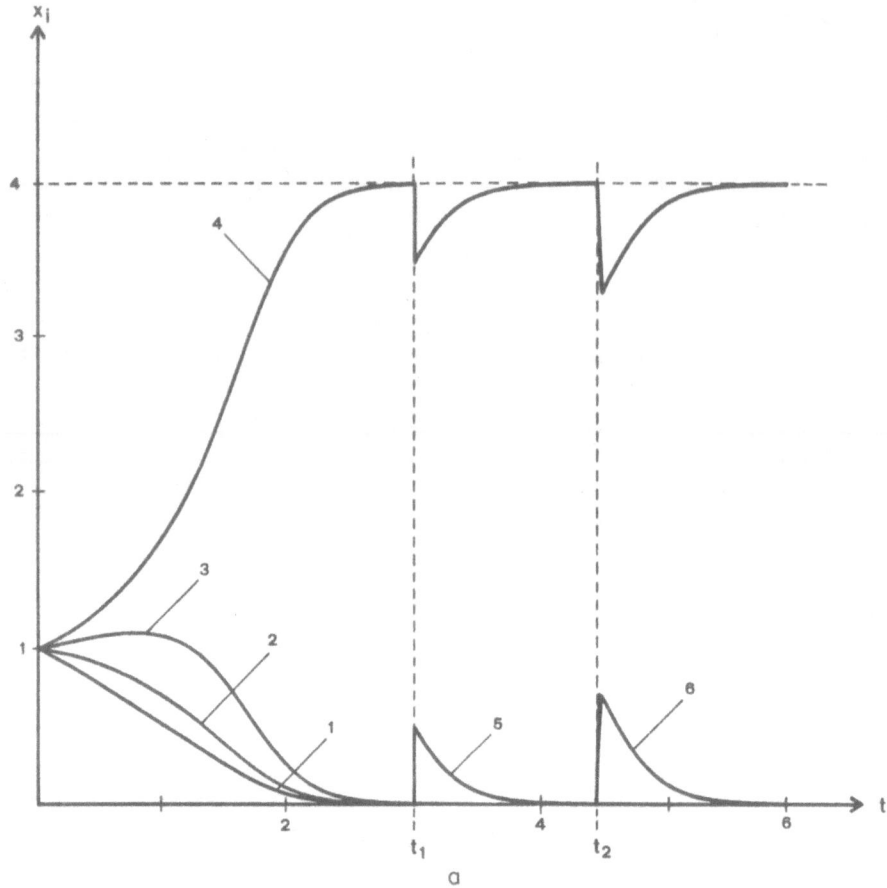

Fig. 17. Selection in an ensemble of systems growing according to
a quadratic growth law under the constraint of constant
organization. We present solution curves of the differential equation (20) for m = 6. The values chosen for the
rate constants are k_1 = 0.3 $[t^{-1} c^{-1}]$, k_2 = 0.6 $[t^{-1} c^{-1}]$,
k_3 = 0.9 $[t^{-1} c^{-1}]$, k_4 = 1.1 $[t^{-1} c^{-1}]$, k_5 = 1.2 $[t^{-1}
c^{-1}]$ and k_6 = 1.3 $[t^{-1} c^{-1}]$. The initial conditions are
$x_1(0) = x_2(0) = x_3(0) = x_4(0) = 1$, $x_5(0) = x_6(0) = 0$.
In the first phase of the development the system with the
largest value of k_i is selected. After one species reached
a stage of dominance it is almost impossible for a favourable mutant to replace this selected species. Even extre-
mely large fluctuations (t_1 = 3$[t]$, Δx_5 = 0.5$[c]$, t_2 = 4.5$[t]$
Δx_6 = 0.7$|c|$) die out in rather short time. Thus, no
optimization takes place after a first process of selection has been completed.

cular level. (2) The origin of the genetic code and the molecular
translation machinery allows to incorporate instructed proteins into
the selfreplicating system. The system is able to design catalysts,
in particular, catalysts for its own replication. (3) The origin
of selfinstructed compartment formation creates the first individual organisms or protocells. This step enables the replicating
unit to make optimal use of its own improvements. Straightforward
evolution of the property obtained in step (2) by means of natural
selection is possible from now on.

Prebiotic chemistry dealing with consecutive reactions which
lead from an ancient reducing atmosphere of the Earth to primitive

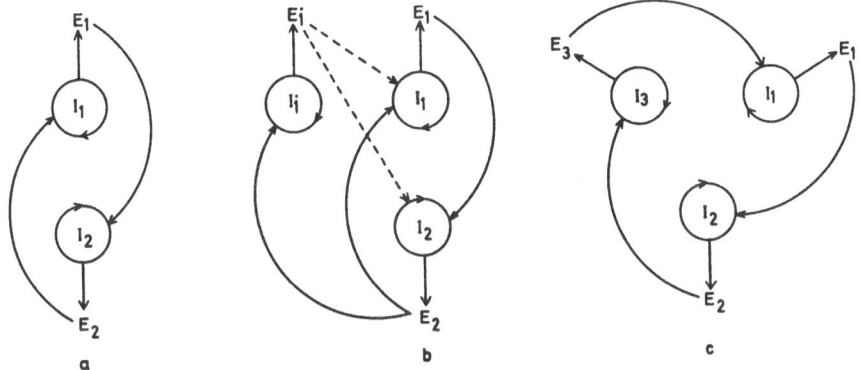

Fig. 18. A mutation–incorporation mechanism for stepwise extension
of hypercycles, a: a two–membered hypercycle with translation, b: a mutant of I_1 denoted by $I_1^{'}$ appears. It has
a certain catalytic effect on the replication of I_1 and
I_2; c: extension of the hypercycle by one member occurs
if the mutant has the following two properties: (1) $I_1^{'}$
(= I_3) is recognized better by E_2 then I_1 is and (2) the
translation product of $I_1^{'}$ (= I_3), the protein $E_1^{'}$ (= E_3)
is a better catalyst for the replication of I_1 than E_2 is.
In other words, the new catalytic coupling terms have to
fulfil the criterion: mutual enhancement along the cycle
$1 \rightarrow 2 \rightarrow 3 \rightarrow 1$ prevails over selfenhancement.

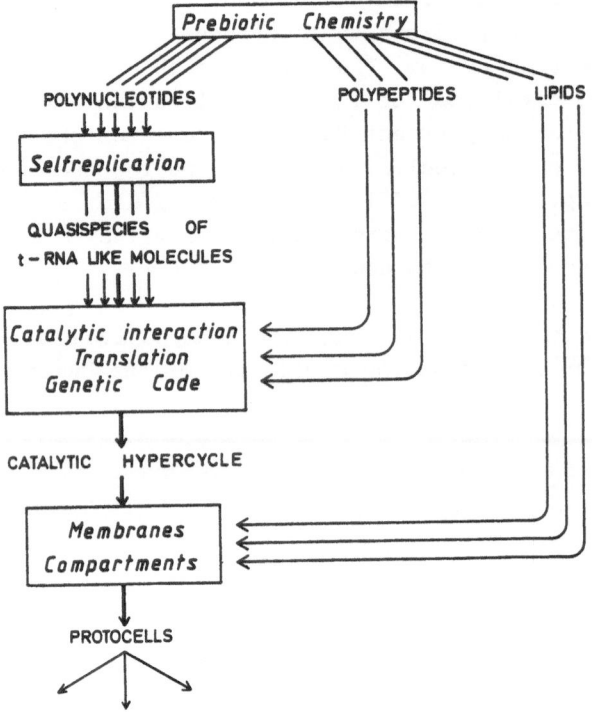

Fig. 19. A scheme for the logical sequence of steps in prebiotic
evolution leading from unorganized macromolecules to
protocells. The stream on the left-hand side of the diagram
represents the course of evolution instructed by polynucleotides. At the beginning instruction is limited to
polynucleotide replication. Later on polynucleotides
brought polypeptides and membranes under their control
through the development of mechanisms for translation and
instructed compartment formation or cell division.

biopolymers of polypeptide and polynucleotide type has been reviewed frequently. We refer here to the book by Miller and Orgel
(1974) as well as to two recent articles (Schwartz, 1981; Schuster,
1981).

Step (1) and conclusive experiments dealing with the primary
issues of polynucleotide replication have been discussed extensively in the previous sections. Replication is subjected to an error

propagation problem. Our analysis centered on the calculation of a maximum length or degree of polymerization which is determined by the accuracy of the replication process. Under prebiotic conditions (Lohrmann et al., 1980) this error threshold sets a limit to the lengths of polynucleotides which is close to the size of the present-day t-RNA´s. Here we can visualize the error threshold also from a different point of view. Within the quasispecies the master copy is selected against its own less efficient mutants. This selection breaks down in case the maximum chain length is exceeded.

Most biologists engaged in molecular evolution share the opinion that t-RNA molecules belong to the oldest class of biomolecules in living organisms. Attempts have been made to reconstruct phylogenetic trees from t-RNA´s in different organisms and to learn about the degree of relationship between different t-RNA´s in the same organisms. We refer to a comprehensive summary (Cedergren et al., 1981) and a recent study (Eigen and Winkler-Oswatitsch, 1981) which aims at the reconstruction of a possible common ancestor of our t-RNA´s. This hypothetical polynucleotide has a number of relevant properties: high GC content, a clear preference for a repetitive -RNY-, in particular -GNC-, pattern of period three[*]. The extrapolated sequence equally allows the formation of a cloverleaf or hairpin structure. The existence of complementary symmetry at the 3´- and 5´-ends of the polynucleotide is of particular importance for replication. The minus strand then has an initial sequence identical with that of the plus strands which facilitates complementary replication since both strands carry the same sequence which is relevant for recognition (see e.g. the case of Qβ discussed above where we found two -CCC- triplets as characteristic recognition sites). All properties of the polynucleotides which are relevant for enzyme free replication are readily optimized by a Darwinian type mechanism. The length of the sequences, however, remains restricted because of the error threshold.

How could the evolution of t-RNA like molecules overcome this dead end? Every possibility of further development points towards an increase in the content of information that could be stored in the sequence of bases. One way to achieve this goal is to develop

---

[*] By R we denote one of the two purine bases A or G, by Y one of the pyrimidine basis U or C and, finally, by N any of the four bases.

a more accurate mechanism of replication. Alternatively, cooperation
of selfreplicating molecules, each one present in a number of copies, would allow also to increase the total length of the transferable genetic information. For the second purpose competition
among the cooperating elements had to be avoided. At the same time
it remains necessary to keep the property of selection against
unfavourable mutants and other competitors. In the preceeding section we saw a formal way to fulfil the two requirements by means
of cyclic catalytic coupling. Any solution of the practical problem seems to combine both concepts for the increase of information:
the t-RNA-like molecules can improve the accuracy of replication
only in case they cooperate and start to design specific catalysts
for their own needs. These specific catalysts are the proteins as
we know from the actual outcome of prebiotic evolution. Thus, co-
operation of small polynucleotides, the origin of translation and
the structure of the genetic code are closely related problems.
Various models for primitive translation have been proposed but
none of them has been verified so far by conclusive experiments.
We mention here one of the most plausible models which was published
more recently and which continues earlier ideas (Crick et al.,
1976). We have modified this model slightly (Eigen and Schuster,
1979). Since an extensive discussion can be found there, we dispense
here with all details. The essential features of our model are
several predictions which are meaningful in the context of prebiotic
chemistry and thus increase the plausibility of this approach:
(1) In order to guarantee sufficient stability of messenger RNA-
t-RNA complexes the primitive codons and anticodons would have to
be rich in G and C. (2) In order to avoid sliding of the t-RNA
along the messenger during ribosome-free translation one needs a
repetitive pattern of period three (Crick et al., 1976). As we mentioned above, a pattern of the type-RNY- has been found in compa-
rative studies of the known sequences of t-RNA molecules (Eigen
and Winkler-Oswatitsch, 1981) as well as in the DNA of some viruses,
bacteria and eucaryotes (Sheperd, 1981a,b). This pattern, as the
author suggests, is a remnant of regularities in primordial genes.
It seems that mutations did not remove this pattern completely over
billions of years and it does still exist as a kind of "background
noise" in present day polynucleotides. (3) Combining (1) and (2)
in a straightforward way we obtain four codons which are presently
used for the amino acids

glycine = GGC,   alanine = GCC,

aspartic acid = GAC   and   valine = GUC.

These four amino acids were the most abundant under prebiotic conditions. The corresponding adaptors, the t-RNA's with complementary anticodons are the four most closely related t-RNA molecules.
They may well be descendants of the same quasispecies.

Our knowledge of the physics of polynucleotide-polypeptide
interaction is fragmentary. Therefore, all the details of the processes and properties which might have played major roles in early
translation are not known yet. The model considerations mentioned
may have their most important impact on experimental research just
in helping to ask relevant questiones about the origin of translation. For our further considerations let us take it for granted
that the polynucleotides managed somehow to organize a primitive
machinery for translation.

Using the primitive machinery for translation, polynucleotides
were able to interfere in the replication of other polynucleotides
by means of their translation products, the corresponding polypeptides. They might have exerted positive or negative influence,
e.g., by coding proteins with specific replicase or nuclease activities. Indirect coupling via translation products is a realistic
model for catalytic interaction in second order selfreplication.
Cyclic coupling by means of proteins will yield the same dynamic
properties of the system as we have discussed for direct coupling.
Kinetic models for hypercycles with translation have been proposed
and analyzed before (Eigen and Schuster, 1979; Eigen et al., 1980).
Moreover, we were able to prove that the property of cooperative
interaction does not depend on a concrete algebraic expression for
the non-linearity (Hofbauer et al., 1981). The only necessary requirement is a plus sign of all the terms in the feed-back loop.

A translation machinery of present-day's complexity is an
enormous involved chemical factory. In early evolution there was
no need for such high perfection. In principle, some of the molecules could act two parts at the beginning: the early transmitters
or precursors of our t-RNA's might have been early genes as well
(Eigen and Winkler-Oswatitsch, 1981). The first aminoacylsynthetases
the enzymes which nowadays attach the correct amino acids to the
corresponding t-RNA's and consequently carry a specific recognition
site for a single polynucleotide, might have acted as specific po-

lymerases as well (Biebricher, unpublished). These are just suggestions to indicate that an early translation using a primitive
highly redundant code for a handful of amino acids could form from
replicating polynucleotides. Admittedly, there is no sufficient
experimental background available yet. We do not know precisely how
probable such an event actually was, which brought the necessary
parts together.

Let us assume that a first, primitive translation machinery
was formed by integration of information stored in a number of
structurally independent genes which might well have originated
from a single quasispecies. Their relative concentrations are controled by hypercyclic organization. The first system that succeeded
in completing its design had an incredible advantage. It increased
the rate of polynucleotide synthesis many times over. At the same
time the accuracy of the replication process was increased as well.
Longer genes became possible which in turn could code for better
enzymes. The dynamical properties of such a replication - translation machinery in solution follow a higher order replication kinetics and do not allow mutants to grow even in case they are somewhat more efficient. Perfection can be achieved by enlargement of
the feedback cycle, i.e. by admission of new members. More and more
amino acids were incorporated into the encoded ensemble. Thereby,
the catalysts were improved and ultimately replication became so
accurate that gene multiplication was no longer necessary. Eventually, the independent genes were ligated to form a genome consisting of a single molecule.

Sooner or later, an inherent disadvantage started to over-
weight the advantages of homogeneous solution. Any favourable genotypic mutation (Fig. 20) is accessible to every beneficiary in the
surrounding solution no matter whether it contributes an honest
share to the common prosperity or not. Parasites may profit from
the achievements of the integrated members. A powerful counter-
action against the dilution of favourable properties through diffusion is spatial separation. This separation may range from very
weak to complete, from hindered diffusion to packing into compartments. Now, the initially mentioned logical step (3) comes into
play. Compartments, which are free of parasites are more prosperous
and multiply faster than their companions which have to feed bene-
ficiaries, will spread and finally win the competition. The positive and negative features of the isolation of genes in compartment

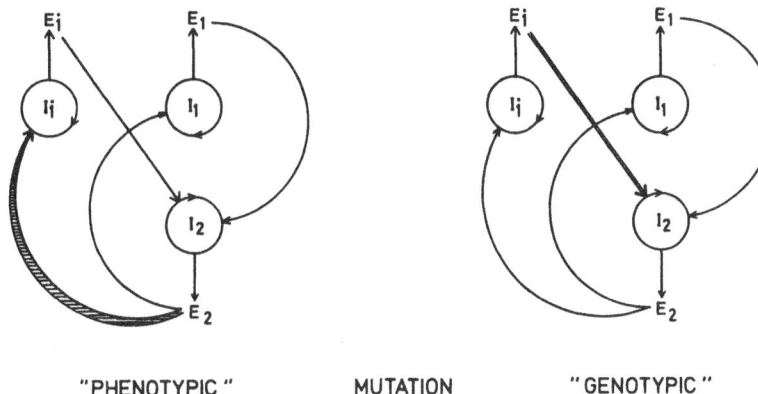

"PHENOTYPIC"   MUTATION   "GENOTYPIC"

Fig. 20. Two idealized classes of mutations in primitive replication
         - translation systems. The "phenotypic" mutations lead to
         mutants ($I_1^\prime$) which are better targets for the specific re-
         plicase, whereas the properties of their translation pro-
         ducts ($E_1^\prime$) are about the same as in the wild type (E_1).
         The "genotypic" mutant, in contrary, is characterized by
         a better translation product but roughly unchanged reco-
         gnition by the replicase.

have been summarized recently (Schuster, 1981; Eigen et al., 1981).
There one can find also a suggestion how a set of genes may exert
control on compartment formation by coding for and synthesizing a
primitive membrane protein. Here, we are mentioning the mechanistic
aspect only. Packaging into compartments brings the system back to
first order replication kinetics since the compartment multiplies
as an entity. Darwinian selection is restored. During the intermediate period of non-linear selfreplication the system was shifted
from the level of individual molecules to the level of protocells
which can be understood also as the level of first primitive but
already integrated and individualized organisms. From now on the
diversity of environments may act on the individuals and thereby
create the enormous variability in finish as we observe it in our
studies on bacteria and blue-green algae.

     Polypeptides or lipids (Fig. 19) presumably played already
an important role in prebiotic chemistry before they were incorporated into the selfreplicating machinery. Then, they are present

as environmental factors of polynucleotide selfreplication, as
general catalysts like mineral surfaces or other inorganic materials.
No direct modification of their catalytic properties by the replicating molecules is possible at this stage. After a translation
machinery has been developed we find a new class of polypeptides
which are part of the replicating unit. The polynucleotides now
instruct the synthesis of these polypeptides which act as enzyme-
like catalysts. Any modification of the primitive "gene" is trans-
lated into the polypeptide sequence - apart from the degeneracies
of the genetic code. Mutants can be evaluated by selection according
to their catalytic properties. Arguments similar to those applied
to the prebiotic role of polypeptides may be used in case of lipid-
like compounds, bilayers, membrane fragments or unicells. Presumably they were formed under prebiotic conditions and were present
in appreciable amounts in the primordial soup. They exercised catalytic power on prebiotic chemistry by forming hydrophobic phases
and interphases. Eventually, they formed primitive compartments.
Sphericles made from proteinoids and hot water (Fox et al., 1959)
may have played an analogous role. These particles, from our point
of view, are environmental factors again, important but not yet
under the influence of replicating system. The situation changes
substantially when the replicating unit succeeds to achieve control
on compartment or unicell formation. From now on the advantages of
spatial isolation are subjected to selection. Any improvement of
the catalytic efficiency of such a "protocell" is accessible only
to the molecules inside the more prosperous compartments and to
its progeny. These protocells represent in a way the first individuals or spatially separated organisms. Here we concentrate on the
properties which are controled by the selfreplicating unit. The
role of proteins and spherules as environmental factor has been
discussed extensively in the past (Fox and Dose, 1972).

Let us summarize our suggestions for prebiotic evolution by
the presentation of a time scale as provided by the fossil records
(Fig. 21). The time available for the development of the first
protocells appears to be rather short compared to other epochs of
biological evolution, in particular compared to that of procaryotic
life. Thus, natural selection had to be a very powerful tool for
evolution at the beginning already. This selection sets in at the
level of polynucleotides as soon as template-induced replication
becomes important. Integration of selfreplicating elements into a
functional unit requires higher order kinetics in order to suppress

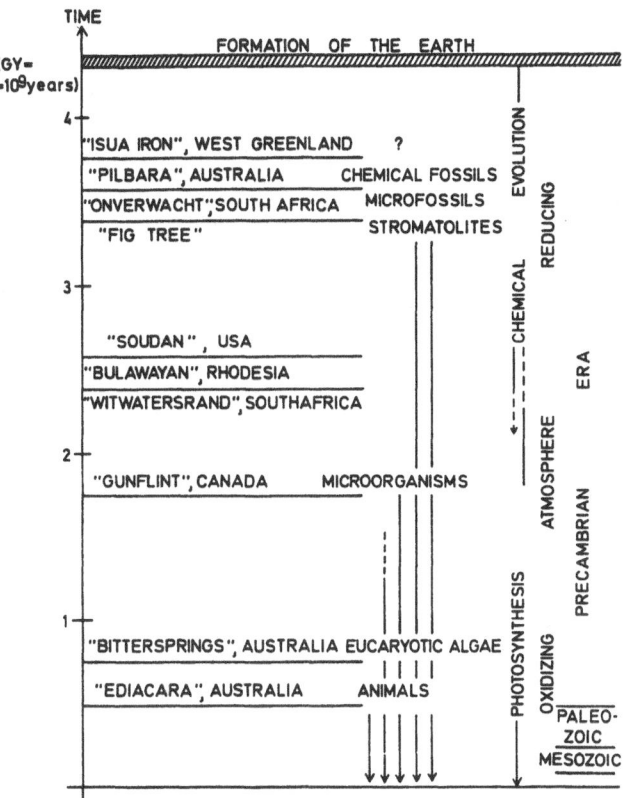

Fig. 21. The time scale of evolution as provided by the fossil re-
         cords. Traces of life on Earth are rather old. The first
         hints are found in Australian sediments which were laid
         down about $3.8 \cdot 10^9$ years ago.

competition between the individual members. This higher order kinetics is not compatible with Darwinian evolution. During such a
phase of "non-linear" development we do not expect optimization
of the newly achieved properties. We expect rather to find "frozen
accidents" or "once for ever" decisions. Some primitive polynucleotides joined their information content and formed together the
coding capacity for a primitive translation machinery. Then, later
in evolution, spatial isolation by compartment formation restored
Darwinian behaviour. From now on variation may start again. But,
the translation machinery as a whole is conserved as an entity and
no longer subject to any serious change.

In principle, a formally similar mechanism may have governed also other bursts in the evolution of the biosphere. The formation of eucaryotic organisms could be a result of similar kind of symbiosis. Some procaryotes had to share their metabolic properties and their cosing capacities in order to create a machinery for organizing the complex genome and the apparatus for mitosis and meiosis observed with all higher organisms. Again it seems very likely that the common features observed with eucaryotes are a kind of "frozen accident", a remnant of a period lacking optimization.

We could try to push analogies further and enter the realm of pure speculations. This, however, is not the aim of this contribution which was essentially directed towards evolution of polynucleotides. There is also another reason which forces us to stop the molecular description at a certain level, because it becomes the more difficult to discover principles of development, the higher organized the biological entities are. Jacob (1977) has formulated this problem so precisely that we need only to refer to his article: "evolution and tinkering". There is no reason for simplicity or intellectual elegance in nature. She does not design with the eyes of an engineer. The only thing that counts, and that is selected for is functional efficiency. Moreover, selection can act only on the things that are there at a certain moment. The basic principles therefore are very hard to detect and well hidden.

ACKNOWLEDGEMENTS

Financial support of the work performed in Vienna was provided by the Austrian "Fonds zur Förderung der wissenschaftlichen Forschung", Projects no 3502 and no 4506. Computer time was generously supplied by the "Interuniversitäres Rechenzentrum Wien". Typing the manuscript by Mrs. J. Jakubetz and drawing of the illustrations by Mr. J. König is gratefully acknowledged.

REFERENCES

Atkins, J.F., Steitz, J.A., Anderson, C.W., and Model, P., 1979, Cell, 18:247.
Batschelet, E., Domingo, E., and Weissmann,C., 1976, Gene, 1:27.
Beremand, M.W., and Blumenthal, T., 1979, Cell, 18:257.

Biebricher, C.K., Diekmann, S., and Luce, R., 1982, J. Mol. Biol., in press.

Biebricher, C.K., Eigen, M., and Luce, R., 1981, J. Mol. Biol., 148:369; Ibid., 148:391.

Cedergren, R.J., Sankoff, D., LaRue, B., and Grosjean, H., 1981, Reviews in Biochemistry, 35.

Crick, F.H.C., Brenner, S., Klug, A., and Pieczenik, G., 1976, Origins of Life, 7:389.

Darwin, C., 1967, "The Origin of Species". Reprinted by J.M. Dent and Sons, Everyman's Library, London, p. 83.

Domingo, E., Davila, M., and Ortin, J., 1980, Gene, 11:333.

Domingo, E., Flavell, R.A., and Weissmann, C., 1976, Gene, 1:3.

Domingo, E., Sabo, D., Taniguchi, T., and Weissmann, C., 1978, Cell, 13:735.

Eigen, M., 1971, Naturwissenschaften, 58:465.

Eigen, M., Gardiner, W., Schuster, P., and Winkler-Oswatisch, R., 1981, Sci. Am.,244:88.

Eigen, M., and Schuster, P., 1979, "The Hypercycle - a Principle of Natural Self-Organization", Springer, Berlin.

Eigen, M., Schuster, P., Sigmund, K., and Wolff, R., 1980, Biosystems, 13:1.

Eigen, M., and Winkler-Oswatitsch, R., 1981, Naturwissenschaften, 68:217.

Eoyang, L., and August, J.T., 1974, Reproduction of RNA bacteriophages, in: "Comprehensive Virology", Vol. 2, H. Fraenkel-Conrat and R.R. Wagner, eds., p. 1.

Fields, S., and Winter, G., 1981, Gene, 15:207.

Fiers, W., Contreras, R., Duerinck, F., Haegman, G., Iserentant, D., Merregaert, J., Min Jou, W., Molemans, F., Raemaekers, A., Van der Berghe, A., Volckaert, G., and Ysebaert, M., 1976, Nature, 260:500.

Fox, S.W., and Dose, K., 1972, "Molecular Evolution and the Origin of Life", Freeman, San Francisco.

Fox, S.W., Haranda, K., and Kendrick,J., 1959, Science, 129:1221.

Gassner, B., and Schuster, P., 1982, Mh. Chem., 113:237.

Gould, S.J., 1978, "Ever Since Darwin", Burnett Books, London, p. 39.

Heinrichs, M., and Schneider, F.W., 1980, Ber. Bunsenges. Phys. Chem., 84:857.

Hofbauer, J., Schuster, P., and Sigmund, K., 1980, SIAM J. Appl. Math. C., 38:282.

Hofbauer, J., Schuster, P., and Sigmund, K., 1981, J. Math. Biol.,
    11:115.
Jacob, B., 1977, Science, 196:1161.
Jones, B.L., Enns, R.H., and Rangnekar, S.S., 1976, Bull. Math.
    Biol., 38:15.
Kunkel, T.A., and Loeb, L.A., 1979, J. Biol. Chem., 254:5718.
Kunkel, T.A., Meyer, R.R., and Loeb, L.A., 1979, Proc. Natl. Acad.
    Sci. USA, 76:6331.
Küppers, B., and Sumper, M., 1975, Proc. Natl. Acad. Sci. USA,
    72:2640.
Lohrmann, R., Bridson, P.K., and Orgel, L.E., 1980, Science,
    208:1464.
Miller, S.L., and Orgel, L.E., 1974, "The Origins of Life on the
    Earth", Prentice Hall Inc., Engelwood Cliffs, New Jersey.
Mills, D.R., Dobkin, C., and Kramer, F.R., 1978, Cell, 15:541.
Ortin, J., Najera, R., Lopez, C., Davila, M., and Domingo, E., 1980,
    Gene, 11:319.
Schneider, F.W., Neuser, D., and Heinrichs, 1979, in: "Molecular
    Mechanisms of Biological Recognition", M. Balaban, ed., El-
    sevier, Amsterdam, p. 241.
Schuster, P., 1981, Prebiotic evolution, in: "Biochemical Evolution",
    H. Gutfreund, ed., Cambridge University Press, Cambridge, U.K.,
    p. 15.
Schuster, P., Sigmund,K., and Wolff, R., 1978, Bull. Math. Biol.,
    40:743.
Schuster, P., Sigmund, K., and Wolff, R., 1979, J. Diff. Equations,
    32:357.
Schuster, P., Sigmund, K., and Wolff, R., 1980, J. Math. Anal.
    Appl., 78:88.
Schwartz, A.W., 1981, Chemical evolution - the genesis of the first
    organic compounds, in: "Marine Organic Chemistry", E.K. Duursma
    and R. Dawson, eds., Elsevier, Amsterdam, p. 7.
Sheperd, J.C.W., 1981a, Proc. Natl. Acad. Sci. USA, 78:1587.
Sheperd, J.C.W., 1981b, J. Mol. Evol., 17:94.
Spiegelman ,S., 1971, Quart. Rev. Biophys., 4:213.
Sumper, M., and Luce, R., 1975, Proc Natl. Acad. Sci. USA, 72:162.
Svetina, J., and Schuster, P., 1982, Selfreplication with errors -
    a model for polynucleotide replication, preprint.
Thomspon, C.J., and McBride, J.L., 1974, Math. Bioscience, 21:127.

LIST OF CONTRIBUTORS

V. Crescenzi, Institute of Physical Chemistry, University of Rome,
   Rome, Italy

F. Franks, Department of Botany, University of Cambridge,
   Cambridge, United Kingdom

A.M. Gotto, Jr., Department of Medicine, Baylor College of Medicine
   and The Methodist Hospital, Houston, Texas 77030, USA

P. Laggner, EMBL Hamburg Outstation, Hamburg, Fed. Rep. Germany,
   and Institut für Röntgenfeinstrukturforschung der Österre-
   ichischen Akademie der Wissenschaften, Graz, Austria

M. Leijonmarck, Institute of Molecular Biology, Uppsala University,
   Uppsala, Sweden

S. Lifson, Department of Chemical Physics, Weizmann Institute of
   Science, Rehovot, Israel

A. Liljas, Institute of Molecular Biology, Uppsala University,
   Uppsala, Sweden

B. Massey, Department of Medicine, Baylor College of Medicine
   and The Methodist Hospital, Houston, Texas 77030, USA

D. Marsh, Max-Planck-Institute für Biophysikalische Chemie,
   Göttingen, Fed. Rep. Germany

H.J. Pownall, Department of Medicine, Baylor College of Medicine
   and The Methodist Hospital, Houston, Texas 77030, USA

H.A. Scheraga, Baker Laboratory of Chemistry, Cornell University,
     Ithaca, New York 24853, USA

P. Schuster, Institut für Theoretische Chemie und Strahlenchemie,
     Universität Wien, A 1090 Wien, Austria

L.C. Smith, Department of Medicine, Baylor College of Medicine
     and The Methodist Hospital, Houston, Texas 77030, USA

J.T. Sparrow, Department of Medicine, Baylor College of Medicine
     and The Methodist Hospital, Houston, Texas 77030, USA

S. Svetina, Institute of Biophysics, Medical Faculty and "J.
     Stefan" Institute, E. Kardelj University, Ljubljana,
     Yugoslavia

INDEX

Adair equation
  for oxygen binding, 281–283
    four-parameter model , 281–
      282
    generalized, 282–283, 286–
      288
Agarose, 88
Alginates, 72, 74, 88
Amides
  consistent force field of,
      35–36
  nonbonded interactions in, 38
Amylopectin, 70
Amylose, 70
Apolipoproteins, (see also Lipo-
      proteins) 179, 190, 208,
      213, 217, 224
  composition of, 215
  distribution of, 214
  functional regions in, 219
  metabolic roles of, 216
  phospholipid association with
      227, 228
  transfer activities of, 218
  and topology of surface com-
      ponents, 209
Autocatalyst
  growth of, 311
    critical concentration for,
      314–315
    exponential, 311

Autocatalyst (continued)
  growth of (continued)
    one-step reaction, 311
    stationary states of, 513–315
    three-step reaction, 342–316
  selection of, 316–317

Bacteriorhodopsin, 137
  three-dimensional structure of,
      137
Bending potential, 9–10
Bond potential, 7–8
Born–Oppenheimer approximation,
      4–6
Born–Oppenheimer energy surface,
      5–7
  empirical description of, 6
  from molecular vibrations, 27
  quantum mechanical description
      of, 8
Bovine pancreatic trypsine in-
      hibitor
  folding of, 50–53

Carboxylic acids
  consistent force field of, 37
  nonbonded interactions in, 38
Carrageenans, 72, 75, 78, 79, 82,
      88
Cellulose, 67, 84, 84
  mesophases of, 84

Cellulose (continued)
  microfibrils in, 68
  structural arrangements of, 69
Cellobiose, 116
Chitin, 67
Chitosan, 68
Cholesterol, 128, 186, 189, 196,
        208, 224, 227
Cholesterol esters, 189, 190,
        191, 193, 197, 208, 216,
        224
Chylomicrons, 206, 213, 218
Conformation
  of amino-acid residues, 48-50
  helix-coil transition, 47-49
  of terminally-blocked amino-
        acid residue, 46
Conjugate gradient method, 24
Consistent force field, 30-32
  of alkenes, 32
  of amides, 32, 38
  of carboxylic acids, 32, 37-39
  for nonbonded interactions, 38
Co-operation,
  in hypercycle, 341-343
  of self-replicating elements,
        339-343
    positive feed-back in, 339,
        341, 349
Coulomb potential, see Electro-
        static potential
Crystal lattice
  energy of, 22

Dextran sulfate, 75, 79
Differentiation
  of murine erythroleukemia
        cells, 299-300
    commitment kinetics of, 299-
        301
    cordycepin perturbed, 300-
        301

Differentiation (continued)
  theoretical modeling of, 294-
        303
    generalized two-state model,
        301
    two-state model, 295-298
Disaccharides, 60, 75, 82, 116
  conformations of, 65
  hydration of, 116
  optical rotation of, 117
Dispersion force, 12-14

Elastic energy
  of membrane, 290
Electrostatic energy
  of membrane in electrolyte,
        290-291
Electrostatic interactions,
  in α-helical polyamino acids,
        47
Electrostatic potential, 3, 16-18
Elongation factors, 246, 247, 254
  binding site for, 261
  crystal forms of, 247
  EF-G, 251
  EF-Ts, 251
  EF-Tu, 247, 248, 251
    comformations of, 251
    domain structure of, 249
    binding of alosteric factors
        to, 248, 251
    "Rossman" fold in, 248, 250
  proteolitic degradation of, 247
Evolution,
  prebiotic, 343-354
    protocell formation, 352
    role of polynucleotides in,
        347-352
    steps in, 343-346
    time scale of, 353
Evolution reactor, 329

Fluorescence photobleaching
    method, 142
Folding
  of bovine pancreatic trypsin
      inhibitor, 50-53
  pathway, 53
  of ribonuclease A, 54-56
    from reduced protein, 55
    thermally induced, 54
Freeze fracture electron micro-
    scopy, 148, 197
Fungal polysaccharides, 76
Furanose, 110
  ring flexibility, 111

Geminal interactions, 11-12
Glucose, 111
  hydration of, 112
Glycoproteins, 61
Glycophorin A, 135
  amino acid sequence of, 136
Glycosaminoglycans, 75
Growth function,
  of autocatalyst, 311-316
  for co-operation, 339-340
  for replication in evolution
      reactor, 334
  for replication with errors,
      334
Guaran, 72-89

HDL, see High density lipopro-
    teins
Hemoglobin
  oxygen binding to, 277-279
    cooperativity of, 283-285
    pH dependence of, 278-279
    thermodynamic description
        of, 280-283
  structure-function relation-
      ship, 286-288
Heparin, 65, 79

Heparin (continued)
  and association with plasma
      protein, 76
High density lipoproteins, 206,
    209, 213
  molecular arrangement in, 190
  subclases of, 206
    HDL_2, 186, 210
    HDL_3, 186, 208, 209
  surface structure of, 190
Hill parameter, 277
  of hemoglobin, 277, 279
  of mutant hemoglobin, 279
Hydration, 95-97
  of disaccharides (see Disac-
      charides), 116
  of ions (see also Lyotropic
      series)
    shell geometry, 102, 103
  of polymers, 118
  of polysaccharides (see Fura-
      nose, Glucose, Ribose)
  of proteins, 119
Hydrophobic effect, 105, 106, 110
  thermodynamics of, 105, 108
Hydrogen bond, 96, 99, 106, 110,
    119, 159
  in amides, 35
  in carboxylic acids, 37
  nature of, 34-39
Hydroxypropylcellulose, 86
Hypercycle, 339-343
  organization of, 342
  selection of, 343
Hyperlipoproteinemia, 207

IDL, see Intermediate density
    lipoproteins
Intermediate density lipoproteins,
    206, 214

LCAT, see Lecithin:cholesterol

LCAT (continued)
        acyltransferase
LDL, see Low density lipoproteins
Lecitin:cholesterol acyltrans-
        ferase, 197, 206, 215,
        231
Lennard-Jones potential, 15
    for alkanes, 33
    in hydrogen bond, 35-40
Lipoprotein lipase, 11, 13, 206,
        215, 217
    the interactions of, 218
Lipoproteins, 179, 182, 186, 205,
        209
    chemical composition of, 212
    classification of, 180
    concentrations of in humans,
        211
    distribution of, 212
    lacking cholesteryl esters
        LpX, 197
    lipid-protein interactions,
        224
    lipid transfer between, 190,
        232
    metabolism of, 205, 224
    physical properties of, 180,
        210
    structure of, 180, 208
London potential, 13
Low density lipoproteins, 190,
        197, 206, 213
    order-disorder transition in,
        192
    physical parameters of, 195
    structural arrangement of
        phospholipids in, 194
    surface structure of, 193
Lp(a), 218
Lyotropic series of ions, 104

Mannan, 69, 72

Membranes, 38, 127, 164
    composition of, 127
    elastic energy of, 290
    electrostatic energy of, 290-
        291
    fluidity, 146
    lipids, 128 (see also Phospho-
        lipids, Cholesterol,
        Natural Lipids)
    and dynamics of chains, 144
    restricted motion of, 165
    rotational motions, 137
    spin labeling of, 165
    structure of, 129
    translational motions, 139
    overall structure of, 131
    proteins (extrinsic), 130, 131
    proteins integral, 7, 15, 130,
        133, 141
    amino acid sequence of, 134
    lateral diffusion, 142, 143
    rotational diffusion, 139,
        141
    rotational relaxation times,
        140, 141
    three-dimensional structure
        of, 137
    stoichiometry of lipid/pro-
        tein interactions in, 164
    viscosity, 141
α-Methyl-maltoside
    optical rotation of, 116
Microbial polysaccharides, 76
Molecular structure
    calculation of, 22-24
    representation of, 20-21
Monosaccharides, 60, 61
Morse potential, 9
    in hydrogen bond, 35, 37
MS2 phage
    life cycle of, 322
    structure of, 322

Mutation
  in Q β phage, 332-334

Neutral lipids, 128, 196
Neutron scattering, 181, 186, 191
Normal modes, 25-27

Oligosaccharides, 63, 66
  structure of, 64, 66

Phosphatidylcholine, 208
Phospholipids, 127, 128, 186,
      191, 196, 224
  amphiphilic nature of, 29
  apolipoproteins association
      with, 227, 228
  bilayer permeability of, 226
  bilayers, 147
    effect of ions and salts,
      156, 158
    and interaction with water,
      121, 149, 154, 159
    termodynamic properties of,
      150, 151
    thermotropic phase transi-
      tions in, 147
    transition temperature
      shifts in, 156
    X-ray diffraction of, 153
  crystal structure of, 132
  fatty acid composition of, 129
  and water, 234
  non-bilayers
  (inverted hexagonal phase), 160
Phosphorescence method, 139
  rotational relaxation times by,
      140
Pitzer potential, see Torsional
      potential
Polydispersity, 185
Polysaccharides, 61, 57, 70, 87
  conformations of, 62

Polysaccharides (continued)
  solutions of, 78, 83
    counterions binding in, 79
    gel formation in, 87
    mesophase formation in, 84
    microcalorimetry of, 80
Polysaccharide gums, 72
Potential energy function, 2
  additivity of, 2
  ECEPP, 45, 46
  minimization of, 22-24, 50-53
  parameters of, 29
  transferability of, 19
Protein biosynthesis, 96, 127,
      245, 246, 263, 264

Q β phage
  life cycle of, 322
  point mutations in, 332
  variants of, 333-334

Recycling reactions
  in catalyst growth, 312, 316
Red blood cell goast
  free energy of, 290-291
  shape of, 291, 292
  volume change of, 292-294
Rhodopsin, 139
Ribonuclease
  contributions to stability of,
      119
  folding of, 54-56
  unfolding of, 54, 55
Ribose, 112, 116
  conformational equilibria in,
      114
  second virial coefficient, 116
Ribosome, 246, 251, 252, 253,
      255
  components, 245, 246, 255
    crystallized proteins as, 256
    sequence determination of, 245

Ribosome (continued)
  particles
    crystallization of, 253
  preparational care, 246
  proteins
    binding of solvent ions to, 261
    structure of L7/L12, 258
    L10, 260
    topology of, 254
RNA replicase
  of Q β phage, 323, 333
RNA replication
  of bacteriophage, 323–327
    of Q β phage, 324–327
    selection of, 326
  enzyme catalyzed, 317
    exponential growth of, 318–320
    linear growth of, 318–320
    selection of, 320–321
  with errors, 334–338
    error threshold, 336–337
    single digit accuracy of, 335–338

Schizophylan, 82
Selection
  between two autoacatalysts, 316–317
  of enzyme catalyzed polynucleotide, 320–321
  in evolution reactor, 331
  of hypercycle, 343
  in serial transfer experiment, 324–326
Sialic acid, 217
Starch, 70
Structural polysaccharides, 67
  (see Polysaccharides)
Sugars, 60
  structure diversity of, 60

Three-body interaction potential, 16
Torsional potential, 10, 11
  optical rotation of, 22
Triglyceride:rich chylomicrons, 211
Triglyceride:rich lipoproteins, 214, 217
Triglycerides, 192, 198, 208, 211, 214, 216, 224
Trisaccharides, 60
t-RNA
  and prebiotic evolution, 347–349

Unfolding,
  of ribonuclease A, 54, 55

Very low density lipoproteins, 206, 209, 214, 221
VLDL, see Very low density lipoproteins
Vibrational energy
  of crystals, 27
  of molecules, 25–27
  and thermodynamic functions, 28–29

Water, 95, 99, 106
  bound, 121 (see also Unfreezable)
  geometry of, 99
  in heterogeneous systems, 122
  physical properties of, 97
    density of, 100
    (radial) distribution functions of liquid, 98
    vibrational modes of, 99
  in polysaccharides, 18, 20
    (see also Glucose, Ribose)
  in protein crystals, 95, 118
    (see also Hydration of

Water (continued)
  in protein crystals (continued)
        proteins)
  undercooled, 98
    density of, 100
    specific heat of, 100
  unfreezable, 121, 123

Xanthan, 77, 89
X-ray small angle scattering,
        181
  contrast variation, 183
  in lipoproteins, 182, 191
Xylan, 69, 71, 83
Xylose
  second virial coefficient of,
        116

Zimm-Bragg parameters, 47